Allied Railways of the Western Front

NARROW GAUGE in the YPRES SECTOR
Before, during and after the First World War

Front Cover: **0.1** The B6 line at the Menin Gate, autumn 1917. (*Watercolour by Jonathan Clay*)

0.2 The station at Bollezeele, junction of the SE line from Herzeele with that from Bergues to St-Momelin. The buildings on the right are the depot for repairs and maintenance for these lines. This was a base for the *10ème Section* of the French *Chemins de Fer de Campagne* during the First World War, and from January to early May 1918 was the base for part of the 85th (Canadian) Train Crews Company, and latterly the 6th (formerly 60th) Australian Broad Gauge Operating Company. These companies operated to Poperinghe and Crombeke in Belgium for the British Army (see Chapters Seven and Eight). Undated postcard. (*Authors' collection*)

Allied Railways of the Western Front

NARROW GAUGE in the YPRES SECTOR
Before, during and after the First World War

Martin & Joan Farebrother

AN IMPRINT OF PEN & SWORD BOOKS LTD.
YORKSHIRE – PHILADELPHIA

First published in Great Britain in 2024 by
Pen and Sword Transport
An imprint of
Pen & Sword Books Ltd.
Yorkshire - Philadelphia

Copyright © Martin and Joan Farebrother, 2024

ISBN 978 1 52678 881 8

The right of Martin and Joan Farebrother to be identified as Authors of this work has been asserted by them in accordance with the Copyright, Designs and Patents Act 1988.

A CIP catalogue record for this book is available from the British Library.

All rights reserved. No part of this book may be reproduced or transmitted in any form or by any means, electronic or mechanical including photocopying, recording or by any information storage and retrieval system, without permission from the Publisher in writing.

Typeset in Palatino by SJmagic DESIGN SERVICES, India.
Printed and bound by Printworks Global Ltd, London/Hong Kong.

Pen & Sword Books Ltd incorporates the imprints of Pen & Sword Books Archaeology, Atlas, Aviation, Battleground, Discovery, Family History, History, Maritime, Military, Naval, Politics, Railways, Select, Transport, True Crime, Fiction, Frontline Books, Leo Cooper, Praetorian Press, Seaforth Publishing, Wharncliffe and White Owl.

For a complete list of Pen & Sword titles please contact

PEN & SWORD BOOKS LIMITED
George House, Units 12 & 13, Beevor Street, Off Pontefract Road,
Barnsley, South Yorkshire, S71 1HN, England
E-mail: enquiries@pen-and-sword.co.uk
Website: www.pen-and-sword.co.uk

or

PEN AND SWORD BOOKS
1950 Lawrence Rd, Havertown, PA 19083, USA
E-mail: uspen-and-sword@casematepublishers.com
Website: www.penandswordbooks.com

Contents

List of Figures		vi
List of Tables		vii
Acknowledgements		ix
Abbreviations		x
Introduction		xii
Chapter One	Introducing the Ypres sector and its railways	1
Chapter Two	The metre gauge railways of French Flanders and related lines and tramways 1894 to 1914	10
Chapter Three	Some metre gauge tramways of Belgian West Flanders and related lines 1885 to 1914	31
Chapter Four	Railways and light railways (60cm gauge) 1914 to 1918	45
Chapter Five	Railways of the Ypres battlefields 3 August 1914 to 25 May 1915 (end of the Second Battle of Ypres)	58
Chapter Six	Railways of the Ypres battlefields 26 May 1915 to 6 June 1917	65
Chapter Seven	Railways of the Ypres battlefields 7 June 1917 to 8 April 1918 Messines and the Third Battle of Ypres, to the final German offensive in the north	86
Chapter Eight	Railways of the Ypres battlefields 9 April 1918 to 11 November 1918	163
Chapter Nine	Light railways and some standard gauge railways after 11 November 1918	196
Chapter Ten	The metre gauge railways of French Flanders and related lines and tramways 1919 to 1954	202
Chapter Eleven	Metre gauge tramways of Belgian West Flanders 1919 to 2022	211
Chapter Twelve	Things to see and do now	221
Bibliography		243
Index		244

List of Figures (pages 137 to 162)

Introduction
0.1 Sectors and Armies on the northern part of the Western Front 1914–1918

Chapter One
1.1 Railways in the Ypres Sector August 1914

Chapter Two
2.1 Metre Gauge and Other Railways in French Flanders, August 1914
2.2 The tramway from Armentières to Halluin 1914

Chapter Three
3.1 Metre Gauge Tramways and Other Railways in Belgian West Flanders, August 1914

Chapter Five
5.1 Allied Railways in the Ypres Sector. End of the Second Battle of Ypres, 25 May 1915

Chapter Six
6.1 Allied Railways in the Ypres Sector, early June 1917. Before the Battle of Messines
6.2 The Proven–Boesinghe & Westonhoek light railway systems (A, B9 & B systems). Early June 1917–before the Battle of Messines
6.3 The Busseboom & Ouderdom light railway systems (D, F & K systems), early June 1917–before the Battle of Messines
6.4 Allied Railways on the Messines Front, early June 1917–before the Battle of Messines

Chapter Seven
7.1 Allied Railways in the Ypres Sector. Early April 1918–before the Battle of the Lys (Fourth Ypres), after the Battles of Messines and Third Ypres
7.2 Allied railways on the Messines Front. Early April 1918–before the Battle of the Lys (Fourth Ypres), after the Battles of Messines and Third Ypres
7.3 The De Kennebak yards on the Douve Valley Railway. Early April 1918
7.4 The Proven–Boesinghe & Westonhoek light railway systems (A, B9 & B systems). Early April 1918–before the Battle of the Lys (Fourth Ypres)
7.5 The Westonhoek, Oakhanger, Edwaarthoek and Peselhoek Yards and surrounding areas. Early April 1918
7.6 The Busseboom & Ouderdom light railway systems (D, F, K, C, V, P & W systems). Early April 1918–before the Battle of the Lys (Fourth Ypres)
7.7 Railways of the British Fourth Army. Belgian Coast, June to November 1917

Chapter Eight
8.1 Allied Railways in the Ypres Sector Battle of the River Lys (Fourth Battle of Ypres) 9–30 April 1918
8.2 Tournehem–St-Momelin metre gauge line, and the Watten–Socx standard gauge line, September 1918
8.3 The Yards at Ferme Bleue (St-Momelin). Tournehem–St-Momelin metre gauge line, Watten–Socx standard gauge line September 1918
8.4 The L System of Light Railways, end August 1918
8.5 Allied Railways – Breakout from the Ypres Salient. 28 September–11 November 1918

Chapter Nine
9.1 Railways in the Ypres Area in 1921

Chapter Ten
10.1 Hazebrouck *Compagnie du Nord* station in 1933, shared with MG line to Hondschoote & Bergues
10.2 Bergues SNCF station 1948, shared with the MG lines to St-Momelin and to Rexpoëde

Chapter Twelve
12.1 A walk in Ieper (Ypres)

List of Tables

Introduction
0.1 French and Dutch place names

Chapter One
1.1 Standard gauge lines in the Ypres sector

Chapter Two
2.1 Hazebrouch–Hondschoote–Bergues Stations and other stops, 1894–1914
2.2 Hazebrouck–Hondschoote & Hondschoote–Bergues Timetable from May 1914
2.3 Connections at Hondschoote and Rexpoëde May 1914
2.4 Bray-Dunes–Hondschoote Stations and other stops, 1903 to 1914
2.5 Bray-Dunes–Hondschoote Timetable from May 1914
2.6 Bergues–Bollezeele & Esquelbecq–St-Momelin Bridges
2.7 Bergues–St-Momelin (from 1914) Herzeele–Bollezeele (from 1910–1912) Stations and other stops 1914
2.8 Armentières–Halluin Stations and other stops, 1895 to 1914
2.9 Tramway Armentières–Halluin Timetable from May 1914

Chapter Three
3.1 Metre gauge tramways in Belgium in the Ypres Sector up to August 1914
3.2 Station stops for SNCV lines in the Ypres Sector south from Furnes and Dixmude up to 1914

Chapter Four
4.1 Ypres Sector 1915-1919. Allocation of Railway Construction Engineers (RCEs) to British Armies
4.2 Technical details of British and French Army Light Railway Steam Locomotives 1914–18
4.3 Technical details of British Army Light Railway Petrol Tractors 1916–18
4.4 Technical details of British Army Light Railway Wagons 1916–18
4.5 British Army Light Railway route miles and tonnage 1917 and 1918, Western Front (Belgium and France)

Chapter Five
5.1. Operation of metre gauge lines in the Allied part of the Ypres Sector in Belgium, October 1914 to 25 May 1915
5.2 Bergues–St-Momelin & Herzeele–Bollezeele. Timetable from August 1914 or soon after

Chapter Six
6.1 Standard gauge railway works in the Ypres Sector June 1915–May 1917
6.2 Metre gauge railway works in the Ypres Sector June 1915–6 June 1917
6.3 Metre gauge locomotives in the Ypres Sector June 1915–May 1917
6.4 Technical details of metre gauge locomotives ordered for the British & French Armies 1916–17
6.5 Light railway works in the Ypres Sector 1916–May 1917

Chapter Seven
7.1 Ypres Sector Standard Gauge summary timetables from 1 July 1917
7.2 Hazebrouck–Hondschoote & Rexpoëde–Bergues Timetable July 1917
7.3 Bergues–St-Momelin & Herzeele–Bollezeele Timetable from 1 July 1917
7.4 Standard gauge railway works on the Messines front 7 June 1917–8 April 1918
7.5 Light railway works on the Messines front 7 June 1917–8 April 1918
7.6 Standard gauge railway works in the Ypres Sector 7 June 1917–8 April 1918
7.7 Light railway works in the Ypres Salient areas 7 June 1917–8 April 1918
7.8 Goods and personnel carried by Fifth Army Light Railways week ending 17 August 1917
7.9 85th Canadian Engine Crews Company metre gauge activity based at Bollezeele, January to March 1918
7.10 Light railway works on the Belgian Coast 20 June 1917–28 November 1917

Chapter Eight
8.1 Standard gauge railways in the Ypres Sector lost or taken out of use, 9–30 April 1918
8.2 Light railway (60cm) lines captured or taken out of use in the Ypres Sector 9–30 April 1918

8.3 Standard gauge railway works in the Ypres Sector April–11 November 1918
8.4 Metre gauge railway works in the Ypres Sector April–September 1918
8.5 Light railway works in the Ypres Sector 9 April–11 November 1918

Chapter Nine
9.1 Light railway works in the Ypres Sector December 1918–July 1919
9.2 Organisation of salvage in the Ypres Sector December 1918–July 1919

Chapter Ten
10.1 Hazebrouck–Hondschoote & Hondschoote–Bergues Timetable from 1929
10.2 Hazebrouck–Hondschoote & Hondschoote–Bergues Timetable from October 1936
10.3 Hazebrouck–Hondschoote & Hondschoote–Bergues Summary Timetable April 1952
10.4 Bergues–St-Momelin & Herzeele–Bollezeele Summary timetable from October 1936
10.5 Bergues–Bollezeele & Herzeele–St-Momelin (St-Omer) Summary timetable from October 1947
10.6 Bray-Dunes–Hondschoote Timetable June 1921

Chapter Eleven
11.1 Metre gauge tramways in Belgium in the Ypres Sector 1919 to 2022

Chapter Twelve
12.1 Location and present status of metre gauge stations in Belgian West Flanders and in French Flanders
12.2 Present location of relevant metre or 60cm gauge motive power rolling stock

Acknowledgements

We wish to thank the following, without whose help this book could not have been produced:

Jonathan Clay, for the watercolour on the jacket; the National Archives at Kew; the Imperial War Museum map archive at Duxford, for help with finding other information on the First World War; the Royal Engineers Museum and Library, Brompton Barracks, Gillingham, Kent for access to War Diaries of the Royal Engineers Railway Companies; Sue Jenkins, for access to the private war diary of her grandfather, Leonard Atkins, who served with the 1st LROC in the First World War; Kim Winter and Ian Hughes, War Office Locomotive Trust; William Shelford, Archivist, Leighton Buzzard Narrow Gauge Railway; Tony Nicholson, Lynton and Barnstaple Railway Trust; Philip Pacey, West Lancashire Light Railway; Adrian Gray, Honorary Archivist, Ffestiniog Railway Company; Simon Lomax and Gareth Roberts, Moseley Railway Trust; Gerry Cork, Amberley Museum and Heritage centre; Tim Kershaw. The Imperial War Museum, London, the Australian War Memorial, Canberra, and the Royal Engineers Museum Library and Archive, for permission to reproduce photographs from their collections. Also, for permission to use photographs from their collections, and for other help, we thank Yves Artur (collection André Artur dcd), William Aves, Sandra Gittins, Robert Jeanfils and Siegrid Vereertbrugghen (Tramsite Schepdaal), Didier Oberlin, Bernard Rozé, Tony Stratford, and Jean Willig. Finally, John Scott-Morgan, for help with the production, and for the use of a photograph from his collection; to Pen & Sword Transport, for the use of a photograph from their archives, and for the production; and to Carol Trow for the editing.

Images and permissions

We have attempted to contact all possible copyright holders. If any have been missed, the copyright holder should contact the publisher. We have done our best to make this book as accurate as possible. We are responsible for any errors and would be pleased to hear about them.

Abbreviations

Railways and Railway Companies, past and present, including heritage organizations

AMHC	Amberley Museum and Heritage Centre
AMTP	*Association du Musée des Transports de Pithiviers*
CdN	*Compagnie du Nord*
CEN	*Compagnie des Chemins de Fer Économiques du Nord*
CF	*Compagnie des Chemins de fer des Flandres*
CFCD	*Chemin de Fer Cappy Dompierre (P'tit train de la Haute Somme)*
CGL	see VFIL
CMV	*Société Anonyme pour l'Exploitation du CFV de Courtrai–Menin–Werwicq*
IC	*Société Anonyme Intercommunale Courtrai*
LBNGRS	Leighton Buzzard Narrow Gauge Railway Society Ltd.
MRT	Moseley Railway Trust
NF	*Compagnie des Chemins de Fer d'intérêt local du Nord de la France*
NMBS	*Nationale Maatschappij der Belgische Spoorwegen* (in French SNCB)
NMVB	*Nationale Maatschappij van Buurtspoorwegen* (in French SNCV)
OB	*CF Électrique d'Ostende - Blankenberghe et Extensions*
ODI	*NM voor de Uitbating der Buurtspoorwegen van den Omtrek Diksmuide - Ieper*
SE	*Société Générale des Chemins de Fer Économiques*
SELVOP	*Société pour l'Exploitation des Lignes Vicinales d'Ostende et des Plages Belges*
SNCB	*Société Nationale des Chemins de Fer Belges*
SNCF	*Société Nationale des Chemins de Fer Français*
SNCV	*Société Nationale des Chemins de Fer Vicinaux (Belge)* (in Dutch NMVB)
TEOL	*Tramways Electriques d'Ostende - Littoral*
TPT	*Tramway de Pithiviers à Toury*
VFIL (or CGL)	*Compagnie Générale des Voie Ferrées d'Intérêt Local*
WHHR	Welsh Highland Heritage Railway

British and Dominion Armies, First World War

ADLR	Assistant Director of Light Railways (with Roman numerals for each army)
Anzac	Australia & New Zealand Army Corps
APB	Australian Pioneer Battalion
ATC	Army Tramway Company (RE)
AT&FC	Army Tramway and Forward Company (RE)
Aus LROC	Australian Light Railway Operating Company
BEF	British Expeditionary Force
BGROC	Broad Gauge Railway Operating Company
CORCC	Canadian Overseas Railway Construction Company
CRCE	Chief Railway Construction Engineer (GHQ)
CRT	Canadian Railway Troops
DGT	Director General of Transportation
DLR	Director of Light Railways
DR	Director of Railways
DRT	Director of Railway Transport
ECC	Engine Crews Company (standard gauge)
GHQ	General Headquarters
LRCE	Light Railway Construction Engineer
LRFC	Light Railway Forward Company
LROC	Light Railway Operating Company
NF	Northumberland Fusiliers
QMG	Quartermaster General
RARE	Royal Anglesey special reserve Company (RE)
RC	Railway Company (RE)
RCE	(with Roman Numerals) Railway Construction Engineer (RE), with HQ & staff each with an RCC (with

Abbreviations

	Roman numerals) Group of Railway Construction Companies (RE)	BREB	Belgian Railway Engineers Battalion
RE	Royal Engineers	CWGC	Commonwealth War Graves Commission
RLC	Railway Labour Company (RE)	DFB	*Deutsche Feldbahnen* (German field railways)
RMRE	Royal Monmouth special reserve Company (RE)	LR	Light Railway
ROD	Railway Operating Division (RE)	MG	Metre gauge
SLR	Superintendent of Light Railways	PE	Petrol electric
TCC	Train Crews Company (light railways, RE)	SACM	*Société Alsacienne de Construction Mécanique*
WD	War Department	SCFC	*Section des Chemins de Fer de Campagne* (Section Field Railway Engineers – French)

Other abbreviations

ACNF	*Ateliers de Construction du Nord de la France (de Blanc Misseron)*	SVCFC	*Section Vicinale des Chemins de Fer en Campagne* (Vicinal Field Railways Section – Belgian)
Alco	American Locomotive Company		
ARP	Ammunition Refill Point	SG	Standard gauge

Introduction

In our previous books, *Narrow Gauge in the Arras Sector – before, during and after the First World War* (Pen & Sword Transport, Barnsley, 2015), and *Narrow Gauge in the Somme Sector – before, during and after the First World War* (Pen & Sword Transport, Barnsley, 2018), we described the history of railways of less than standard gauge in the Arras and Somme sectors of the Western Front respectively. These were put into the context of the existing and new standard gauge lines. In addition, the narrow gauge history of those areas was followed from the beginnings, through the First World War, and then on into their subsequent use. In this book we have addressed these subjects for the Ypres sector of the Western Front.

Many in France and Belgium regard the term 'narrow gauge' as not including the metre gauge but only gauges less than this. This was also the view of the British army in the First World War, to whom the metre gauge was coupled with the standard gauge as 'Broad Gauge Railways'. 'Light Railways' were, with very few exceptions, of 60cm gauge. However, we have concentrated here on all railways of less than standard gauge in the Ypres sector. There is also enough information on standard gauge railways to set the scene for this.

Those who have read our previous books will realise that there are parts of this book which are the same. We have done this so that this book can stand alone, and new readers are not disadvantaged. All the photographs are new, and all the figures with the exception of Figure 0.1, which has been modified to correct some inaccuracies, and for the dates to fit better with the key events in the Ypres sector.

Throughout this book the term 'British army' means the British and British Empire (now Commonwealth) units which were under the command of the British Expeditionary Force (BEF) General Headquarters (GHQ). This included important armies from the Dominions of Canada, Australia, New Zealand, and South Africa, and from India.

The British sectors and approximate army positions from November 1914 to September 1918 are shown in Figure 0.1. The Ypres, Arras and Somme Sectors are also defined on this Figure.

It was during the Battle of the Somme in the summer of 1916 that grave deficiencies in the supply lines became most apparent. As part of general mobilisation at the beginning of the war, the French government had placed all railways in the country under military control, although the railways generally continued to be run by the regular personnel. Belgian railways made similar arrangements for the small unoccupied part of the country. In 1916, Sir Eric Geddes, an experienced railway manager who had previously worked in London with Lloyd George when the latter was Secretary of State for War, was appointed Director General for Transportation, France, and given the rank of Lt Colonel in the Engineering and Railway Staff Corps. He was given responsibility for standard gauge and 'light' railways, roads, and canals.

This recognised the importance of transport, and especially railway transport, in the effort to win the war. In 1915 it was agreed that the British army could operate on some French and Belgian standard gauge lines, and in December 1916 there was a further agreement for the British army to import more locomotives and wagons for these lines. The British army Railway Operating Division (ROD) operated extensively on the standard and metre gauge railways in Belgium and the north of France from 1915. The agreement also led to the building and operating of military standard gauge railways exclusive to the British army, which we have called 'British army lines'. An increase in the rate of railway building and upgrading followed, culminating in an enormous effort in 1917 and 1918.

The railway infrastructure before 1914 was key to the developments during the war. Afterwards these railways played a major part in the recovery and reconstruction. However, the war had a major effect on the later development of railways in these parts of France and Belgium. If the war had not occurred, railways, particularly secondary lines (those of *Intérêt Local* in France), might have been more extensive, and might have been in a much better position to compete against road transport. The war, and financial problems afterwards, impaired infrastructure investment, including in particular electrification.

We have made all the necessary translations from French. We have followed the British convention for the configuration of wheels on locomotives. The French convention is only to count the wheels on one side of the locomotive and not to use hyphens, so that a locomotive that is to the British a 2-6-0 is to the French a 130.

Introduction

Times and timetables
In summary timetables, all times have been given in the 24 hour clock. All these tables are abstracted from the originals. All of the original timetables are in the 24 hour clock, except for some in the nineteenth and early twentieth centuries which are in the 12 hour clock. In the text we have given all times in the 12 hour clock with am or pm.

Languages and Place names
Most of the Ypres Sector as defined for this book which is in Belgium is in the Province of West Flanders. However, there is in the south a small detached part of the province of Hainaut. This extends along the north of the Lys River, which is the French border, from just north of Armentières, east to Comines. As such, it includes a small but important part of the First World War front line, around Ploegsteert Wood.

Before and during the First World War, the French language was dominant in Belgium. Only French names appear on maps, and are those universally used in records, including those of the British army. Since then, there has been a resurgence in Flemish identity and culture. The Dutch language is now universally used in the Provinces of the Flemish Region, with the use of Flemish names. West Flanders (West-Vlaanderen) is a Flemish Province and modern maps show the Dutch names, although French maps tend to show the French alternatives for major places in brackets. Hainaut, including the detached part near Armentières, is a Province of the Walloon Region, which is French speaking.

In this book, the names in use for the period being described have been used, as they were at the time. In particular during the First World War French names are used. Where necessary, particularly for descriptions of things which still exist, both are used. An example is for the descriptions of the railway networks in Chapter One. The French and Dutch names for important places for this book are shown in Table 0.1. Occasionally places also have an English version for the name (Dunkirk, Ostend); these have been avoided in the main text but are included in Table 0.1. And (of course) for the British soldier Ypres (Ieper) was always 'Wipers'.

Units of measurement

Length and distance
1 metre = 100 centimetres (cm) = 1,000 millimetres (mm) = 3.28 feet = 3 feet $3\frac{3}{8}$ inches

1 kilometre (km) = 1,000 metres = 0.62 miles = approx. $\frac{5}{8}$ mile

1 mile = 1.609 kilometres

Weight
1 kilogram (kg) = 1,000 grams (gm) = 2.2046 pounds (lb)

1 tonne (metric ton) = 1,000 kilograms = 2,204.6 pounds

1 ton (Imperial ton) = 2,240 pounds

Volume
1 litre = 1.7598 pints

1 Gallon = 4.546 litres

In this book most units are metric, except where the originals are in imperial units, as in most British army documents from the First World War. Some important distances and heights are given in both.

Layout of the book
For production reasons the Figures are bound together in the centre of the book (pages 137 to 162). However, they are in the order in which they are first referenced in the book, and the number before the decimal point indicates the chapter to which they wholly or mainly relate. Photographs are in the text close to the relevant position, but all are in black and white even when the original was in colour.

Narrow Gauge in the Ypres Sector

Table 0.1 French and Dutch place names

	Dutch	French	English
Country	België	Belgique	Belgium
Province	West-Vlaanderen	Flandres occidentale	West Flanders
Towns & villages	Abele	Abeele	
	Bikschote	Bixschoote	
	Boezinge	Boesinghe	
	Brugge	Bruges	
	De Panne	la Panne	
	De Seule	le Seau	
	Dikkebus	Dickebusch	
	Diksmuide	Dixmude	
	Elverdinge	Elverdinghe	
	Gent	Gand	Ghent
	Ieper	Ypres	
	Koksijde	Coxyde	
	Kortrijk	Courtrai	
	Langemark	Langemarck	
	Menen	Menin	
	Merkem	Merckem	
	Mesen	Messines	
	Nieuwkerke	Neuve-Église	
	Nieuwpoort	Nieuport	
	Oostduinkerke	Oost-Dunkerke	
	Oostende	Ostende	Ostend
	Passendale	Passchendaele	
	Poelkapelle	Poelcappelle	
	Poperinge	Poperinghe	
	Rousbrugge	Rousbrugge	
	Roeselare	Roulers	
	Sint-Jan	St-Jean	
	Sint-Juliaan	St-Julien	
	Torhout	Thourout	
	Veurne	Furnes	
	Vlamertinge	Vlamertinghe	
	Wijtschate	Wytschaete	
Rivers	Leie	Lys	
	IJzer	Yser	
Other words used with or in place names			
	bad	bains	bathing place, beach resort
	dijk	digue	dyke
	dorp	village	village
	stad	ville	town
	French	**Dutch**	
Province	Hainaut	Henegouwen	
Towns & villages	Comines	Komen	
	Mons	Bergen	
	Mouscron	Moeskroen	
	Warneton	Waasten	
Country	France	Frankrijk	France
Towns	Dunkerque	Duinkerke	Dunkirk
	Lille	Rijsel	

Chapter One

Introducing the Ypres sector and its railways

We have defined the Ypres Sector for the purposes of this book as the part of France and Belgium north from Armentières (France) to Nieuport (Nieuwpoort) (Belgium), which is on the North Sea coast. A look at Figure 0.1 or any map of western Belgium and northern France, shows that Ypres (Ieper) is in a corner of western Belgium. France is 20km (12½ miles) to the west, and only 12km (7½ miles) to the south-west and the south-east. Armentières, just on the French side of the border, is 20km due south. For most of the war, the front line ran north from Armentières to Nieuport, with a bulge east around Ypres (the Salient).

Reference to Figures 0.1 and 1.1 also shows that the Ypres sector was smaller than the Arras Sector or the Somme Sector, which we have described in our previous books. The straight line distance from Armentières to the sea at Nieuport is only 55km (34 miles). In addition, there was relatively less movement of the front line in this sector between November 1914 and September 1918. The Salient around Ypres to the east became bigger and smaller, but the greatest movement was the German advance south of Ypres into the Lys Pocket in April 1918. It was this mostly static state of the front line which led to the enormous concentration of camps, supply dumps, batteries, other military facilities, dressing stations and hospitals, which reached its peak in 1917 and 1918. Not least among these facilities was a particularly dense network of railways, of standard gauge, 60cm gauge (light railways), and to a lesser extent of metre gauge. This was possibly the densest area of railway development ever seen. It was in contrast with the situation before August 1914. As Figure 1.1 shows, in that era there was only one standard gauge railway within the lines linking Hazebrouck to Dunkerque and Armentières, that linking Dunkerque east into Belgium, and the later front line north from Armentières to the sea at Nieuport. This was the single track line linking Hazebrouck with Ypres via Poperinghe.

From autumn 1914 to the end of the war in November 1918, part of this front, usually a large part, was held by the British. The Belgians held the front line from north of Ypres to the coast. This was especially important to them, because this was the only part of Belgium not in German occupation, and their army was led personally by King Albert. The French also played a major part, most notably during the defence of the area in autumn 1914, during the Third Battle of Ypres in 1917 (see Chapter Seven), and in the final battles of autumn 1918 (see Chapter Eight). Two American divisions were involved in 1918, but this was never a major front for the US army.

Figure 1.1 also shows how, although the front line north from Armentières to Nieuport (Nieuwpoort) was almost entirely in Belgium, major parts of the back and support areas were in France. We have defined the southern boundary of the Ypres Sector as the standard gauge main line railway east from St-Omer to Lille via Hazebrouck and Armentières. The western boundary is the canalised Aa river north from St-Omer to the North Sea at Gravelines. This is also the boundary between the Nord *département* of France and the Pas-de-Calais *département*. West of this were important supply lines and facilities in the northern part of Pas-de-Calais, including at the ports of Calais and Boulogne, and we have included some information on the support lines and bases in the northern Pas-de-Calais *département*.

The main area covered by this book can best be described as Flanders. Today it is an area divided between France and Belgium. In France it is the northern part of the Nord *département* while in Belgium it is the western part of the Province of West-Vlaanderen (West Flanders), with the small detached part of the Province of Hainaut just north and east of Armentiéres. The rest of Belgian Flanders is beyond the scope of this book.

Geography

The coast of this part of Flanders runs in a curve eastwards, bending towards the north and straightening east of the Belgian border, with sandy beaches and a rampart of dunes. In the distant past the land behind the dunes was marshland with numerous water channels but over the centuries these have been drained to form the Polders, less extensive than those in Holland but still very characteristic of this area. Drainage has revealed an alluvial soil allowing agriculture to flourish. The Polders run from 6 to 10 miles inland.

Flanders has three major rivers, two rise in France before crossing Belgium, the third stays in France. The longest is the Lys (Leie) which rises in the Pas de Calais, and north-east from Armentières for 20km

(12 miles) forms the French-Belgian border. Crossing Hainaut, West and East Flanders it joins the Scheldt in Ghent. Its length is 202km (126 miles) and it was used commercially from the Middle Ages. Severe floods in the seventeenth century led to major works. Six locks and weirs were constructed in an attempt to control the river. It was, until quite recently, severely polluted by the extensive Flemish flax industry along its banks. The next in length is the French one, the Aa. The origin of the name is the word for 'water' in Old Dutch. This rises in the Artois Hills and its total course is 89km (55 miles). It has two distinct sections. In the first, longer, part it is simply a chalk stream running through the Artois Hills down to St-Omer. The second part is the 29km (18 miles) towards the sea and here it has been navigable for centuries. It connects with the Canal de Calais and the Canal de Bourbourg leading to Dunkerque before reaching the sea at Gravelines. The low-lying nature of Flanders can be shown dramatically in the Aa's prehistory. St-Omer formerly lay at the head of the Aa's estuary. Calais was on its Western edge and Bergues was on its Eastern with Dunkerque developing on the dunes across the marshes. The land of the estuary was reclaimed as some of the earliest Polders. Our final river is the Yser (IJzer) which rises in the Nord *département*, on the Northern flanks of Mont Cassel. It crosses Flanders and enters the sea at Nieuwpoort. It runs for 78km (48 miles) of which 30km (19 miles) are in France. Like the Aa, its nature has changed over the centuries. Until the tenth century its estuary ran for several kilometres inland. When the Polders were created with sluices at Nieuport, the level of the river could be controlled. It was the opening of these sluices in 1914 that stopped the German advance.

There were also many canals. The most important for this book is the Yser Canal, linking Ypres (Ieper) north 16km (10 miles) to the Yser (IJzer) river. From Ypres there was a narrower canal south to Comines. Another canal linked the Yser to Furnes (Veurne) and the east-west canal near the coast from Ostend to Dunkerque. Waterways remained commercially important well into the twentieth century, but now many see only leisure boating.

Although Flanders has a predominately flat landscape it does have a string of low hills. There are ten peaks, ranging in height from Mont Cassel at 176m (577ft) to Mont Kokerel at 110m (361ft). The town of Cassel on Mont Cassel was important, not least as the headquarters of the British Second Army. In the Cretaceous period the area was under water and when the sea retreated it left a plateau. Erosion over the ensuing centuries left these isolated hills of the harder rocks, formed of limestone, some bearing iron and giving it a yellow colour.

The hills run in a line from France east into Belgium, the furthest east being Mont Kemmel (Kemmelberg), 159m (522ft) high. East of Mont Kemmel is Messines (Mesen) Ridge, 80m (260ft) high at Wytschaete (Wijtschate). From here, a low ridge curves round south and then east of Ypres, remaining above 50m (165ft) high to north of Passchendaele (Passendale). Holding this ridge for much of the war allowed the Germans to observe the salient, hence the struggles to take it in 1917 (Third Battle of Ypres, see Chapter Seven).

History

The area covered in this book is one which has been fought over for centuries with powerful neighbours vying for control. To the east the powerbase shifted gradually from the German and Holy Roman Empires to the Duchy of Burgundy and finally to Spain. To the west, France always had an eye on shifting her borders east. Hence it can be best described as a 'frontier zone'.

During the Middle Ages, the urban elites of rich merchants in the powerful cities sought to make their towns top of the league, thereby increasing their riches and their power. The land-owning aristocracy tended to be based on their estates and looked to the French for support. The towns however tended to look to England for support since they needed English cooperation in the supply of the wool that kept their industry alive and the money flowing in. These tensions were exacerbated by religious differences. The Low Countries had been receptive to the new religious ideas of Martin Luther and John Calvin. This was especially true of the northern parts which largely make up today's Netherlands. The southern parts, which largely make up today's Belgium, remained predominantly Catholic. By the sixteenth century under the Spain of Phillip II, an attempt was made to abolish all Protestantism in the Spanish empire. This led to a mass revolt and years of wars, massacres and destruction. During the seventeenth century, Spain had been forced to acknowledge the unyielding resistance of the northern Protestants and in 1648 Spain recognised the independence of the Protestant provinces as the United Provinces.

France now enters the scene. In 1678 Louis XIV takes Flanders and Hainaut in the name of his Spanish wife. By the eighteenth century the picture has become more complicated as Austria enters the field. Charles II of Spain has died childless and Austria and France have a claim. War ensues and Austria gains the Netherlands. There is much unrest and calls for independence from Austria. The Austrians are temporarily defeated and the United States of Belgium comes into being but the tension between the wealthy merchants and the smaller crafts increases and the Austrians regain power. France is by now a Republic and it wins two battles in 1792 and 1794 that expel the Austrians. France now creates the nine *départements* that are still reflected in today's Belgian provinces. Flanders was under French control until 1814 but following the defeat of Napoleon, Belgium and the Netherlands were joined as the United Kingdom of the Netherlands. However by 1830 pressure for Belgian independence led to the creation of a separate Belgian state.

We now enter the final stage of the story up to 1914. In the nineteenth century, Belgium had serious economic difficulties, with a severe famine in Flanders from 1845 to 1848, and was slow to make progress in the Industrial Revolution. Otherwise, the nineteenth century was one of peace and increasing industrialisation. However, in 1914 Belgium found herself once again on the frontline in a major war.

Perhaps the complex history of Flanders can best be brought into focus by looking at events in four of the powerful towns of the region, two in Belgium and two in France.

Ypres - Ieper

The first glimpse we have of Ypres (Ieper) is when it suffered an attack by the Romans in the first century BC. Its name is supposedly taken from its river *Ieperlee*. This naming dates from the eleventh century. By adopting skilful and efficient methods of cloth making the city prospered and by 1200 had 40,000 inhabitants. There were close ties to England as the producer of raw materials. Ypres and Bruges (Brugge) were the founder members of the Flemish Hansa, a trading organisation set up to trade with England. There were rich merchants from fifteen other Flemish towns in the organisation. They traded not only with England but sold their wares as far away as Russia. There is a record of cloth sales in the twelfth century to Novgorod where the German Hansa, a merchant group which predated the Hanseatic League, had a warehouse, the *Peteryard*. However, the good times did not last. In 1241 a fire destroyed most of the old city and during the rest of the thirteenth century Ypres' prosperity declined. In 1388, Ypres found itself facing an attack from yet one more of its powerful neighbours, the English. This was a time of religious unrest and schism in the Catholic Church. There were two popes, one in Rome and one in Avignon. The latter was Clement VI (or VII according to the French). Louis, Count of Flanders was a supporter of Clement. Although 'officially' called a 'Crusade' by the English Parliament against a supporter of the 'Anti-Pope', the attack was really an excuse to gain power in Flanders to protect their commercial interests. The invasion was called the Norwich Crusade since it was led by the Bishop of Norwich, Henry le Despenser. It was badly supplied and poorly led but the English besieged Ypres for four months before the city was relieved by the French and the inhabitants could try and build their economy again.

1.1 The Cloth Hall at Ypres before 1914. (*Authors' collection*)

Moving to the sixteenth century, wars can occasionally bring benefits to cities. After Thérouanne in France was destroyed by Charles V in 1553, its bishopric was moved to Ypres in 1561. The church of St. Martin became the cathedral. In the sixteenth and seventeenth centuries we begin to see most clearly the results of being a frontier town. It was besieged by the Spanish in 1583. Starvation was the winner and the Spanish increased the town's fortifications. The French attacked three times in 1644, 1648 and 1658, finally taking it in 1678. Within two weeks Louis XIV had set his military architect, the brilliant Sébastian le Prestre de Vauban, to build defences around the city. In 1697 the city was handed back to Spain. The eighteenth century saw another war, this time between France and Austria, with England backing Austria. The English were led by John Churchill (later Duke of Marlborough), who was looking for cities to capture. In 1709 he considered Ypres as a target but dismissed it on account of its swampy, disease-ridden site. Instead he took Tournai (Doornik). The end of that war saw Ypres change from Spanish to Austrian rule in 1714. It was the Austrians who, in 1782, decided to demolish the walls to increase trade. It didn't have any effect and so parts were left. However, enough had gone to allow the French to besiege and then capture it in 1794. Some of the ramparts were demolished in 1852 but the earthworks and water provided an attractive leisure area. With the defeat of Napoleon in 1815, Ypres settled down as a Flemish city and Belgium had a century of peace until it became one of the most famous sites of the First World War, with its complex railway network that is the core of this book.

Bergues

The first of our French examples is Bergues, sitting just 7 miles (11km) from the Belgian border. Its position always made it a frontier town and looking through its history it seems to have been captured by the French, English, Spanish and Dutch at some time or other. Its origins are a matter of some debate but they clearly involve a seventh century monk, Winnoc, who was a Celt either from Brittany or Wales. He was based in St-Omer and here the stories vary. Some say he only came from St-Omer to convert the heathen, others say he retired there to find religious peace. Winnoc was canonised and became the local saint. This may have been the doing of the first Count of Flanders, Baldwin (Baudouin) who found numerous 'relics' to improve the status of the towns he was fortifying. Bergues' turn was in 1022. At this time Bergues was a port and became a textile trading centre. It was a member of the Flemish Hansa, mentioned above under Ypres. It was fortified against French attacks and held by Spain until

1.2 The *Porte de Cassel* (Cassel Gate) at Bergues before 1914. (*Authors' collection*)

it was handed to France in 1688 in one of the many deals that closed the multiple wars of the seventeenth century. As at Ypres, Vauban was sent to improve the town fortifications. He strengthened and improved the medieval walls. Louis XVI, however, decided to promote Dunkirk and Bergues began its steady decline as its access to the sea gradually silted up. In both World Wars Bergues was badly damaged and like Ypres was rebuilt in the same pattern.

Dunkerque

Dunkerque is a close neighbour of Bergues and their histories are connected. It was a small fishing settlement on the coastal marshes when St. Eloi settled there in 645 and built a church known as 'The Church on the Dunes'. It suffered in the same way as Bergues, being sacked six times in the Middle Ages by the same powers. By the mid-seventeenth century, as most of Flanders, it was under Spanish control. However here Dunkerque stars in a rather strange piece of power-play. England, under the Protectorate of Oliver Cromwell, and France, under the rule of Louis XIV, were allied against Spain. They won the Battle of the Dunes in 1658 under the command of the great French general Turenne, and by a treaty of the previous year, England was given Dunkerque and Mardyck in 1659. It was garrisoned by a curious mixture of soldiers from Cromwell's New Model Army and Royalists who had served Charles II in exile, Puritans and Catholics! Cromwell died that year and in the chaos that followed Charles II was restored to the English throne in 1660. Charles was chronically short of money and in 1662 he agreed to sell Dunkerque to Louis for 5 million livres. As a French frontier town, it joined the string of those fortified by Vauban. There was an even more pressing need for security in Dunkerque because it was the base from which Jean Bart, the French pirate, operated and Louis was keen to protect his profitable piracy. The walls were destroyed during the eighteenth century and the town expanded, becoming ultimately the third largest port in France. It was badly damaged in both World Wars. Unlike Ypres and Bergues it was reborn as a modern port.

Furnes - Veurne

The other Belgian example is Veurne (Furnes), a small town not far from the coast. It was founded in 870 by Baudouin (Baldwin) I, known as Iron-Arm, who was the first Count of Flanders. Looking at his story, we see again how important Flanders was to the balance of power in Northern Europe at this time. He was a member of the court of the Frankish king, Charles the Bald, and in 861 he eloped with the king's

1.3 The station at Dunkerque before 1914, with an electric standard gauge tram in the forecourt. (*Authors' collection*)

daughter, Judith. She had already been married for two years each to two kings of Wessex, a father and son, Æthelwulf and Æthelbald. After intercession by the Pope, Charles forgave them. They were welcomed back to court, officially married and sent to rule in Flanders, protecting France from Viking attack. His successors were successful in this and beat the Vikings soundly in 890. Veurne was another of the many Flemish towns which became prosperous from trade with England. When this failed in the late thirteenth century the town went into a decline. France was always an enemy and in 1297 there was a battle at Furnes, with the French victorious. For the next three centuries Veurne suffered the power struggles we have seen in our other examples and succumbed to the same fate under Louis XIV. Vauban improved the town's fortifications. When Austria gained control in the eighteenth century the walls suffered a similar fate to those at Ypres, but in the case of Furnes the job was completed. All was quiet from Waterloo (1815) until the First World War except for being the first Belgian town to welcome Leopold I, first King of the Belgians, in 1831. In the First World War Furnes was central to the Yser Pocket, the area where the Belgians held out, and it was an important hospital centre. For this in 1920 it was given the Croix de Guerre by M. Poincaré, the French President.

Development of railways in the Ypres sector

The main network of standard and metre gauge lines in the Ypres sector, in France and Belgium, is shown in Figure 1.1 as it was by August 1914. This includes all the major battlefield areas of the war in this sector, but not some of the supply lines further west, in the northern part of the Pas-de-Calais *département*. Figure 1.1 does not show all of the communities in this area, only the larger ones, and those of most relevance to the railway network. In addition, the urban tramways of Dunkerque, Ostende (Oostende), Armentières, and the northern part of the Lille conurbation (Roubaix and Tourcoing) are not shown. For more detail of the metre gauge railways in French Flanders see Chapter Two, and for those in Belgium see Chapter Three.

In Table 1.1 the standard gauge lines in the area of interest of this book are listed, up to 1914. The exception is the line from Adinkerke (later De Panne) to Poperinghe (Poperinge). This was built by Belgian and British Military Engineers in 1915. From 1921 it was opened for civilian traffic. This line is included because it was the only military line which was retained after the war for civilian use. Also, it was the only substantial new civilian standard gauge line in this area from the end of the war until the opening of the *Ligne de Grande Vitesse Nord* from Calais and the Channel Tunnel to Lille in 1993. In Figure 1.1, standard gauge lines leaving the Ypres Sector are also shown, even though they are not listed in Table 1.1.

The three sections of Table 1.1 relate to the next three sections of text, dealing with standard gauge lines in France, lines across the French-Belgian border, and lines in Belgium. The national line numbers are given with each line. For the lines across the border, the French numbers are given first.

Standard gauge lines in France (all of *Intérêt Général*)

In France, the definitions of lines of *Intérêt Général* and those of *Intérêt Local* were codified in the *Loi Migneret* of 1865. Lines of *Intérêt Général* were those of sufficient length, importance, or strategic worth to be at least partially a charge on the State and were administered by the Ministry of Public Works in Paris. Those of *Intérêt Local* were administered by the *département* concerned and were (and are still) a responsibility of the *Préfet* (Prefect). The Chief Engineer of *Ponts et Chaussées* (bridges and highways) for the *département* reported to the *Préfet* and was also responsible for railways. Local engineers were based in the chief towns. Unless otherwise qualified, the terms 'Chief Engineer', 'Local Engineer' or 'Engineer' in this book refer to these departmental employees, not to employees of the operating companies. Although the right to build and operate the railways was conceded to companies, the State and the *départements* kept very firm control over them.

The Freyciney Plan in 1879–80 further encouraged the development of lines of *Intérêt Local*. Within this category, tramways were also defined. These were lines built at least 70 per cent in roads, or on the verges of roads. For a full discussion of all these decisions and their effects we recommend Chapter One of *Minor Railways of France* (W.J.K. Davies, Plateway Press, 2000).

A businessman or a company would usually be the originator of plans for railways of *Intérêt Local*, frequently with local encouragement. If found suitable after public and other enquiries, the proposed line would be declared *d'utilité publique* ('in the public interest'), and the concession to build and operate the line would be granted to the businessman or company for a fixed number of years. A *Compagnie* or *Société Anonyme* (limited company) would be formed at or before this stage. At the end of its life, local agreement was sufficient to close a line, but the formal decree of *déclassement* (declassification) had to come from the office of the President of France and would be published in the official journal of the Republic.

Introducing the Ypres sector and its railways

Table 1.1 Standard gauge lines in the Ypres sector (where no date of closure given, lines are still open)

Line & line number	via	type	opened	closed pass.	closed goods	electrified	length (km)	notes
Lines in France (all of *Intérêt Général*)								
1 Paris - Lille 272	Amiens, Arras	DT	1846			1958	251	now SNCF
2 Lille - Calais 295	Hazebrouck, St-Omer	DT	1848 (1)			Y(2)	107	now SNCF
3 Hazebrouck - Dunkerque 301	Cassel, Bergues	DT	1848			1962	41	now SNCF
4 Dunkerque - Calais 304	Bourbourg	ST	1876			2014	47	now SNCF
5 Watten - Bourbourg 303		ST	1888	1939	1958 (3)	N	15	*déclassée* 1954/1967
6 Tourcoing - Halluin 268		ST	1879	1971		N	11	*déclassée* 1986
7 Lille (La Madeleine) - Comines (F) 296		ST	1876	2019		N	16	(4)
Lines across the French-Belgian frontier								
8 Lille - Courtrai (Kortijk) 278/75	Tourcoing (F), Mouscron (B)	DT	1842			1982	30	now SNCF/Infrabel SNCB
9 Halluin (F) - Menin (Menen) (B) 268/71		ST	1879	1939		N	3	extension of 6 (above) (5)
10 Comines (F) - Comines (Komen) (B) 296/70		ST	1876	NK		N	2	extension of 7 (above)
11 Armentières (F) - Comines (B) 298/67	Houplines (F)	ST	1870	1988		N	13	*déclassée* 1991 (6)
12 Hazebrouck - Poperinghe 299/69	Godewaersvelde (F), Abeele (B)	ST	1870	1954	1990	N	16	(7)
13 Dunkerque - Adinkerke (De Panne) 300/73	Bray-Dunes-Ghyvelde (F)	ST	1870	2003	2003	N	18	(8)
Lines in Belgium								
14 Courtrai (Kortijk) - Bruges (Brugge) 66	Roulers (Roeselare), Thourout (Torhout)	DT (9)	1846-47			Y (date NK)	53	now Infrabel/SNCB
15 Bruges (Brugge) - Ostende 50A		DT	1838			1954	22	now Infrabel/SNCB
16 Thourout (Torhout) - Ostende 62		DT (10)	1868	1963	1963		25	*déclassée* 1984
17 Menin (Menen) - Roulers (Roeselare) 65		ST	1889	NK	NK		44	
18 Poperinghe - Courtrai (Kortrijk) 69	Ypres, Comines, Menin	ST/DT	1853/54			1987-92		now Infrabel/SNCB
19 Ypres - Roulers (Roeselare) 64		ST	1868	1953	1953		32	(11)
20 Ypres - Thourout (Torhout) 63	Staden	ST	1873	1955	2003		87	
21 Adinkerke - Gent (Gand) 73 & 72	Dixmude, Thielt (Tielt)	DT/ST	1855-80			1996		now Infrabel/SNCB
22 Dixmude - Nieuport 74	Pervyse	ST	1868	1952			18	*déclassée* 1974
23 Adinkerke (De Panne) - Poperinghe 76	Rousbrugge	ST	1920 (12)	1934	1939		34	taken up 1942

DT double track
ST single track - where both stated there are sections of each, predominant stated first
NK not known

(1) To provisional terminus at St-Pierre-lès-Calais, extended to Calais-Ville in 1888-9.
(2) Electrified Lille - Hazebrouck 1964, Hazebrouck - Calais 1993
(3) Closed for goods 1940 between Watten-Éperlecques and St-Pierre-Brouck
(4) May be re-opened as a tramway
(5) *déclassée* 1975
(6) Joined Line from Poperinghe to Courtrai (18) at Comines (B)
(7) Closed for local passengers Poperinghe - Abeele 1950. Closed goods Caëstre - Poperinghe 1970, Hazebrouck - Caëstre 1990 or shortly before.
(8) Closed Bray-Dunes -Adinkerke 1958-1960, Dunkerque - Bray-Dunes 1994, re-opened for passengers and goods 1999-2003
(9) Originally ST, date doubled not known
(10) ST until 1895
(11) Closed in part in 1950
(12) Opened as military line 1915

SNCF *Société Nationale des Chemins de Fer Français*
SNCB *Société Nationale des Chemins de Fer Belges*

1.4 The *Compagnie du Nord* station at Hazebrouck, viewed from the west approach. The metre gauge line from Hondschoote and Bergues is on the left (see Figure 10.1). The standard gauge lines towards the photographer are to St-Omer and Calais, and to Bergues and Dunkerque. Postcard postmarked 1905. (*Authors' collection*)

Usually, standard gauge lines were of *Intérêt Général*, and those of metre gauge and narrower gauges were of *Intérêt Local*. The *Compagnie des chemins de fer du Nord*, formed in 1845, was eventually responsible for almost all the standard gauge lines in the area of France covered by this book. In 1937, French railways were nationalised under the *Société Nationale des Chemins de fer Français* (SNCF).

Standard gauge lines across the French - Belgian Frontier

Those relevant to the Ypres Sector are listed in Table 1.1. The line from Ypres in Belgium via Poperinghe to Hazebrouck, of particular importance during hostilities, was built and operated by the Belgian *Flandres Occidentale* Company. It opened on 10 June 1870.

1.5 The track side of the standard gauge station at Ypres before 1914. (*Authors' collection*)

1.6 The road side of the standard gauge station at Furnes (Veurne) before 1914. A metre gauge SNCV train in the forecourt is headed towards La Panne (line 115) or Nieuport and Ostende (line 2). The SNCV depot is to the right at the end of the forecourt. (*Authors' collection*)

Standard gauge lines in Belgium

On 5 May 1835, Belgium opened the first railway in continental Europe, from Brussels to Mechelen (Malines), using locomotives imported from the Stephenson Company of Darlington, UK. By 1840, the main lines to the west extended to Ostend and were partly completed to Courtrai.

Initially, the lines were built and opened by the *Chemins de fer de l'État Belge* (Belgian State railways), but soon many private companies were formed. Table 1.1 includes the lines in Belgium (14 to 23 in the Table) relevant to this book. They were all constructed before 1914, with the exception of the line from Adinkerke (later renamed De Panne) to Poperinghe (number 23).

In 1926, all Belgian standard gauge lines were nationalised under the *Société Nationale des Chemin de Fer Belges* (SNCB), in Dutch the *Nationale Maatschappij de Belgische Spoorwegen* (NMBS). Some previously single track lines were doubled. Electrification started in 1935 and the majority of remaining lines are now electrified. However, after the opening of the Adinkerke to Poperinghe line for civilian use in 1920, there were no new long distance standard gauge lines in the area of this book until the opening of the *Ligne de Grande Vitesse* from Paris to Lille and Calais in 1993, extended from Lille to Brussels in 1997.

Metre Gauge lines

In this book we have only examined in detail the metre gauge lines in or closely involved with the battlefields of the Ypres Sector. This includes those in the forward area, and in the back and support areas. It does not include details of lines in the supply areas from the Channel Ports, except for connections. Information on other metre gauge lines in the Nord and Pas-de-Calais *départements* of France, and in the West Flanders Province of Belgium, can be found in books in the bibliography. Metre gauge lines in the supply areas of the Pas-de-Calais *département* are covered in detail in *Tortillards of Artois. The Metre Gauge railways and tramways of the Western Pas-de-Calais* (Oakwood Press, 2008), by the present authors.

Before 1914 there was one metre gauge line which crossed the border. The Belgian line south from Ypres via Kemmel and Neuve Église (Nieuwkerke) ran into France for less than 2km (about 1 mile) to terminate at Steenwerck station. This is actually at La Crèche, on the main line between Armentières and Hazebrouck. This line is described in Chapter Three.

Chapters Two and Three deal with the metre gauge lines of the Ypres sector up to 1914, those in France in Chapter Two and those in Belgium in Chapter Three. The metre gauge lines during the war are described in Chapters Five to Eight, with the light railways. In Chapters Ten and Eleven the story of the metre gauge railways of these areas of France and Belgium is followed from 1919 to closure.

Chapter Two

The metre gauge railways of French Flanders and related lines and tramways 1894 to 1914

The railways of French Flanders, in August 1914, are shown in Figure 2.1. This shows the three main metre gauge lines of this area, Hazebrouck to Hondschoote, with Rexpoëde to Bergues; Hondschoote to Bray-Dunes-Plage; and Herzeele to St-Momelin with Bergues to Bollezeele. The electrified tramway at Cassel is also shown. The Drincham to Bourbourg line, under construction at the outbreak of hostilities but never completed, is shown. The Armentières to Halluin line is discussed later in the Chapter and is shown in Figure 2.2.

The potential importance of the extension of the lines originating from Herzeele and Bergues to St-Omer, which were never built, is discussed on page 18, with the wrangling between operating companies which delayed them. For this reason, the proposed connecting lines from St-Momelin to St-Omer, St-Omer to Tournehem, and Tournehem to Audruicq are also shown in Figure 2.1. Tournehem was on the line from Anvin to Calais, opened in 1881-82. The link from Tournehem to the St-Momelin line was built in 1918, as a military line following a different route (see Chapter Eight, page 177).

Hazebrouck to Hondschoote, and Rexpoëde to Bergues

The lines from Hazebrouck to Hondschoote via Herzeele and Rexpoëde and from Bergues to Hondschoote via Rexpoëde were run by the *Compagnie des Chemins de fer des Flandres* (CF) owned by Alfred Lambert. This company was, from 1919, part of the *Compagnie Général des Voies Ferrées d'Intérêt Local* (CGL-VFIL). The story began in 1883, when the Nord and Pas-de-Calais *départements* wanted to concede to the *Société d'Entreprises Générales* a narrow gauge line of *Intérêt Local* from Lille to Dunkerque with branches to Bergues and Hondschoote. The state, perhaps influenced by the *Compagnie du Nord* (CdN), rejected that proposal and undertook new less ambitious studies. After CEN (*Compagnie des Chemins de Fer Économique du Nord*) declined the project, it was Alfred Lambert who obtained the concession in 1890 for a line from Hazebrouck to Hondschoote with a branch to Bergues, declared *d'utilité publique* in 1891. The *Compagnie des Chemins de fer des Flandres* (CF) was formed to construct and operate the line for the 99 years of the concession, with the assistance and support of the Anvin-Calais company.

2.1 The 'Lambert' type station at Steenvoorde, with a train from Hazebrouck, probably a mixed train, headed towards Rexpoëde and Hondschoote. Postcard postmarked 1905. *(Collection Jean Willig)*

Building and opening

The line from Hazebrouck to Hondschoote, and the branch from Rexpoëde to Bergues (Porte de Cassel), were opened on 8 September 1894. At Bergues, the closeness to the fortified walls required talks with the military authorities and the *Compagnie du Nord*, which delayed the opening of the last section to Bergues-Gare until 1 July 1897. A special train ran for the ceremonial 'opening' on 17 August 1894 with a 'commission of reception'. After this, on 4 September, the Prefect approved the opening from Berques Porte de Cassel to Hazebrouck, with a branch to Hondschoote. This *service assuré* was run by the Anvin-Calais Company and the arrangement was regularised on 24 February 1895.

Description of the lines

The line from Hazebrouck to Hondschoote began at the CdN station in Hazebrouck on the main line from Lille to Calais. Branching off to the north, it served Steenvoorde, halfway between the Mont de Cassel and the Belgian border, and then Herzeele where, after 1910, the line of the *Société Générale des Chemins de Fer Économiques* (SE) to Esquelbecq branched off. After crossing the Yser the line arrived at Rexpoëde, the junction to Bergues. Going on to Hondschoote, it connected with trains to Bray-Dunes run by the *Compagnie des Chemins de Fer d'intérêt local du Nord de la France* (NF). The length of the line from Hazebrouck to Hondschoote was 34km.

The branch from Rexpoëde to Bergues was 9km long. It ran east–west and served a *huilerie* before the station at Warhem. Near the Porte-de-Cassel at Bergues the ditches were crossed by two metal bridges. The line then ran alongside the standard gauge line from Lille to Dunkerque and into Bergues CdN station. From 1914 the line to Bollezeele and St-Momelin of the SE left from the other side of the station but there was no connection between the two lines.

Engineering

The line ran in its own path throughout, with steel Vignoles rails of 20kg/m and sleepers at 90cm intervals. The maximum gradient was 2.5 per cent, with curves of minimum 100m radius. The line crossed the Yser river just south of Bambecque on a *tablier métallique* bridge. Near Bergues Porte-de-Cassel the ditches surrounding the fortifications were crossed by two metal bridges. The remains of one of these are shown in the picture on page 229 (Chapter Twelve).

2.2 The 'Lambert' type station at Rexpoëde, with goods hall, from the road side. Undated postcard. This was the junction of the line from Bergues with that between Hazebrouck and Hondschoote. (*Authors' collection*)

Stations and other stops

These are shown in Table 2.1. The distances are from the *Chaix* timetable of 1914. These only show completed kilometres, and for this reason sometimes the distance is the same if the stops are less than one kilometre apart. Likewise, the distances between stops are sometimes apparently different, depending on the point of origin from which the distance is measured. This applies to all the station lists in this chapter.

Buildings

The stations are similar to those on other lines associated with M. Lambert – the Aire to Berck line in the Pas-de-Calais *département*, and the lines from Noyon to Ham and to Montdidier, and from Milly to Formerie, in the Somme and Oise *départements*. The passenger buildings were of two storeys, with three openings, doors or windows, on each side on the ground floor, and three windows on each side on the first. They were built of brick with minimal corner decoration. This was echoed in the decoration under the end eaves. There were tall thin rectangular windows in the end walls at attic level but no other end windows or doors. The goods buildings were attached to one end of the passenger buildings, with overhanging roofs, and loading platforms on both the track and the road side. There were separate lavatory blocks of brick also with ridged tiled roofs.

Track Layouts

At Hondschoote there were five platform lines. The centre one was used for the service to Bray-Dunes run by the NF Company. We have seen a partial track plan of the station at Bergues by 1914. This showed standard gauge lines with the SE line (opened 1914) to Bollezeele and St-Momelin on the other side of the station. The layout in 1933 of the station shared with the CdN at Hazebrouck is shown in Figure 10.1, and that at Bergues in 1948 in Figure 10.2.

Depots

The depot and workshops were at Hondeghem which was 4km from Hazebrouck. In February 1894, CF requested the CdN to provide inspection and maintenance for their rolling stock. For this, CdN put a third rail on one track at their depot at Hazebrouck.

Table 2.1 Hazebrouck - Hondschoote - Bergues
Stations and other stops, 1894 to 1914
Compagnie des Chemins de fer des Flandres (CF).

Name	Type	Distance km		Altitude m	(ft)
		From Chaix 1914			
		fr Hazebrouck	fr Hondschoote		
Hazebrouck	CdN shared	0	34	29	(95)
SG lines Lille - Calais, Paris - Arras - Béthune - Dunkerque, Hazebrouck - Merville (IL), Hazebrouck - Poperinghe (Belgium)					
Le Pont-Rommel	arrêt (1)	0	34		
Wecke-Meule-W	arrêt (1)	3	31		
Hondeghem	station	4	30	42	
Le Korten-Loop	arrêt (1)	6	28		
St-Sylvestre-Cappel	station	8	26	55	(180)
Terdeghem	arrêt (1)	11	24	33	
Steenvoorde	station	12	22	26	
Winnezeele	station	16	18	25	
Herzeele	station	22	12	14	
junction with line to Esquelbecq (from 1910) & St-Momelin (from 1912) (SE)					
Bambecque	station	24	10	8	
Rexpoëde	station	29	6	21	
Killem	station	31	3	10	
Hondschoote	station	34	0	6	(20)
origin of MG line to Bray-Dunes (from 1903) (NF)					
Rexpoëde	station	29	6	21	(69)
Le Rattekot	arrêt (1)	29	8		
Warhem	station	33	10	15	
La Maison-Rouge	arrêt (1)	33	12		
Bergues	arrêt (1)	37	14		
Bergues (2)	CdN shared	37	15	2	(7)
SG line Hazebrouck - Dunkerque					
Origin of MG line to St-Momelin SE (from 1914)					

CdN	Compagnie du Nord	SE	Société générale des Chemins de fer Économiques
IL	Intérêt Local	NF	Compagnie des Chemins de fer d'intérêt local du Nord de la France

(1) *arrêt* only open for the service of passengers without baggage
(2) from 1897

2.3 The track side of the station at Rexpoëde, with a train arriving from either Hazebrouck or from Bergues. Picture taken before 1914. Postcard written on 3 July 1915, by a Frenchman based at the station, possibly a military engineer. The locomotive is no. 32, 0-6-2T Corpet constructed in 1890 or 1891, and put in service on this line in 1894. (*Authors' collection*)

Industrial links

There was a link to a *distillerie* at Rexpoëde and the branch from Rexpoëde to Bergues served a *huilerie* near the station at Warhem.

Rolling stock

The rolling stock was provided with a single central buffer and had a coupling below. There were continuous vacuum brakes for locomotives and passenger carriages only.

2.4 The 'Lambert' type station at Hondschoote. Picture taken before 1914. Postcard written on 16 September 1915. The locomotive on the right is no. 31, 0-6-2T Corpet constructed in 1890 or 1891, and put in service on this line in 1894. The number of the similar locomotive on the left is obscured by a man. These trains are headed for Hazebrouck and Bergues, on the CF lines. The train in the middle, of which the rear is seen, is probably headed for Bray-Dunes-Plage on the NF line, opened from here in 1903. This is probably the scene in mid-morning, preparatory for the departures to all these destinations between 10.10 and 11.10 am (Table 2.3). (*Authors' collection*)

2.5 The 'Lambert' type station at Warhem, on the line between Rexpoëde and Bergues, with a passenger train headed by locomotive no. 01. This 0-6-2T Corpet-Louvet of 1890 had been transferred in 1912 from the Estrées-Froissy line in the Oise *département*, a line also associated with M. Alfred Lambert. (*Authors' collection*)

Steam locomotives

The line opened in 1894 with 5 light Corpet-Louvet 0-6-2Ts. They were 16 tonnes empty, 21 tonnes in running order. They were 2.25m wide and 3.20m high. They were all constructed between 1890 and 1891 for the Aire to Berck line but they were insufficiently powerful and were transferred to Flanders at various dates. Numbers 31 to 33, works numbers 523 to 525, were manufactured in 1890, and were delivered in 1894 after a brief spell on the Aire to Berck line. Number 34, works number 528, also came from the Aire to Berck line but was immediately sent to the Oise network, another Lambert concession. In 1912 the Estrées to Froissy line, part of this network, sent Number 01 (works number 509 of 1890) to Flanders. Also in 1912, Number 36, works number 530, was transferred from the Aire to Berck line.

Passenger carriages

The line originally had three mixed first and second class bogie coaches plus three, two axle, four wheel coaches for second class. Two more bogie coaches were delivered in 1912.

Goods wagons

The original stock of wagons of 1893 comprised:

- 5 *fourgons* for baggage
- 14 covered wagons
- 49 open wagons
- 24 flat wagons with light balustrades, either open or solid sided. Some of these were specially adapted for the transport of sugar beet.
- 4 flat wagons with mobile struts or crosspieces.

These were supplemented in 1911-12 to a total of 101 wagons.

Operations

Services were run between Hondschoote and Hazebrouck, and between Hondschoote and Bergues. This provided a double service between Hondschoote and Rexpoëde. The *Cahier des charges* shows a minimum of three trains each way per day throughout in two classes. Prior to 1905 there were three trains each way per day on each line. From 1905 to 1914 there were four trains each way daily from Hazebrouck to Hondschoote and four from Bergues to Hondschoote, plus one shuttle from Rexpoëde to Hondschoote.

Timetables

The timetable for May 1914 is shown in Table 2.2. This confirms that by May 1914 there were four trains each way per day over both routes. Most trains started from and ended at Hondschoote, but one appears to originate at Rexpoëde. There was a locomotive shed at Hondschoote, which was probably the main base for

The metre gauge railways of French Flanders and related lines and tramways 1894 to 1914

Table 2.2 Hazebrouck - Hondschoote & Hondschoote - Bergues
Timetable from May 1914 (main stops only, excludes *arrêts*)
***Compagnie des Chemins de fer des Flandres* (CF)**

Hazebrouck (CdN)		08.29	12.45	15.51	19.01
SG lines Lille - Calais, Paris - Arras - Béthune - Dunkerque, Hazebrouck - Merville (IL), Hazebrouck - Poperinghe (Belgium)					
Hondeghem		08.41	12.57	16.04	19.14
St-Sylvestre-Cappel		08.50	13.06	16.16	19.25
Steenvoorde		09.04	13.21	16.33	19.41
Winnezeele		09.14	13.31	16.45	19.52
Herzeele		09.25	13.45	17.06	20.04
junction with line to Esquelbecq (from 1910) & St-Momelin (from 1912) (SE)					
Bambecque		09.32	13.51	17.12	20.10
Rexpoëde (1)		09.45	14.03	17.26	20.25
Killem		09.52	14.10	17.34	20.33
Hondschoote		09.58	14.16	17.40	20.39
origin of MG line to Bray-Dunes (Plage) (NF)					
Hondschoote		07.23	11.10	14.30	18.00
Killem		07.30	11.17	14.37	18.07
Rexpoëde (1)		07.36	11.28	14.45	18.18
Warhem		07.53	11.43	14.58	18.33
Bergues (CdN)		08.08	11.58	15.12	18.47
SG line Hazebrouck - Dunkerque, origin of MG line to St-Momelin (from August 1914) (SE)					

		(2)	(3)			
Bergues (CdN)		08.28	10.16	12.40	15.30	19.52
Warhem		08.44	10.32	12.55	15.45	20.07
Rexpoëde (1)	06.40	09.00	10.45	13.10	15.58	20.25
Killem	06.48	09.07	10.52	13.17	16.05	
Hondschoote	06.54	09.13	10.58	13.23	16.11	
origin of MG line to Bray-Dunes (Plage) (NF)						
Hondschoote	06.00		10.10	13.46		16.30
Killem	06.07		10.17	13.53		16.37
Rexpoëde (1)	06.18		10.26	14.05		16.46
Bambecque	06.30		10.38	14.17		16.58
Herzeele	06.37		10.44	14.23		17.09
junction with line to Esquelbecq (from 1910) & St-Momelin (from 1912) (SE)						
Winnezeele	06.52		10.56	14.35		17.30
Steenvoorde	07.08		11.08	14.50		17.39
St-Sylvestre-Cappel	07.22		11.19	15.01		17.46
Hondeghem	07.36		11.28	15.10		17.55
Hazebrouck (CdN)	07.48		11.39	15.21		18.06
SG lines Lille - Calais, Paris - Arras - Béthune - Dunkerque, Hazebrouck - Merville (IL), Hazebrouck - Poperinghe (Belgium)						

CdN   *Compagnie du Nord*
SG   Standard Gauge
NF   *Compagnie des Chemins de fer d'intérêt local du Nord de la France*
SE   *Société générale des Chemins de fer Économiques*

(1) Junction of lines to Hazebrouck - Hondschoote and Hondschoote - Bergues
(2) Except Mondays (Monday market day in Bergues)
(3) Mondays only (Monday market day in Bergues)

(departure times shown, except at end of journey, sometimes arrivals a few minutes earlier)

operating locomotives, and the depot at Hondeghem was most likely a repair and maintenance facility rather than an operating base.

The journey time between Hazebrouck and Hondschoote was about an hour and a half except for one run which took twenty minutes more. We assume that this was a mixed passenger and goods train and the extra time was needed to pick up and drop wagons. A similar pattern is seen on the return journey from Hondschoote to Hazebrouck. The best average speed was 22.4kph (14.3mph). The journey time between Hondschoote and Bergues was about 45 minutes, and the best average speed was 21.4kph (13.3mph).

Connections

These are shown on the list of stations (Table 2.1). The connection times at Hondschoote and Rexpoëde in May 1914 are shown in Table 2.3. Passengers to and from Hazebrouck or Bergues towards Bray-Dunes would change at Hondschoote. Passengers between Bergues and Hazebrouck could just as well change at Rexpoëde. Table 2.3 shows good connections mid-morning in all

Narrow Gauge in the Ypres Sector

Table 2.3 Connections at Hondschoote and Rexpoëde May 1914
Compagnie des Chemins de fer des Flandres (CF)
Compagnie des Chemins de fer d'intérêt local du Nord de la France (NF)

Hondschoote								
from Bray-Dunes (Plage) NF		07.08	10.06			16.14 (1)	18.06 (2)	20.35 (3)
from Hazebrouck CF			09.58		14.16		17.40	20.39
from Bergues CF	06.54		09.13 (4)	10.58 (5)	13.23	16.11		
to Bergues CF		07.23		11.10	14.30		18.00	
to Hazebrouck CF	06.00			10.10	13.46	16.30		
to Bray-Dunes (Plage) NF		07.20		10.37		17.50 (1)	18.11 (2)	20.50 (3)
Rexpoëde								
from Hazebrouck CF			09.45		14.03		17.26	20.25
from Bergues CF			09.00 (4)	10.45 (5)	13.10	15.58		20.25
to Bergues CF		07.36		11.28	14.45		18.18	
to Hazebrouck CF		06.18		10.26	14.05	16.46		

(1) Except Saturday, and Friday if Saturday is a public holiday
(2) Saturday only, and Friday if Saturday is a public holiday
(3) Sundays and festivals from 1 July to 15 September 1914 (Friday market day in Hondschoote)
(4) Except Mondays (Monday market day in Bergues)
(5) Mondays only (Monday market day in Bergues)

directions at Hondschoote and at Rexpoëde. The only exception to this was on Mondays, which was market day in Bergues, when the train ran later and passengers from Bergues missed all the connections. There were reasonable connections between the Hazebrouck to Hondschoote and Bergues to Hondschoote lines in early afternoon at either Hondschoote or Rexpoëde, and between all lines in either late afternoon or early evening.

Fares

The fares from Hazebrouck to Hondschoote were first class 2fr 65 cent, second class 1fr 95 cent. The fares from Hazebrouck to Bergues were first class 2fr 85 centimes, second class 2fr 10 centimes.

Hondschoote to Bray-Dunes-Plage

When in 1892 the establishment of a line from Hondschoote to Bray-Dunes was proposed, it was natural to look to Lambert, already concessionaire of the Flanders network under construction as far as Hondschoote. But he could not obtain the necessary financial guarantees and the *département* turned to another entrepreneur of secondary railways, Mathieu Michon. He received the concession for this line, declared *d'utilité publique* in 1902. The line was run by the NF. They ran other lines of *intérêt locale* in Aisne and Somme and a standard gauge line in Nord. Unusually this line was assigned to the concessionaire M. Michon personally. When it opened, Michon asked for NF to be substituted, on the grounds that they operated the other lines. Curiously, this simple formality was not completed, and the public authorities only accepted this 27 years later, in 1930, after the line had closed. The line opened on 29 August 1903.

Description of the line

The line was 15.169km long. From its terminus, the CF station at Hondschoote, it served the *cartonnerie* Cartiaux, at Pont-aux-Cerfs, crossing the canal on a bridge, and from there the line crossed the marshy area of les Moëres, the next station. The line reached the hamlet of Ghyvelde, another station, on 10 August 1903. From there it crossed the main road and the canal from Dunkerque to Furnes. Next it met the CdN standard gauge line from Dunkerque to Furnes at the station of Ghyvelde–Bray-Dune, its last in France. The Hondschoote to Bray-Dune line shared this station, stopping on the north side. It then turned north to a terminus near the beach at Bray-Dune-Plage.

Engineering

The line crossed the standard gauge Dunkerque to Furnes line on the level, at the east end of Ghyvelde-Bray-Dunes CdN station. There was a bridge over the canal de Basse-Colme at the Pont-aux-Cerfs *cartonnerie*. and a bridge over the canal from Dunkerque to Furnes.

Stations and other stops

These are shown in Table 2.4

Buildings

There was a grandiose terminus building at Bray-Dunes-Plage. Other stations were two storey, but very simple.

Industrial link

At Pont-aux-Cerfs, the *cartonnerie* Cartiaux, built on the side of the canal de Basse-Colme, was served by the line. In the area of les Moëres there was a branch to a *distillerie* only a few hundred metres from the Belgian frontier.

The metre gauge railways of French Flanders and related lines and tramways 1894 to 1914

Table 2.4 Bray-Dunes - Hondschoote
Stations and other stops, 1903 to 1914
Compagnie des Chemins de fer d'interêt local du Nord de la France (NF)

Name	Type	Distance km		Altitude m	(ft)
		From Chaix 1914			
		fr Bray-Dunes	fr Hondschoote		
Bray-Dunes (Plage)	special station	0	16	8	(26)
Ghyvelde (Bray-Dunes) (Nord)	CdN shared	3	14	7	
SG line Dunkerque - Dixmude (Belgium)					
Ghyvelde (Ville)	station	4	12	4	
Les Moëres	station	11	5	0	
Hondschoote	CF station	16	0	6	(20)
origin of MG line to Rexpoëde, Hazebrouck and Bergues CF (from 1894)					

CdN *Compagnie du Nord*
CF *Compagnie des Chemins de fer des Flandres*

Rolling stock

This was chosen to be compatible with the neighbouring network of CF. It had single central buffers with couplings below and continuous air braking. The line had two Corpet-Louvet 0-6-0T steam locomotives. No. 1 was delivered new in 1903 and it had a works number 808. The line also had No. 2, which apparently came from another network. There were two mixed first and second class bogie carriages, one *fourgon* for baggage, and 25 goods wagons.

Operations

There were up to 1914 three trains each way per day, with a fourth one on Sunday evening.

2.6 A train consisting of a passenger carriage and *fourgon* at the terminus at Bray-Dunes-Plage waiting to leave for Hondschoote. The locomotive is 0-6-0T Corpet-Louvet no. 2, acquired from another network. (*Collection Jean Willig*)

Timetables
These are shown in Table 2.5. The journey from Bray-Dunes-Plage to Hondschoote took about an hour except one on Sunday which took 35 minutes. The trains ran more frequently in summer. The best average speed was 26kph (15mph).

Connections
These are shown in Table 2.4 (stations). Connections to and from other metre gauge services at Hondschoote have already been discussed under the Hazebrouck to Hondschoote and Bergues lines (CF), see Table 2.3.

Fares
The fares were first class 1fr 25 centimes, second class 90 centimes for the whole length of the line

Herzeele - St-Momelin, Bergues - Bollezeele
This third and most recent network of the French Flanders metre gauge system was never fully completed because of the war. The project was studied from 1892, but negotiations held over about 15 years with potential candidates were unfruitful. An initial concession was made with CEN in 1904 but they were too heavily committed to works elsewhere in France. Therefore, in 1905 the *département* granted the concession to the SE on 27 July 1907. It was to be part of the northern network of SE.

Building and opening
The line from Herzeele to Esquelbecq was opened for passengers on 23 June 1910 and for goods on 15 January 1911. The line from Esquelbecq to St-Momelin opened on 21 October 1912. The line from Bergues to Bollezeele was opened on 3 August 1914, the day war was declared between France and Germany. A branch to Bourbourg leaving the Bergues to Bollezeele line at Drincham was under construction at this time. Works were never resumed after the war.

Description of the lines
The line from Herzeele to St-Momelin ran for 30km. From the junction at Herzeele with the Hazebrouck to Hondschoote line of CF, the line travelled west through the large village of Wormhoudt and crossed the Standard Gauge line from Hazebrouck to Dunkerque at Esquelbecq. The line went on to Bollezeele where it had its technical centre. On its way to St-Momelin the line ran partly on its own track and partly on the verge of the road.

The section from Bergues to Bollezeele ran for 20km. It started at the CdN station at Bergues, on the opposite side to the CF line from Rexpoëde. Transhipment facilities with CdN existed but there was no link between the two networks. The line went south across an area of canals and marshes to Drincham where the line to Bourbourg should have branched off. It then ran to Bollezeele to meet the line from Herzeele.

Proposed extension to St-Omer
At St-Momelin, the track continued to the Aa canal, but the line never crossed the canal to reach St-Omer. A line had been proposed to run from St-Omer to Tournehem in the Pas-de-Calais. The concessionaire was M. Émile

Table 2.5 Bray-Dunes - Hondschoote
Timetable from May 1914
Compagnie des Chemins de fer d'intérêt local du Nord de la France (NF)

			(1)	(2)	(3)
Bray-Dunes (Plage)	06.12	09.18	15.10	17.16	20.00
Ghyvelde (Nord)	06.26	09.31	15.30	17.28	20.05
SG line Dunkerque - Dixmude (Belgium)					
Ghyvelde (Ville)	06.35	09.39	15.39	17.35	20.12
Les Moëres	06.57	09.56	16.02	17.54	20.25
Hondschoote	07.08	10.06	16.14	18.06	20.35
origin of MG line to Rexpoëde, Hazebrouck and Bergues (CF)					
			(1)	(2)	(3)
Hondschoote	07.20	10.37	17.50	18.11	20.50
origin of MG line to Rexpoëde, Hazebrouck and Bergues (CF)					
Les Moëres	07.32	10.52	18.03	18.21	21.01
Ghyvelde (Ville)	07.49	11.14	18.26	18.43	21.17
Ghyvelde (Nord)	07.58	11.31	18.37	18.53	21.30
SG line Dunkerque - Dixmude (Belgium)					
Bray-Dunes (Plage)	08.01	11.34	18.40	18.56	21.33

SG Standard Gauge
MG Metre Gauge
CF *Compagnie des Chemins de fer des Flandres*

(1) Except Saturday, and Friday if Saturday is a public holiday
(2) Saturday only, and Friday if Saturday is a public holiday
(3) Sundays and festivals from 1 July to 15 September 1914

(Friday market day in Hondschoote)

2.7 The station at Wormhoudt (or Wormhout) on the SE line from Bollezeele to Herzeele. The section from Herzeele to Esquelbecq through Wormhoudt opened in June 1910, and this photograph was taken before May 1911. The pale bricks are characteristic of the Flanders SE lines. Postcard written on 1 March 1915 by a French or Belgian man from Crombeke in Belgium. The train headed towards Herzeele is hauled by locomotive 3.751. (*Authors' collection*)

Level, owner of the Anvin to Calais line. This line was to have been part of Pas-de-Calais proposals for the 'third network' which had been planned at least since 1910. The proposed line would have met the Herzeele to St-Momelin line at the proposed station of Salperwick between St-Momelin and St-Omer. The issue was how should the crucial connexion between this line and the SE line, when it was extended from St-Momelin to St-Omer, be handled. Both these lines are shown in Figure 2.1.

At the *Conseil Générale* of Pas-de-Calais in September 1913 the vote was to proceed with the line, and to regard this as independent of the line from Salperwick to St-Momelin. It was urged that the two lines must be regarded as separate and that the latter should only go ahead when the inter-departmental problems had been ironed out. This decision led to a letter from M. Level in Paris to the Chief Engineer of the Pas-de-Calais *département*, on 20 March 1914. He stated that the St-Omer to Tournehem line is the principal line, and the line to St-Momelin a branch, and therefore the trains from Bergues via St-Momelin must terminate at Salperwick. Alternatively, they must pay to run in to St-Omer, an option he did not favour. Clearly this would have seriously compromised the business case for extending the line from Bergues. In the event the war intervened and neither line was built then or later. The 1918 military line from St-Momelin to Tournehem took a different route (see Chapter Eight).

Engineering

The line was built with Vignoles rails (20kg/m). The maximum gradient was 2.8 per cent, and the minimum radius of curvature 100 metres.

Bridges

There was a bridge over the CdN Hazebrouck to Dunkerque line at Esquelbecq. This was to the north of the station. We have more details of this bridge and other bridges for the lines from Esqulbecq to St-Momelin and from Bergues to Bollezeele from the demolition list drawn up by the 2nd Battalion Canadian Railway

2.8 The station at Wormhout (Wormhoudt) before 1914, with a train headed towards Esquelbecq, Bollezeele or St-Momelin. *(Collection Jean Willig)*

Troops (2CRT) in 1918 (see Table 2.6). Particularly striking are the number of long bridges required in the flat country west of Bergues, with multiple drainage ditches.

Stations and other stops

These are shown in Table 2.7. The distances are from the timetables of 1936 and 1947, which give them to the nearest whole kilometre. In this table, the distances to and from Heerzeele, St-Momelin and Bollezeele are all shown. When the line was fully open from 3 August 1914, the services were between Bergues and St-Momelin via Bollezeele, and between Bollezeele and Herzeele. At some time after the First World War (see Chapter Ten), the services were between Herzeele and St-Momelin via Bollezeele, and between Bollezeele and Bergues; hence the need for all the distances. At Bollezeele and St-Momelin there were 18ft (5.5m)

Table 2.6 Bergues - Bollezeele & Esquelbecq - St-Momelin Bridges (from demolition list July-August 1918).

| *Esquelbecq to St-Momelin* | | | | | |
km (from Herzeele)	location	bridge span ft in	m	type	notes
10.39	Esquelbecq	15'	4.6	brick arch	2 15ft arches
	"	25'	7.6	lattice girder	over standard gauge railway Hazebrouck - Bergues
	"	9'	2.7	rolled steel girder	
11.95	Zeggers-Cappel	3'3"	1.0	brick culvert	
16.83	1km east of Bollezeele	3'3"	1.0	brick culvert	
19.24	near Merckeghem	3'3"	1.0	brick culvert	
21.70	Volkerinckhove	5'0"	1.5	brick culvert	
Bergues to Bollezeele					
km (from Bergues)	location	bridge span ft in	m	type	notes
1.55	Bierne	16'	4.9	2 steel girder bridges, brick abutments	over Nouveau Bieren Dyck 50° 57' 42.80" N 2° 25' 06.61" E Abutments of 1 present over old canal route
1.90	"	16'	4.9	steel girder bridge, brick abutments	over Hout Gracht 50° 57' 41.59" N 2° 24' 44.72" E'
4.01		3'9"	1.1	brick culvert	
4.50		10'	3.0	steel girder bridge, brick abutments	over Canal de Steene 50° 57' 40.27" N 2° 22' 35.97" E' abutments still present
9.60	Pitgam Halt	16'	4.9	steel girder bridge, brick abutments	over Hout Gracht 50° 56' 15.44" N 2° 19' 12.49" E' a bridge still present
12.02		16'	4.9	steel girder bridge, brick abutments	over Deullert Gracht 50° 55' 09.92" N 2° 19' 22.46" E' a bridge still present

The metre gauge railways of French Flanders and related lines and tramways 1894 to 1914

Table 2.7 Bergues - St-Momelin (from 1914)
Herzeele - Bollezeele (from 1910-1912)
Stations and other stops 1914
Société générale des Chemins de fer Economiques (SE)

Name	Type	opened	Distance km			Altitude m	(ft)
			From TTs 1936 and 1947 (with decimals 2CRT 1918)				
			fr Bergues	fr Bollezeele	fr St-Momelin		
Bergues	CdN shared	1914	0	20	32	2	(7)
SG line Hazebrouck - Dunkerque							
Origin of MG line to Rexpoëde, Hazebrouck & Hondschoote CF (from 1897)							
Bierne	station	1914	2.50	17	29	4	
Steene	station	1914	6.40	13	25	3	
(Grand-Mille-Brugge)							
Pitgam	halt	1914				1	
Pitgam	station	1914	10.90	9	21	9	
Drincham	station	1914	14.00	5	17	11	
Eringhem	*arrêt*	1914	17	3	15		
Bollezeele	station	1912	19.00	0	12	47	(154)
Junction with MG line to Herzeele (also SE)							
Volkerinckhove	station	1912	22.80		8	28	
Lederzeele	station	1912	24.90		6	32	
St-Momelin	halt	1912				15	
St-Momelin	station	1912	30.50		0	14	(46)
(St-Momelin wharf)			(30.90)			3	(10)
			fr Herzeele	fr Bollezeele	fr St-Momelin		
Herzeele	station (CF)	1894 (for CF line) 1910 (this line)	0	18	30	14	(46)
junction with MG line Hazebrouck - Rexpoide - Hondschoote & Bergues CF							
Wormhoudt	station	1910	6	12	24	19	
Esquelbecq	halt					18	
Esquelbecq	CdN shared	1910	10.39	8	20	21	
SG line Hazebrouck - Dunkerque							
Zeggers-Cappel	station	1912	12.20	5	17	26	
Bollezeele	station	1912	18.00	0	12	47	(154)
Junction with MG line to Bergues - St-Momelin (also SE)							
Volkerinckhove	station	1912	21.80		8	28	
Lederzeele	station	1912	23.90		6	32	
St-Momelin	halt	1912				15	
St-Momelin	station	1912	29.50		0	14	(46)

CdN	*Compagnie du Nord*		2CRT	2nd Battalion Canadian Railway Troops
CF	*Compagnie des Chemins de fer des Flandres*		SG	Standard Gauge
SE	*Société générale des Chemins de fer Economiques*		MG	Metre Gauge

turntables on a central cast iron bearing. The layout in 1948 of the station shared with the CdN at Bergues is shown in Figure 10.2.

Buildings

These were in a layout different from the normal SE style, but they conformed with the *département's* wish for its technical services to be unified. Overall, these stations do not fit readily into any pattern, having a mixture of elements. Unlike on the SE Somme network, the roof ridge ran parallel to the track. The stations were of pale brick with medium red brick embellishments at the corners and round the windows. They had attractive patterning just below the roof line on the long sides. They were two-storey usually and had a goods hall. On the forecourt side there was a central door with windows either side. There were three windows above. On the track side there were one or two doors on the ground floor with three windows above. The name was on track side.

2.9 A mixed train at Bollezeele, headed towards St-Momelin. Station called Bollezècle on the postcard heading but the station name is clear behind. Postcard postmarked May 1916, but picture taken before 1914. (*Collection Jean Willig*)

We consider these the most attractive metre gauge stations in French Flanders.

Depots
The depot and workshops responsible for maintenance and repairs were at Bollezeele. There was also a locomotive shed with places for nine locomotives (see frontispiece, page ii).

Industrial links
At Esquelbecq there was a goods transhipment area, a line for the loading of *truck porteurs*, and a branch to a *huilerie*. At St-Momelin there was a branch to a *huilerie*.

Rolling stock
This was typical of SE's normal stock. They had a central buffer, connections below this and vacuum brakes. There were adaptations to allow eventually exchanges with the CF network at Herzeele.

Steam locomotives
There were six 2-6-0T locomotives of 24.6 tonnes constructed by Blanc-Misseron-Tubize in 1909 and having the numbers 3.661 to 3.666. They were delivered on 19 January 1909 by the *Ateliers de Construction du Nord de la France de Blanc-Misseron* (ACNF). Their weight empty was 23 tonnes. These were similar to, and numbered in sequence with, the SE locomotives on the Somme network.

Passenger carriages
The line had 16 passenger carriages constructed by Decauville. They had bogies and wooden bodies with end access platforms. There were six mixed first and second class, and ten second class carriages.

Wagons
There were 139 wagons, all four wheel (two axle). These were originally bought in 1910 but were supplemented in 1935. The stock comprised:

5 Fourgons
26 covered wagons
79 open wagons
23 flat wagons
5 flats with mobile traverses
1 mobile crane, moved from the Avesnes-Solesmes line

Operations
Before 1914 there were three trains each way per day between Herzeele and St-Momelin. Note that this was before the line from Bergues to Bollezeele was open.

Timetables
The section from Bergues to Bollezeele was opened on 3 August 1914, the day war was declared. Because of the requirements of military traffic, the civilian service on the existing line and on the newly opened section became two slow mixed trains per day between Herzeele and Bollezeele and two between Bergues and St-Momelin. A timetable from early in the war, probably August 1914, is shown in Table 5.2 (Chapter Five). This was labelled 'May 1914' but cannot be from then, because the Bergues to Bollezeele section is shown as open, and the service on the whole network is reduced to the wartime two trains per day.

Journey times were as follows:

Bergues to St-Momelin 1hr 41mins, or 1hr 31mins
St-Momelin to Bergues 1hr 45mins, 1hr 48mins, or 1hr 36mins

2.10 A train heading towards Esquelbecq crosses the Grande Place at Wormhout (Wormhoudt). This line was opened for passenger traffic in June 1910, and the postcard is postmarked October 1910. (*Collection Jean Willig*)

The fastest average speed was 20kph
Herzeele to Bollezeele 1hr 45mins
Bollezeele to Herzeele 1hr 53mins
The fastest average speed was 15kph

The slower speeds were probably related to wartime service. These may have been mixed passenger and goods trains on lines that were also being used by military trains (see Chapter Five).

Connections

There were dwell times at Esquelbecq, probably for CdN standard gauge connections. There was also a standard gauge connection at Bergues. We do not know the nature of the connections at Herzeele with CF service Hazebrouck to Hondschoote. However, the times of the connections at Bollezeele were as follows:

From St-Momelin	07.11 (1)	08.11 (2)	15.15
From Bergues	12.00		20.25
From Herzeele	11.00		17.00
To Herzeele	13.05		19.22
To Bergues	07.11 (1)	08.11 (2)	15.15
To St-Momelin	12.00		20.25

(1) Mondays only, market day in Bergues
(2) Except Mondays

As can be seen, connections at Bollezeele were very poor. It should also be noted that the Bergues to Bollezeele line only opened on 3 August 1914. This meant that pre-war the only service at Bollezeele was that through between Herzeele and St-Momelin.

Lines proposed but never built

The branch from Drincham to Bourbourg, which was under construction at the declaration of war, was never finished. As we noted above, from St-Momelin the line should have crossed the Aa canal into the *département* of Pas-de-Calais to reach St-Omer, but this connection was never realised. As discussed above, the line from S-Omer to Tournehem was never built. It was planned to have stations or stops at St-Martin-au-Laërt, Salperwick, Tilques, Serques, Morilles, Eperlecques, Nordausques (with a connection to Audruicq), Zouafques (near the junction with the Anvin to Calais line) and Tournehem.

Tramway from Armentières to Halluin

This steam tramway was part of the vast inter-urban network which extended along the frontier between the conurbations of Valenciennes, Lille-Roubaix-Tourcoing, and Armentières. It had strong analogies with the '*vicinals*' of Belgium. The lines opened from 1881. They were built and run by various companies, but they all had a connection with Baron Empain, a

Belgian developer. The authorities knew that these lines operated under a *'provisoire'* title without legal basis. In 1885 the State pronounced them *d'utilité publique* and imposed the retrospective concession on a French company. This was how the CEN was formed. Since the company had its roots in the Belgian *Société National des Chemins de fer Vicinal* (SNCV) it inevitably developed urban and inter-urban steam tramways. Finally, those lines having the characteristics of the *Intérêt Local* network were conceded to other companies (including CF). Only the extensions to the Valenciennes network, and the isolated line from Armentières to Halluin, were conceded to CEN by various decrees between 1890 and 1893. The line from Armentières to Halluin was conceded to Baron Empain and CEN in 1891.

Building and opening

The first line trials were run on 21 March 1895 and the line was opened on 23 May 1895.

Description of the line

The route of the line is shown in Figure 2.2. It was a metre gauge tramway, 26km long running from Armentières to Halluin. Its total length was in France. It ran not far from the south side of the canalised River Lys and was separated from the Lille conurbation. The line started in the *Rue de la Gare* by the CdN station at Armentières and crossed the town, which had its own network of urban electric tramways. There was a section of line shared with the electric Tramways of Armentières (Line 1).

At Houplines it crossed the international Armentières to Comines line on the level. Its path followed the Belgian frontier, which was sometimes only a few metres away. It crossed successively, the international line from Lille to Comines (Komen, Belgium) at Comines-France, and that from Tourcoing to Menin (Menen, Belgium) a little before the terminus at Halluin. At Halluin the line did not join the Belgian line from Menin to Mouscron (*Vicinal* line 85), nor the eventual suburban line of the Tourcoing network, although both were of the same gauge.

Engineering

As a tramway it ran on the verge of the road, and in the streets of conurbations. It ran on Vignoles rails, 20kg/m, on the verges of roads, and on Marsillon rails embedded in the roadway in built up areas. The rails on the verges were strengthened later with the Vignole 30kg/m type. It crossed the river Dêule just to the west of Dêulémont, on a bridge shared with the road that is now the D945.

2.11 A train headed towards Armentières at the *Douane* (customs post) in Halluin, with a double-decker horse bus. The Belgian frontier is immediately behind the photographer, with the SNCV line from Menin to Courtrai via Mouscron not far beyond, but the two lines were never joined (see picture 3.7). About where the photographer is standing, the line to Armentières takes a sharp left turn (right of the photographer) to stay in France. The caption indicates that the horse bus is transferring passengers from and to the SNCV lines in Menin. Postcard written in 1911. (*Authors' collection*)

Stations and other stops
These are listed in Table 2.8. Many of these are *arrêts*, a feature typical of tramway type lines.

Buildings
The brick stations and toilet blocks match those of CEN on the Boulogne to Bonningues and Lens to Frévent lines. They had a central door and two windows on the ground floor on both the road and track sides. The upper storey had three windows on both sides. The ends were different. One had two doors and a single window. The other had one door and a very characteristic blind *oeil de boeuf* (bullseye) window. There was no goods hall, but a lavatory block was often present. The tramline shared the passenger facilities at CdN stations.

Depots
In the steam era the main depot and workshops for CEN in the Nord *département* were established at Raismes. Subsequently there was a depot and repair shop in the Place de la République at Houplines. This survived long after the closure of the line but was demolished in 1992-3 for a supermarket to be built.

Rolling stock
This was typical CEN stock that had one central buffer and coupling below. They had vacuum brakes, of the Smith-Hardy system.

Steam locomotives.
The whole inter-urban network was provided with about 30 locomotives of tramway (*bicabine*) type similar

Table 2.8 Armentières - Halluin
Stations and other stops, 1895 to 1914
Société des Chemins de fer Économiques du Nord (CEN)

Name	Type	Distance km		Altitude m	(ft)
		From Chaix1914 fr Armentières	fr Halluin		
Armentières Gare	CdN	0	25	19	(62)
SG lines Lille - Calais, Armentières - Menin & Courtrai (Belgium), Armentières - Lens & Arras, Armentières - Isbergues					
Armentières rue de Lille	*arrêt*				
Armentières rond-point	*arrêt*				
Houplines Place Chanzy	*arrêt*				
Houplines Octroi	*arrêt*				
Houplines Gare	CdN	3	22	20	
SG line Armentières - Menin & Courtrai (Belgium)					
Houplines Nouvelle	*arrêt*				
Houplines Pont Casier	*arrêt*				
Houplines Rue Brune	*arrêt*				
Houplines La Ruage	*arrêt*				
Frelinghien Brasserie Lutun	*arrêt*				
Frelinghien Place	*arrêt*				
Frelinghien Gare	station 1a	7	18	15	
Deûlémont Pont Rouge	*arrêt*				
Deûlémont Gare	station 1a	9	16	16	
Warneton Sainte-Barbe	*arrêt*				
Comines Le Hel	*arrêt*				
Comines Rue de la République	*arrêt*				
Comines (France) Gare	CdN	15	10	18	
SG line Lille - Ypres (Belgium)					
Comines Trois Ballots	*arrêt*				
Wervicq-Sud (France) Gare	station 1a	18	7	15	
Bousbecque Chemin des Vaches	*arrêt*				
Bousbecque Gare	station 1a	20	5	19	
Halluin Le Malplaquet	*arrêt*				
Halluin Rue Varna	*arrêt*				
Halluin Frontière	*arrêt*				
Halluin Place	*arrêt*				
Halluin Gare	CdN	25	0	15	(49)
SG line Lille - Menin (Belgium)					

CdN *Compagnie du Nord*
SG   Standard Gauge

to those used on the Belgian *vicinal* networks. The tare was 10 to 12.5 tonnes. The oldest was constructed by the Belgian firm Tubize, the later ones by ACNF in collaboration with Tubize. As part of the general stock of CEN, the numbering was designed to fit in with the systems of their other networks, those of the Pas-de-Calais, Isère and Haute-Savoie. This line had four locomotives for the opening in 1895. They were *bicabine* 0-6-0Ts from a series of seven, numbered 15, 16, 19 and 21. Number 15 was later sent to the Hellèmes line.

Passenger carriages
In the steam era, the network had about 100 carriages, of two axle four wheel type, about 65 for the Valenciennes network, 13 for the Hellèmes line and 21 for Armentières–Halluin. Those ordered by CEN in 1886, which would include those for this line, had a central platform. The last two were delivered in the 1920s to replace those which disappeared in the war.

Wagons
The whole network had almost 150 wagons. Most were open wagons used for coal and sugar beet, but they also had covered wagons and flat wagons. Of these, twenty-four were allocated to the Armentières–Halluin line.

Operations
There were two classes, first and second. The maximum length of trains allowed was 60m, that is 10 vehicles at most, and the maximum speed allowed was 20kph (12mph).

Timetables
A timetable for May 1914 is shown in Table 2.9. It took 1hr 45mins to travel the whole line except for one run from Armentières to Halluin which was faster at 1hr 33mins. The fastest average speed for the whole line was 16.8kph. The line usually ran at 14.9kph.

Fares
A ticket for the whole line one way was 1fr 95 centimes for first class, 1fr 25 centimes for second class.

Tramways of Armentières
The metre gauge electrified town tramways in Armentières were conceded in 1899 to the *Compagnie des tramways d'Armentières*, subsidiary of *l'Omnium lyonnais de chemins de fer et tramways* and declared *d'utilité publique* on 29 January 1900. In 1914 it was ceded to *La Compagnie de l'Électrique Lille Roubaix Tourcoing*. The lines opened in 1901 and are shown in Figure 2.2.

Description of the lines
There were two lines. Line 1 ran from Armentières station to Le Bizet. It started outside the CdN main station and went north. It terminated at Armentières Église St-Joseph, about 500 metres short of the border at Le Bizet. Line 2 began at the level crossing at la Chappelle-d'Armentières, at the eastern end of the station, and crossed the first at the junction of the Rue de Lille with the Rue Gambetta (D933). Then it followed the D933 north and then north-west to Nieppe Place. The total length of both lines was 6.5km (4 miles). A section of line was shared with the steam tramway from Armentières to Halluin (see figure 2.2). There was a depot, just off line 1 at the *Rue des Fusillés*, at the end of the *Quai de la Dérivation*, by a canal. There were 14 stops on line 1, and 18 on line 2.

Rolling stock
The lines had ten four-wheel two axle power cars with *archet* (single loop) overhead pick-ups.

2.12 The Grande Place at Armentières before 1914, with the *Hôtel de Ville* (Town Hall), and a tram on line 1 headed towards Armentières standard gauge station. (*Authors' collection*)

The metre gauge railways of French Flanders and related lines and tramways 1894 to 1914

Table 2.9 Tramway Armentières - Halluin
Timetable from May 1914 (main stations only)

	(1)(2)	(1)	(3)	(2)										(2)(4)	(3)	(4)(5)
Armentières CdN station	05.14	05.15	06.00	06.15	07.30	08.58	10.10	11.25	12.43	14.07	15.29	16.40	17.53	19.20	20.48	20.55	21.55
SG lines Lille - Calais, Armentières - Menin & Courtrai (Belgium), Armentières - Lens & Arras, Armentières - Isbergues
Houplines		05.16	05.17	06.16	06.31	07.46	09.14	10.26	11.41	12.59	14.23	15.45	16.56	18.09	19.36	20.54	21.09	22.01
SG line Armentières - Menin & Courtrai (Belgium)
Frelinghien	05.33	05.34	06.33	06.48	08.03	09.31	10.43	11.58	13.16	14.40	16.02	17.13	18.26	19.53	20.54
Deûlemont		05.42	06.41		08.11		10.51		13.24		16.10		18.34
Comines		06.04			08.33		11.13		13.46		16.32		18.56 (5)
SG line Lille - Ypres (Belgium)
Wervicq		06.14			08.43		11.23		13.56		16.42		19.06 (5)
Bousbecques		06.24			08.53		11.33		14.06		16.52		19.16 (5)
Halluin		06.48			09.15		11.55		14.28		17.14		19.38 (5)
SG line Lille - Menin (Belgium)

	(1)(3)	(2)	(3)	(2)							(2)	(3)	(5)
Halluin				07.07		09.31		12.20		14.50	17.30	20.05
SG line Lille - Menin (Belgium)
Bousbecques				07.29		09.53		12.42		15.12	17.52	20.27
Wervicq				07.39		10.03		12.52		15.22	18.02	20.37
Comines				07.49		10.13		13.02		15.32	18.12	20.47
SG line Lille - Ypres (Belgium)
Deûlemont			06.42	08.11		10.35		13.24		15.54	18.34	20.03	21.09
Frelinghien	05.35		06.50	08.19	09.33	10.43	12.05	13.32	14.42	16.02	17.15	18.42	20.00	20.11	21.17
Houplines	05.52		07.07	08.36	09.50	11.00	12.22	13.49	14.50	16.19	17.32	18.59	20.17	20.28	21.34
SG line Armentières - Menin & Courtrai (Belgium)
Armentières CdN station	05.54	06.08	07.23	08.52	10.06	11.16	12.38	14.05	15.15	16.35	17.48	19.15	20.33	20.44	21.50
SG lines Lille - Calais, Armentières - Menin & Courtrai (Belgium), Armentières - Lens & Arras, Armentières - Isbergues

Underlined - terminates, other than at end of line
SG Standard Gauge

(1) starts from the depot at Houplines at the stated time
(2) weekdays (Monday to Saturday)
(3) Sundays and public holidays
(4) to depot at Houplines
(5) daily 1 April to 1 October, and Sundays and feast days at other times

Closure

The tramways ceased operation at the outbreak of war. During the battles for possession of Armentières in autumn 1914, the lines were severely damaged, especially the overhead electric lines. The tramways were never reconstructed.

Tramway de Cassel

The town of Cassel stands on one of the *Monts de Flandres* and dominates the surrounding plain from a height of 176 metres. Naturally the standard gauge station had been built on the plain at the foot of the hill. Various projects to facilitate access to the station for the town's inhabitants were investigated from 1893. In 1899 the *Compagnie du Tramway de Cassel* obtained the concession, and the line was declared *d'utilité publique*. From 1910 the company expanded to also produce town gas, and became the *Compagnie du Tramway électrique de Cassel et d'éclairage par le gaz*. They built an electric tramway from the standard gauge station to the town centre on the hill. It was opened on 20 July 1900.

Description of the line

It was classified as a tramway with a total length of 3.36km (2.1 miles). At the station there was a small shed for rolling stock and a transhipment siding for goods. The line left the forecourt of the station and followed the verge of the RN 42 to the entrance to the town, and then continued in the road through the town to the terminus in the Grande Place. At the town entrance there was a branch to a depot and workshops and to a goods hall.

Engineering

It had an electric overhead connection to deliver a 600 volt DC current. This was carried by a simple single overhead wire supported by wooden posts, or in the town by cross suspension from the buildings. At the beginning, electricity generated was produced by a gas plant in the forecourt of the station. This was replaced in 1912 by a more substantial installation which also supplied the town gas. The track was metre gauge with Vignole 24kg/m rails for the road verge and Broca 36kg/m rails where the track was embedded in the roadway. The maximum gradient was 6.75 per cent and minimum radius of curvature was 35m.

Rolling stock
Power cars

Four were constructed in 1900 by *Société Alsacienne de Construction Mécanique* (SACM) at Belfort. They had two 25hp motors taking current through an

2.13 Electric tramway of Cassel. Two passenger power cars and a power *fourgon* in the Grande Place at Cassel on a market day. Picture taken before 1914. Postcard written on 10 May 1918. (*Authors' collection*)

archet (single loop) pickup. Three were equipped for passengers. Access was through a platform at each end. This was later closed in on the passenger cars. The fourth car was equipped as a *fourgon* for the carriage of packets and small parcels. The cars had bars with towing/coupling connections and originally only had hand brakes. Later they were fitted with compressed air brakes.

Trailers and wagons

Also, for passengers, there were two open carriages and one enclosed passenger trailer with a central platform. There were four small wagons with open network sides for the transport of larger goods and bulk goods.

Operations

A typical train had a power car with *fourgon* for baggage, a closed trailer for passengers, and a wagon for merchandise. Up to 1914 there were 11 shuttles per day, which could be increased at busy times. The journey took 15 minutes up the hill and 10 minutes down. In May 1914, trams ran from 6.20am to 9.05pm from the station, and from 5.50am to 8.35pm from the Grande Place, and the fare was 35 centimes up the hill and 25 centimes down.

Tramway from Dunkerque to St-Pol

On 11 January 1896 the *Société du tramway de St-Pol-sur-Mer à Dunkerque* opened a 60cm gauge street tramway from Dunkerque station to the *Place de la Mairie* in St-Pol. It was horse drawn and ran for about 2km (1.25 miles). It had six carriages with open platforms and a stable of 12 to 20 horses. Following a decree cancelling the concession on 26 December 1912, it was taken over in 1913 by the Electric Tramways of Dunkerque. The line was converted to standard gauge with overhead electrification and became Line E of Dunkerque Tramways.

Tramways of Dunkerque

Some information on these has been included even though they were of standard gauge. In 1873 the Town Council requested two tramway lines, linking the station with the Casino at Malo-les-Bains, and to the station at Rosendaël. The original concessionaire was M. Spilliaerdt-Caymax from Antwerp.

Building and opening

The first horse drawn line to the chapel of Notre-Dame des Dunes opened 13 June 1880. It was 2km (1.25 miles) long and was soon extended to the Casino at Malo-les-Bains. A second line to Rosendaël

2.14 A horse drawn tram on the 60cm gauge tramway from Dunkerque station, in the Rue de la République at St-Pol-sur-Mer. Postcard undated, but the line was converted to standard gauge and electric traction in 1913. (*Collection André Artur dcd, now Yves Artur*).

station opened on 17 July 1898, which linked to the line from Dunkerque station to the Kursaal. After unsuccessful trials of steam trams in 1880, and of trams with batteries, an overhead electrical system was agreed and put into service in 1903, to Malo-Terminus, to Rosendaël station, and to the Basse-Ville. Meanwhile in 1902 the concession was transferred from M. Spilliaerdt-Caymax to the *Société Anonyme des Tramways de Dunkerque et Extensions* (TDE).

Description of the lines
The lines were standard gauge (see image 1.3, Chapter One):

Line A ran from the station to the Kursaal.
Line B ran from rue de l'Église to Rosendaël from 17 July 1898.
Line C was the ligne des Darses, and ran to Darses mole No. 2.
Line D was the ligne de Basse-Ville. It was extended in 1906 to Couderkerque–Branche station.
Line E was the line to Mairie de St-Pol. Prior to 1913 this had been 60cm gauge.

There were also lines outside the territory of the Dunkerque *commune*:

Line No 1 was line B extended to Rosendaël station
Line No 2 went from the Place de la Mairie de Rosendaël to the Place des Kursaal at Malo-les-Bains
Line No 3 'de Malo-Terminus' extended to the Casino in 1906

Engineering
The major work on the lines was the long viaduct-like bridge about 0.5km east of the terminus of Line D at Couderkerque-Branche station. This carried the tram line in two spans over a main road and then the standard gauge railway from Dunkerque to Furnes (Belgium). The lines had overhead electrification at 500-550 volts DC, with a power station from 1903 in the Avenue de la République at Rosendaël.

Depots
After electrification there were two depots, one in the Boulevard de la République at Malo-les-Bains and one near the Place de la République at Rosendaël.

Rolling stock
Some of the early horse-drawn trams and battery-operated trial trams were double deck. After overhead electrification there were 34 power cars numbered 1 to 34. They had open platforms at the ends, were on SACM chasses and had two Jeumont motors of 20hp. They were capable of 20kph. They had mechanical and electric braking. Their capacity was 40 passengers, including 10 standing on the platforms. Trailer cars were numbered 36 to 59. Some were old battery vehicles with the upper deck removed, some were open sided (*baladeuses*), and some closed. They were like the power cars but shorter.

Operations
The agreement to open the lines demanded a minimum daily service as follows:

	Low season 1 Oct - 31 Mar	*Spring season* 1 Apr - 15 June	*High (Summer) season* 16 June - 30 September
Line A Dunkerque Station - Malo-les-Bains Kursaal	80	120	160 trams per day
Lines B & D Dunkerque station - Rosendaël gare or Basse-Ville and Couderkerque-Branche	56	80	80
Line Malo-Kursaal - Malo Terminus	12	14	14
Line Malo-les-Bains - Rosendaël	39	42	42

Services ran from 6.30am to 10.30pm in the high season, slightly shorter hours at other times.

Steenwerck–Ypres
This line was opened from Ypres to Neuve Église on 22 December 1897 and then on to the French border at Le Seau, and to Steenwerck station on the main French standard gauge line from Armentières to Hazebrouck, on 1 May 1909. This station is some way from the centre of Steenwerck, near the village of La Crèche. The line was managed by the Belgian Company SNCV as their Line 75 (see Chapter Three).

Chapter Three

Some metre gauge tramways of Belgian West Flanders and related lines 1885 to 1914

The vast majority of narrow gauge lines in Belgium at this period were metre gauge and were built and run by the *Societé Nationale des Chemins de Fer Vicinaux/ Nationale Maatschappij van Buurtspoorwegen* (SNCV/NMVB) but there were a few exceptions in our area and we will start with those.

La Panne to Adinkerke (60cm gauge)

This line began in 1894 as a horse-drawn bus. It was set up by Mme Terlinck, a local hotel owner. From 14 July 1901 this was replaced by a horse tramway. It is believed that the line was built by Chapel et Pluntz, but it was later controlled by another hotel owner, Van Neufville. It was run by *SA Tramway de la Panne*. The line was 3.7km (2.3 miles) long, starting from Adinkerke station forecourt. It immediately crossed the Dunkerque to Furnes standard gauge line and finished at La Panne in the Avenue de la Mer. There were three passing loops at Adinkerke (Markt), Adinkerke (Veurnestraat) and La Panne (St Pieterskirke). At its seaside terminus there was only a hut for the passengers and a stable for the horse. The line had six or seven one man/one horse vehicles. These were bogie carriages with a clerestory and end platforms with longitudinal seats reportedly so close that knees interlocked! Luggage was piled high on a grid at the back or in high season up to the roof. It is said that the crews had bare feet but wore blue pantaloons, a rust-coloured jumper, and a kepi. The line was taken over by the Allies. After the war the line was commonly known as the 'Adele' (see Chapter Eleven).

3.1 Horse tram on the 60cm gauge tramway from Adinkerke to La Panne, at La Panne. This picture also shows the size of the sand dunes along this coast. Postmarked 1919, but picture taken before the First World War. (*Authors' collection*)

La Panne to St-Idesbald (60cm gauge)
This line ran along the coast from La Panne to the *Hôtel des Dunes* at St-Idesbald. It was 1.7km (1 mile) long. It was built and operated by a development Company, *SA de St-Idesbald*, in about 1910. It opened on 29 August 1910. It became redundant when the SNCV line opened on 1 July 1914, and it closed on 3 August (see Chapter Five).

Coxyde to Coxyde-Bains (Metre Gauge)
This was a single- track horse tramway 3km long from Coxyde village to Coxyde-Bains, a rapidly developing resort. It met the SNCV line 2 running from Nieuport-Ville to Furnes at Coxyde Village. It opened on 24 July 1904 and was operated by Coxyde Council (*Gemeente Koksijde*). It was laid down the centre of an unpaved road, and ended on the beach approach. It seems to have had one closed four wheel carriage with four side windows and end platforms, and probably no others. It was replaced in part by SNCV line 193 which opened on 1 July 1909. This was steam operated until it was electrified in 1920.

SNCV Lines
We will start this section with a general description of SNCV/NMVB and its lines, including engineering and rolling stock. This will be followed by a specific description of each line in the area of this book.

By the early 1870s, development of main line railways had almost reached the limit of what was economically viable. It was also realised that local lines were badly needed. Thus in 1875 the SNCV/NMVB was proposed to co-ordinate a local system. It was to be 'national but not nationalised'. It was intended that SNCV should plan, build and administer the secondary transport system without funding or operating it. This organisation was created and started work in 1884. A Royal Charter was granted after parliamentary approval on 1 July 1885. This was an unusual system, but it worked well, at once a limited company and a non-profit public cooperative. It was independent of the state except for obtaining initial authorisations and for controlling overall tariffs. The network eventually covered the whole country, having a greater length than the standard gauge system. Half the system was eventually electrified. We have indicated what remains to be seen of the system in Chapter Twelve.

How it worked
In general, the idea for a new line came either from SNCV or the provinces themselves. The state would then give tentative approval. SNCV gave the title 'Capital' to each line, with a number. However, we have used the name 'Line' with the number in this book. We have used these numbers in the tables and maps in this chapter and later chapters.

In raising the money, the state would agree its share. The rest was raised from the provinces and communes. Occasionally private interests would become involved. If agreed, detailed plans would be made of prospects, costs, proposed tariffs, probable receipts, and specifications with maps, plans, and technical drawings. These would be submitted to the Department of Agriculture, Industry and Public Works. If approved, often after long negotiations, SNCV would get the funds. The chief advantage was that SNCV provided the lines with equipment which they acquired more cheaply under bulk buying contracts. Another advantage was that all the maintenance was supervised by SNCV.

Building and opening
When state funding was approved, SNCV would put the line out to public tender. These were often offered by specially formed companies, *Société Anonyme pour l'exploitation du CFV de …* (*Naamloze Maatschappij voor …*). Sometimes, SNCV had to operate the line itself. The concession period was usually 30 years.

The network developed rapidly. In 1904, 20 years after start up, 2,680km (1,666 miles) were in full operation with a similar length being considered. Unlike in some other countries, the growth was properly coordinated and controlled. It was mainly metre gauge. In general, they were built alongside roads and through towns and villages as 'tramways'. This reduced the cost and contrasts with the more common French model in which lines ran mostly in their own path. They carried considerable quantities of freight, especially timber and agricultural produce, as well as passengers.

Finance
We will give one example to illustrate how the finance for the tramways was apportioned. The *Liège–Seraing* (LS) Company obtained the concession for the Ostende to Furnes tramway. This (Line 2) opened in 1885-6. The company would receive 70 per cent of receipts up to 4,000 Francs/km per year. This would drop in stages to 65 per cent if receipts exceeded 5,500 Francs/km per year. Because the paths were mainly along roads, building costs were low. £2,300 per mile was quoted for the 1880s.

Engineering
Path and trackbed
Another advantage of building along roads was that it speeded up the construction. They mostly only deviated from roads to bypass a steep gradient or narrow street. However, as shown on Figure 3.1, quite a lot had some 'cross country' sections. Another advantage of building along roads and through villages was that much of the land

was already publicly owned. Running the lines through communities was clearly more convenient for passengers and contrasts with the French lines where stations were often a long way from the community served. However, it was not all positive. Maintenance costs were higher since the network also had responsibility for maintaining the road edge nearest the track. There was also the cost of laying and maintaining the cobbles or tarmac surface so that the rails did not project.

Rails
The offroad track and that running alongside the road consisted of Vignole 23kg/m rails resting on metal bearing plates (chairs) spiked down to hardwood sleepers. The latter were 0.20m wide, 0.1m deep and 1.80m long. There were ten sleepers per nine metres of rail, spaced to give extra support at the joints (fish plates). The track was bedded on 0.20m of gravel or cinder ballast and ballasted to sleeper depth. In paved roads with setts, flat bottomed rail of 31kg/m was used rather than grooved tram rails. It was screwed or bolted to sleepers laid below the surface. Spacer bars were bolted or rivetted to one side for the wheel flange or the track was laid on creosoted wood and the paving formed the edge of the flangeway.

Pointwork and curvatures
The tramway points had special sleepers, with a frog angle of 1:5 or 1:6 or where space allowed 1:8. They were worked by counterweighted levers, or in streets by using a key set between the rails to turn. The points were left set for the through road. The radius of curvature was of minimum 75m for free standing track but could be less in towns. The outer rail was elevated relative to the inner.

Dual gauge and Crossings on the level
There were two dual gauge sections. The first was Nieuport-Ville to Nieuport-Bains, opened in 1889. It was 1.1km (0.69 miles) long and was three rail. The second was Line 2 from Nieuport-Ville to Groenen Dijk. Crossings on the level or standard gauge merges were marked by a red disc or square-board (*carré*) signals.

Level crossings
On electric lines automatic signalling was activated by the pantograph. We presume that on steam lines a whistle would be sounded and the engine would proceed with caution if clear.

Electrification
The limitations of steam power, especially in urban and suburban areas, were quickly recognised. Another problem that had arisen was the steep gradients in some Belgian cities, although this cannot have been much of a problem in the area under discussion here. From 1884 to 1904 they tried light steam railcars, battery-electric railcars in Ghent and Ostende, and petrol-electric railcars in Brussels and Mons. None of these were very successful. The Belgians were pioneers in the use of overhead electrification which was coming into vogue in the 1890s. The first testing ground was the hilly line 11.2km (7 miles) from Brussels to Petite Espinette (an outer suburb), later extended to Waterloo. It was very successful and other developments followed rapidly. An early example was Ostende to Westende-Bains, the first section of which was electrified in 1897. The line east from Ostende along the coast was also electrified. We can gauge the spread of electrification by looking at the data pre-1914 (see bottom of page).

The coastal system with heavy tourist traffic was an early candidate for electrification, since it had a more generous loading gauge. The coast lines took stock 2.40m wide, whereas the norm was 2.00m or 2.20m. The standard current was 600v DC. Higher voltage monophase systems were tried south-west of Mons (Borinage) in 1906-07, but this was later converted to the standard. The systems used overhead single wires. Pick-ups were initially trolley poles, or single loop (bow or *archet*) collectors. Later, pantographs came into widespread use.

Stations and other stops
The stations stood on their own ground where possible. They were usually set off road but occasionally they intruded into the road. It should be noted that many country roads were not made up until the 1920s. Trains were signalled away from the stations by the guard blowing a whistle or a cornet. At all but the most important stations there would be an agent. They would handle any matters arising for the tramway. Often these agents owned a nearby café, bar or hotel which would then be used as a waiting room. They would be paid a retainer by the tramway and of course it was good for business. Where buildings were provided, they were often substantial two storey structures.

Building styles varied and there was an almost complete absence of the French-style goods building.

Electrification up to 1913

Year	% of system	% of traffic	% receipts	total km	(miles)
1904	3.82	22	20	97	(60)
1913	10	42	29.6	409	(254)

Customers themselves were responsible for loading and unloading wagons. However, some places had permanent ramps for cattle or sugar beet. The main provision at the manned stations of larger communities was a parcels office for 'smalls' (*colis* in French).

Other stops
These had standard stop signs, a column with a plaque on top with '*Société Nationale des Chemins de Fer Vicinaux – Arrêt du Train*' on one side and '*Trein Stilstand*' on the other. We assume that most of these required the passenger to signal to the driver, but some may have been scheduled stops.

Depots
All the lines had one or more depots. Each Group had at least one substantial depot for running repairs with multiple lines into a repair shed. In addition to this, each line had at least one more small shed with inspection pit, one or more carriage sheds, a lamp room (for acetylene or paraffin lamps), and a coal store to lock away coal briquettes. There would be a water tower or column usually supplied with a hot air pump.

Rolling stock
From the beginning, the Rolling Stock was ordered centrally by SNCV, who preferred to stick to a few favourite types. Since large numbers would be needed, each type would be ordered from several manufacturers. Many wagons and all the carriages were fitted with handbrakes, operated from the balcony. The country wide remit of SNCV meant no line need be overstocked; a 'central pool' could be kept.

Steam locomotives
From the beginning, SNCV wanted a 'tram engine' of 0-6-0 type and a short wheel base to deal with some fairly sharp curves. These were *bicabines* with driving positions at both ends to allow the driver to always be at the front but avoiding the need for turning facilities. The two outside cylinders and outside couplings were covered, for safe use in towns and villages. Of the twenty-three final classes, including some brought in after the armistice, all but seven were of this type. The first two classes were based on a Tubize design already in existence and were relatively light. The series culminated before 1914 with type 18, of which 127 were built by several manufacturers. They had outside cylinders and Walschaerts valve gear. A reverser, regulator, hand and vacuum brakes and a whistle cord were provided at both ends. There are six preserved examples of these.

British Army locomotives WD 201-250 were in use from 1916 to 1918. Of these, 48 survived to became SNCV type 19 after the war. They were almost identical to Type 18 (see Chapter Eleven).

Conventional metre gauge 0-6-0T locomotives were used mainly for industrial sites and shunting duties. SNCV also had standard gauge conventional and 'tram' type locomotives but as far as we know none was used in the Ypres sector. The first steam locomotive was constructed by the *Société 'La Métallurgique'* of Nivelles. It was a Type 3, weight 12.5 tonnes empty and 18.5 tonnes operating. Also constructed later were two heavy (60 tonne) locomotives of Garratt type for goods trains.

Passenger carriages
Two axle, four wheel carriages were standard. A few bogie coaches were produced before 1914 but they were found to be less flexible in use. First class, first/second composite, second class, second class with luggage composites, and luggage cars were all produced. First class had either upholstered benches arranged along each side, or one longitudinal bench facing four seat (upholstered) bays. Second class had varnished wood seats arranged transversely with a central gangway. There were no toilets. The carriages were heated initially with solid fuel stoves which replaced some seats. Later hot water heating systems were used. Finally, from 1912, Charpentier hot air systems were introduced. Lighting was initially with oil lamps but culminated in battery-powered electric lighting.

The earliest carriages were of low slung (*Surbaissée*) type of 1885 to 1886. There had been concerns that narrow gauge vehicles with higher centre of gravity would prove unstable. This proved unfounded. They had six window saloon bodies and end balconies. These were followed by the classic type, produced in considerable numbers between 1886 and 1911. They were accessed by end balconies and there were minor variations between manufacturers. Some had the first class section upholstered in velour. A further development gave a longer wheelbase to get greater capacity without bogies. These had eight windows on each side. There were 45 built between 1892 and 1899. A final development led to cars with glazed end screens (*à paravents*). They appeared from 1912 and eventually 453 were built. Initially they had wooden bodies and Charpentier underfloor hot air heating, paraffin lamps, and open balconies protected by the glazed screens.

Wagons
The first wagons were of 10 tonnes capacity. Mostly they had no continuous braking since the locomotives had powerful steam brakes. In 1912, the network had 6,011 open wagons, 577 low sided wagons, 1,067 vans,

111 tank wagons (all 2 axle), and specialist wagons of various types. Some of these were for timber transport in the Ardennes, which were more flexible to better allow for curves. Some had *haussettes* (raised sides) for transporting coal, sand and sugar beet. They also had tank wagons for transport of *purin* (slurry), especially in the Antwerp area. Later the network had flat and covered wagons of 20-30 tonnes capacity.

Electric trams

The first power cars appeared on the line Bruxelles-Espinette in 1894. Verkaegen de Malines (Mechelen) made the first 12 which were ordered on 28 September 1893. This rolling stock, 'rapid and without smoke', allowed the slopes of the *Vivier d'Oie*, previously impossible for steam trams, to be opened to public transport.

The earliest trams consisted of a wooden body on a chassis hanging from the axles that carried the driving and braking systems. They had two axles with a short wheelbase of 1.8m and each axle had a small motor of 20hp. These had six equal arched windows on each side, and open-end platforms with a waist high rail. All these cars were adapted by the addition of end screens. Initially, they had drop window ventilation, later these were changed to louvres. End screens were fitted from new from 1906. Six-wheel cars were very uncommon although some were used around Gent.

From 1907, SNCV adopted another general body style. These had panelled bodies with a short clerestory over the passenger saloon and six unequal, usually arched, windows. Their end platforms were partially enclosed with bowed end screens. A further type was introduced just before the war began, called the 'Manage' type. They had five unequal windows. This enabled the interior to be divided into two unequal saloons for first and second class passengers. These were in use from 1914 to the mid-1920s. The coast routes had special vehicles with large windows for passengers to enjoy their sea-views. The Ostende to Blankenberghe line had the most characteristic of these. There were also open sided *balladeuses* for the tourist season. Powered baggage cars were constructed by *Ateliers Franco-Belge* in 1909, for the coastal lines, to take the trunks and cases of summer visitors to the various beaches.

Operations

The speed on non-electric lines was limited to 30kph (18.6mph) in the country and 10kph (6.2mph) in towns. The train crew was responsible for picking up and setting down wagons and selling tickets, helped by the local agent if required. The *Chef de Train* did the paperwork.

Signalling

The signalling for crossings on the level of standard gauge lines was by red disc or square-board (*carré*) signals with red and green lights at night. Otherwise, there was very little signalling on steam lines. Country steam lines were run on timetable and train order.

There were however some fixed warning signs:

A   end of restriction
F   *fluten* or S *siffler* (whistle)
V   slow down
H   (on electrified routes) – three overhead masts away from a stop

Coloured light signalling was introduced later on long inter-urban electric sections.

Activity and financial results

In 1889, five years after the start-up, there were 704km (438 miles) of tramways operating. These were spread over 35 lines throughout Belgium. The overall operating ratio, which is the ratio of costs to receipts, averaged at 73 per cent. On the whole network, only four lines ran a constant deficit and none of these was in the Ypres sector. Ten years after start-up the average operating ratio was 70 per cent. By 1914, most lines were run at a profit or broke even.

Description of the lines in the Ypres Sector

These are shown in Figure 3.1 and listed in Table 3.1, with all the opening dates. The main station stops on the principal rural steam lines south from Furnes and Dixmude and around Ypres are shown in Table 3.2, with their French and Dutch names. The area of West Vlaanderen was very early in promoting tramways. A few large operating companies gradually acquired most concessions originally allocated to small local companies. In West Vlaanderen these were:

Mij tot Uitbating den Buurtspoorwegen van het Norden van Westvlaanderen
Société Anonyme Intercommunale Courtrai (IC)
NM voor de Uitbating der Buurtspoorwegen van den Omtrek Diksmuide - Ieper (ODI)

There was also the private company *Tramways Electriques d'Ostende–Littoral* (TEOL) which ran line 142 from 1897 to 1905 and continued to run it when it was incorporated into SNCV, and the *Société Anonyme des Railways Economiques de Liège–Seriang* which spread throughout Belgium forming a string of subsidiaries. This was part of the complex owned by Baron Empain. The decade from 1904 to 1914 saw a further major expansion.

Table 3.1 Metre Gauge Tramways in Belgium in the Ypres Sector up to August 1914

Littoral Group

Line no.	(Capital) From	to	via	length (total) km	section	length (section) km	opened	electrified	closed for WW1
2	Ostende	Furnes	Nieuport Ville, Coxyde	29.3	Ostende - Middlekerke	8.3	05.7.1885		19.10.1914
					Middelkerke - Nieuport Ville	9.8	15.7.1885		"
					Nieuport Ville - Furnes	11.2	22.7.1886		"
142(6)	Ostende	Westende Bains (1)(2)		14.7	Ostende - Mariakerke digue	4	19.7.1897	19.7.1897	18.10.1914
					Mariakerke digue - Middelkerke Bains	6	31.7.1897	31.7.1897	"
					Middelkerke Bains - Westende Bains	5	29.6.1903	29.6.1903	"
2	Nieuport Ville	Groenen Dijk	Nieuport Bains	3.6	Nieuport Ville - Nieuport Bains	1.1	5.1889		19.10.1914
					Nieuport Bains - Groenen Dijk Bains	1.4	31.5.1903		"
					Groenen Dijk Bains - Groenen Dijk	1.1	"		(4)
193	Coxyde	Coxyde Bains		7.5		3.0 (5)	1.7.1909		19.10.1914
	Coxyde Bains	La Panne	St-Idesbald			4.5	1.7.1914		"

Dixmude Group

Line no.	(Capital) From	to	via	length (total) km	section	length (section) km	opened	electrified	closed for WW1
29	Ypres	Furnes	Alveringhem, Oostvleteren, Elverdinghe	36.9			15.7.1889		16.10.1914
75	Ypres	Steenwerck (F)	Voormezeele	18.2	Ypres - Kemmel	10	22.12.1897		1914
					Kemmel - Neuve Église	4.5	"		"
				le Seau (border)		Neuve Église - Steenwerck (F)	4.5	1.5.1909		yes ?date
	Kemmel	Warneton	Messines	12.0			22.12.1897		"
115	Poperinghe	La Panne	Watou, Rousbrugge	46.2	Poperinghe - Watou	8	24.10.1905		16.10.1914
					Watou - Furnes	31	1.7.1906		"
					Furnes - La Panne	7.2	25.7.1901		"
107	Dixmude	Poperinghe Merckem, Oostvleteren		41.8		31	25.9.1906		16.10.1914
	Merckem	Elverdinghe				10	"		"
132	Dixmude	Ostende	Leke, Steene	26.1			29.6.1907		"
150	Roulers	Woumen (to 107 Ypres - Dixmude)		21	Roulers - Staden	11	15.2.1911		16.10.1914
					Staden - Woumen	10	1.10.1911		"
150	Roulers	Langemarck		14	Roulers - Westroosebeke	8	1.3.1913		"
					Westroosebeke - Langemarck	6	"		"
					(Langemarck - Bixschoote)	5	(planned but not built)		

Some metre gauge tramways of Belgian West Flanders and related lines 1885 to 1914

Courtrai Group

Line	From	To	Via	Distance	Section	km	Opened	Closed
41	Courtrai	Wervicq	Dadizeele, Gheluwe	29.0	Courtrai station - Ledeghem	*12*	13.2.1893	9.11.1914
					Ledeghem - Gheluwe	*8*	8.12.1892	"
					Gheluwe - Wervicq	*5*	13.2.1893	"
					Gheluwe - Menin	*4*	8.12.1892	"
85	Courtrai	Menin		22	Courtrai - Mouscron Station	*12*	15.6.1902	1.7.1917
					Mouscron Station - Menin Grotemarkt	*10*	8.8.1900	"
	Mouscron	Mont-à-Leux		2			4.5.1906	"
121	Ypres	Gheluwe	Gheluvelt, Becelaere	17.7			14.7.1905	9.10.1914
153	Iseghem	Wevelghem	Gulleghem (41)	13.9			11.4.1911	9.11.1914

Swevezeele Group

Line	From	To	Via	Distance	Section	km	Opened	Closed
33	Houglede	Thielt	Roulers	32.8	Houglede - Roulers	*5*	24.12.1889	1914
					Roulers - Thielt	*28*	(not relevant)	
137	Bruges	Leke	Couckelaere	25	Bruges - Couckelaere	*18*	(not relevant)	
					Couckelaere - Leke	*7*	22.3.1910	1914
137	(Couckelaere	Dixmude)		*13*	(132 Ostende - Dixmude)		(planned but not built)	

Distances in italics estimated from maps and other information

(1) electrified from opening
(2) Mainly summers, with limited winter services, 1902 - 1904
(3) 3 rail dual gauge from 1889 to 31 December 1902 to serve Nieuport Bains, with SG line from Dixmude
(4) did not re-open after First World War
(5) reported length of previous horse tramway
(6) line 142 under independent company until 1905 - *Tramways Electriques d'Ostende - Littoral* (TEOL)

Table 3.2 Station stops for SNCV lines in the Ypres Sector south from Furnes and Dixmude up to 1914
French names in first column, Dutch names in second column

Dixmude (Diksmuide) Group   *Courtrai (Kortijk) Group*

Line 115		Line 29		Line 75		Line 107		Line 121	
Furnes	Veurne	Furnes	Veurne	Ypres	Ieper	Dixmude	Diksmuide	Ypres	Ieper
Bulscamp	Bulskamp	Nieuwe-Herberg	Nieuwe-Herberg	Voormezeele	Voormezele	Woumen	Woumen	Gheluvelt	Geluveld
Houthem	Houtem	Alveringhem	Alveringem	Vierstraat	Vierstraat	Kippe	Kippe	Becelaere	Beselare
Leysele	Leisele	Forthem	Fortem	Kemmel	Kemmel	Merckem	Merkem	Gheluwe	Geluwe
Beveren	Beveren	Loo	Lo	Neuve-Église	Nieuwkerke	Luighem	Luigem		
Rousbrugge (Haringhe)	Roesbrugge	Pollinchove	Pollinkhove	le Seau	de Seule	Norsdschoote	Noorschote		
		Linde	Linde	Steenwerck (F)		Reninghe	Reninge		
Proven	Proven	Elsendamme	Elzendamme	(La Crèche)		Ostvleteren	Oostvleteren		
Watou	Watou	Ostvleteren	Oostvleteren			Westvleteren	Westvleteren		
Poperinghe	Poperinge	Woesten	Woesten	**Line 75 branch**		Crombeke	Krombeke		
		Elverdinghe	Elverdinge	Kemmel	Kemmel	Poperinghe	Poperinge		
		Brielen	Brielen	Wytschaete	Wijtschate				
		Ypres	Ieper	Messines	Mesen	**Line 107 branch**			
				Warneton	Waasten	Merckem	Merkem		
						Bixschoote	Bikschote		
						Steenstraat	Steenstraat		
						Zuydschoote	Zuidschote		
						Elverdinghe	Elverdinge		

F - in France

By 1914, the total length of lines in East and West Flanders was 1,061km (660 miles), of which 81 (50 miles) were electrified. At end of 1911 SNCV published a list of major connected systems (*réseaux*) usually with a single concessionaire working from a single HQ which gave the Group its name. In 1914, the major operating groups for the Ypres Sector were:

1. Littoral Group (Three lines in the Ypres Sector)
2. Dixmude Group (Six lines in the Ypres Sector)
3. Courtrai Group (Three lines in the Ypres Sector)
4. Swevezeele Group (Two lines in the Ypres Sector)

Littoral Group
Line 2 Ostende to Furnes
The first SNCV line to open, Ostende to Nieuport Ville, opened 15 July 1885. It was planned from 1875 by *Tramways de la Flandres Orientale* but nothing was done until SNCV took it over in 1884 as Line 2. It was authorised retrospectively in 1886. It opened from Ostende to Middlekerke Dorp on 5 July 1885, but stock shortage delayed public services until 13 July. It was open to Nieuport-Ville by 15 July and from Nieuport-Ville to Furnes by 22 July 1886.

Operation was ceded to the *Société Anonyme des Railways Économiques de Liège–Seriang* and its partner *Compagnie Générale des Railways à Voie Étroite*, both part of the Empain complex. It was transferred in 1905 to its wholly owned subsidiary *CF Électrique d'Ostende–Blankenberghe et Extensions* (OB).

The tramway mostly followed the main road, therefore this was a genuine all-purpose steam worked tramway. It had depots at Ostende, Nieuport and Furnes. There were a few additions. The first was a branch line from Nieuport-Ville to Nieuport-Bains where it operated on a dual gauge track with the standard gauge line from Dixmude. This line operated in the summer from 1889 by agreement with the *Chemin de Fer de l'État Belge*. Some sources say it was a three-rail dual gauge. The dual gauge agreement ended on 31 December 1902. After that, a loop of Line 2 separate from the standard gauge ran from Nieuport-Ville via Nieuport-Bains, on to Groenen-Dijk-Bains, and back to the existing line between Nieuport and Oost-Dunkerke. This opened for the summer season from 31 May 1903.

Line 142 Ostende to Westende-Bains
This tramway was run by TEOL until 1905. This was a company based in Antwerp but had been promoted by an Englishman, Colonel T. North, who lived in Eltham, Kent. It had been formed to run electric tramways in Ostende and along the coast. The Colonel was apparently a friend of King Leopold. Despite resistance from SNCV, they were given the concession on 11 March 1897 to build and operate an electric line from Ostende (Waterhuis) to Middelkerke (Bains) along the Digue (Dijk or seawall). Colonel North had died in 1896 but the company was also known as the *Compagnie North* after its English promoter. It was electrified overhead at 600v DC with trolley pole pickups. This was the only electrified line in the Ypres Sector before 1914. The tramway had fourteen closed and three open sided power cars and twelve trailers.

It opened from Ostende Handelsplein to Mariakerke Dijk on 19 July 1897 and on to Middelkerke on 31 July. The official opening was on 2 August and it ran until 31 October when it closed for the winter.

Some metre gauge tramways of Belgian West Flanders and related lines 1885 to 1914

3.2 A metre gauge train crosses the Nieuport Bridge over the Canal de Furnes, between the standard gauge station area at Furnes and the old town, headed towards Nieuport via Coxyde (SNCV line 2) or La Panne (line 115). Postcard written in November 1914. (*Authors' collection*)

Although it was damaged by storms in November 1897, it reopened on 15 May 1898. From then it usually operated from mid-May to mid-October. It was extended to Westende Bains from 29 June 1903 and had limited winter services between 1902 and 1904. In 1904 the government ordered SNCV to take the line over and make it part of the system, and negotiations to sell to SNCV started from the end of the summer service on 1 November 1904. The sale was agreed on 20 April 1905 and transferred on 29 May as Line 142. The concession was handed to LS who passed it to OB. They ran services from 10 June 1905 and the service ran all through the year. The line was upgraded and partly re-aligned in the winter of 1906/7.

The line ran along the coast road most of the way to Nieuport. It ran in parallel with the existing Line 2 between Ostende and Nieuport, which ran along the main road slightly inland and was steam operated. The Westende route was eventually merged into Line 2 in 1930.

Line 193 Coxyde to La Panne
The metre gauge horse drawn tramway from Coxyde to Coxyde-Bains (see page 32), opened on 24 July 1904. This was replaced by the SNCV steam tramway (Line 193) from 4 July 1909. From 1 July 1914, this was extended along the coast west to St-Idesbald and La Panne. It was intended to go through to Dunkerque, but the outbreak of war stopped that. This, effectively, made the 60cm horse tram from La Panne to St-Idesbald (see page 32) redundant, and it closed from 3 August 1914. Line 193 closed for civilian services on 19 October 1914, and reopened on 1 October 1920.

Dixmude (Diksmuide) Group
This was classed as a group by SNCV because they eventually came under the same concessionaire. In the early days it was related to the Littoral system through the original concessionaire, the LS. It ran Line 29 with the part of Line 115 that linked it to the coast, Furnes to La Panne, and the two branched frontier system south of Ypres, Line 75. Later it was connected to the coastal system and ran trams into Ostende and La Panne.

Line 29 Ypres to Furnes
This tramway was authorised on 22 March 1888 and ran north-west from Ypres along the main road, N8, through Brielen, Elverdinghe and Ostvleteren

3.3 Metre gauge lines in the forecourt of Ypres station, called *la station* not *la gare*, with a *bicabine* tram locomotive in the centre, and four wheel (two axle) carriages right and left. There are standard gauge wagons in the goods area behind the fence left, and part of a passenger train in the station to the right. No date but the dress, and exclusively horse road transport, indicates before 1914. Metre gauge tramways from here went to Furnes, Gheluwe, Warneton, and Steenwerck (France). (*Authors' collection*)

to Linde. There it suddenly turned north-east in a big loop through Pollinchove, Loo and Alveringhem. The section from Pollinchove to Forthem ran along the west bank of the Furnes to Yser canal. After Alveringhem, it came back to the main road for a straight run to Furnes standard gauge station on the north-east side of Furnes. It opened on 15 July 1889. It had a total length of 36.44km. There were depots in Ypres and Furnes. It was soon linked to the coast by a 7.17km section of Line 115, Furnes to La Panne. It was authorised on 23 December 1901 and opened on 25 July 1901. It was operated by LS, which handed it over to its coastal subsidiary OB. From 22 October 1910 this line was taken over by the ODI who were originally called *NM voor de Uitbating der Buurtspoorwegen van de Om trek Dixmude–Yper–Poperinghe*. The renaming indicated that the company ran the whole collection of lines around Dixmude and Ypres.

Line 75 Ypres to Steenwerck and Warneton

There had been plans as early as 1888 to connect Ypres with a French metre gauge line to the south but nothing happened. This line gained authorisation for the parts entirely in Belgium on 24 December 1895. This was Ypres to Neuve Église via Kemmel, and a branch from Kemmel to Messines and Warneton. The line from Ypres to Steenwerck in France opened from Ypres to Neuve Église on 22 December 1897 and extended across the border at Le Seau to Steenwerck in 1909. The extension to Steenwerck was not authorised until 19 March 1904 for the Belgian part and 11 September 1906 for the French part. It opened on 1 May 1909 for passengers only. Its terminus was at Steenwerck station on the main Hazebrouck to Lille line. The station was actually a little way out of the town at La Crèche. It was supposed to connect with an extension of the Armentières to Halluin line (CEN) but this never materialised (see Chapter Two). Line 75 followed minor roads for the main part. The chief depot was at Ypres station. We assume this was the same as that for line 29 to Furnes. The length of both branches was 34 to 35km. From 22 October 1910 this line was also taken over by ODI.

Line 115 Poperinghe to La Panne via Watou and Furnes

The line from Furnes to La Panne had opened on 25 July 1901 but the line was only authorised on 23 December 1901. This section was under the control of OB until 1 July 1914 when it was transferred to ODI.

Some metre gauge tramways of Belgian West Flanders and related lines 1885 to 1914

3.4 A train waiting to depart from the station and depot at Pont-Rouge, Warneton. This was at the end of the branch of line 75 from Kemmel. Undated postcard, dress indicates before 1914. (*Authors' collection*)

3.5 Steam locomotive and train at the terminus of SNCV line 115 at La Panne. Postcard written in 1906. (*Authors' collection*)

The line from Poperinghe to Rousbrugge (Haringhe) opened on 24 October 1905 and the tramway was completed to Furnes on 1 July 1906. Its total length from Poperinghe to La Panne was 46.18km. It had depots at Poperinghe and at Furnes, the latter combined with the depot for the Ypres to Furnes tramway. The line rambled around often in sight of the French frontier. It closed to civilian traffic on 16 October 1914 and opened again on 4 December 1918.

Line 107 Dixmude to Poperinghe and Ypres

This line was authorised on 25 March 1901 and opened on 25 September 1905. It was opened and operated by ODI. It ran south from Dixmude along the Ypres road, taking a small detour at the 'cluster of farmhouses' called Merckem. Here it split with one part going along minor roads to Ostvleteren where it crossed Line 29 before turning south at Crombeke to Poperinghe. The other part rejoined the Dixmude to Ypres road as far as Elverdinghe, where it joined Line 29 with running rights into Ypres. Its total length was 41.80km (26 miles). It was worked in sections which must mean there were no through trains.

Line 132 Dixmude to Ostende

This tramway had been mooted since 1889 but SNCV only obtained the authorisation on 19 March 1904. It opened on the 29 June 1907. It followed roads all the way, running almost north through Beerst to Leke to a junction with Line 137, a long cross country line to Bruges. Line 132 then ran on to Steene, from where it ran over OB lines into Ostende by the old standard gauge station. It had main depots at Dixmude and Steene and was 26.13km (16.2 miles) long.

Line 150 Roulers to Dixmude & Roulers to Langemarck

This tramway had two branches based on Roulers, at the junction with Line 33 Houglede to Thielt. The first branch ran through Roulers town then north-west along the road to Staden. From there it went, sometimes on the road and sometimes across country, to Woumen, a station on Line 107, from where it had running rights into Dixmude. It was authorised on 2 May 1908 and opened from Roulers to Staden on 15 February 1911 and to Woumen on 1 October 1911. It had depots at Roulers and Dixmude and the line ran for 11.97km (7.4 miles).

3.6 A train headed to Elverdinghe at the station at Bixschoote, on the branch of line 107 from Merckem. This was not in Bixschoote Village but at Smiske on the road from Merckem to Steenstraat. The single storey station building is at the back. The building in the foreground is advertising itself as the horse stables at Smiske Station. (*Authors' collection*)

The second branch opened on 1 March 1913 from Roulers to Langemarck. The tramway ran roughly west along minor roads to the standard gauge station at Langemarck. It was intended for the line to carry on to Bixschoote but it was delayed in crossing the standard gauge line and then war broke out. The path was complete and was probably used for military light railways. It was never re-commenced after the war. The total length would have been 19.15km (11.9 miles).

Courtrai (Kortrijk) Group

Only those lines and part lines relevant to this book are mentioned here. The area is partly French speaking and partly Dutch speaking. Prior to 1914 all these tramways were steam powered. Some were electrified after the war.

Line 41 Courtrai to Menin and Wervicq via Gheluwe

This was an early extension of the Courtrai suburban lines. It was authorised on 7 June 1890 after 4 years negotiation and was opened in 1892 to 1893. These lines had their origin in the local *Société Anonyme pour l'Exploitation du CFV de Courtrai–Menin–Werwicq* (CMV) which was formed in December 1892. The line opened from Ledeghem to Gheluwe and Menin on 8 December 1892. The two ends, Courtrai–Ledeghem and Gheluwe–Werwicq, opened on 13 February 1893.

The line ran from Courtrai west to Bisseghem, then in a big loop north to Moorseele before tracking across country to Gulleghem, where it crossed the Menin–Roulers standard gauge line. The tramway then ran west cross country to Dadizeele, and south-west along roads to Gheluwe. There it met Line 121 from Ypres, and then ran on to Wervicq, with a branch running south-east along the Ypres–Menin Road to Menin. This was a very circuitous route. Coutrai to Menin as the crow flies is only 13km (8 miles) but the tramway covered more than twice that distance. However, the line prospered. In 1899, CMV changed its name to *Société Anonyme Intercommunale Courtrai* (IC). All the succeeding lines around Courtrai were allocated to this company.

Line 85 section Courtrai to Menin via Mouscron

This line also ran from Aarsele to Courtrai but that section is not relevant to this book so we are only dealing with the section from Courtrai station to

3.7 The road to Halluin at Menin where it crosses one of the old branches of the River Lys (Leie), with the track of SNCV line 85 to Mouscron and Courtrai in the roadway. There is a footbridge (*passerelle*). The *Douane* (customs post) for the French border at Halluin is nearby (see picture 2.11). (*Authors' collection*)

Menin. The line was approved on 8 September 1897. It opened from Menin to Mouscron on 8 August 1900 and from Mouscron to Courtrai on 15 June 1902.

The line started on the south side of Courtrai station and to reach Menin it took a great loop to the south. It went along the Tournai road and then ran across country to Mouscron station (Luinghe) on the Lille to Courtrai main standard gauge line. Then it turned north-west across country and joined the Mouscron to Menin road. It followed this, approaching the French border at Halluin. Here it was very close to the French CEN line from Armentières which had opened in 1895 (see Chapter Two). The tramway however turned away from France to cross the Lys into Menin where its terminus was in the town square. A spur from Mouscron to the frontier at Mont-à-Leux was approved on 17 September 1904 and opened on 4 May 1906.

Line 121 Ypres to Gheluwe
This line was authorised on 24 June 1902 and opened on 14 July 1905. On leaving Ypres station to the north, it shared track with Line 29. It then curved round north of Ypres following the old wall and ditch, to the beginning of the Menin road proper, slightly east of the Menin gate. It went under the standard gauge line from Ypres to Roulers at what became known later as 'Hell Fire Corner'. The tramway followed the Ypres to Menin road, only detouring to the village of Becelaere. Its length was 17.7km (11 miles), 18.3km with the running rights on Line 29.

Swevezeele Group
As for the Courtrai Group, only those lines of relevance to our area are mentioned here.

Line 33 Houglede to Thielt
This line was authorised on 12 December 1888 and opened on 24 December 1889. It was operated by the *Société Anonyme pour l'Exploitation du CFV Thielt–Hooghlede*. Only the 5km between Roulers and Houglede are relevant to this book. At Roulers it connected with the standard gauge lines to Ostende, Menin, Ypres and Courtrai. It also met Line 150 to Dixmude and Langemarck.

Line 137 Bruges–Leke–Dixmude
This line was authorised on 5 October 1904 and opened on 22 March 1910, however only the section Couckelaere–Leke is relevant to this book. At Leke it joined Line 132 from Ostende to Dixmude. Trains ran through on this line to Dixmude.

Lines proposed but never built
There were two in our area – see Table 3.1. The first was to extend Line 150 Roulers–Langemarck from Langemarck to Bixschoote, on the Merckem to Elverdinghe line which was part of Line 107. The second was a branch of Line 137 (Bruges to Dixmude via Leke) from Couckelaere to Dixmude via Vladsloo.

The situation in summer 1914
In the summer before the outbreak of war, the Belgian SNCV/NMVB system of tramways was still expanding with some lines part constructed. It was almost all Metre Gauge. The system was very efficient and widely admired by other countries. By the end of 1913 there were 3,826km (2,378 miles) of steam operated track, and 409km (254 miles) of electrified track.

Chapter Four

Railways and light railways (60cm gauge) 1914 to 1918

It was inevitable that the railways of the north of France and Belgium would play a vital role in supply and communication for the Allies. As part of general mobilisation at the beginning of the war, the French government had placed all railways in the country under military control.

The Belgian government did the same. As the Germans advanced through Belgium in 1914, the Belgian railways moved rolling stock west, and then destroyed facilities as far as possible to obstruct the German supply lines. A large quantity of Belgian standard gauge rolling stock, and some metre gauge stock, was successfully withdrawn into the small part of Belgium not occupied, and into northern France.

The British front lines from autumn 1914 to 1918 have already been described in the introduction. Figure 0.1 also shows the distribution of the British, French and Belgians at various stages of the War. The British which were involved in the Ypres sector were the Second, Fourth and Fifth. A brief statement of the changes to the British approach to railway transport from 1916 has also been given in the introduction.

Line of railway command

By September 1914, the BEF, commanded by General Sir John French, had set up their GHQ at St-Omer. By the middle of September 1914, a Director of Railway Transport (DRT), Lt. Colonel (later Brigadier General) J.H. Twiss, had been appointed, based at GHQ. The agreement of the French authorities was obtained for the deployment of British railway troops on what was then optimistically called the British 'line of advance' into Belgium. From October 1915 the DRT was sometimes called the Director of Railways (DR). A Chief Railway Construction Engineer (CRCE) was also appointed.

Also by October 1914 Major A.M. Henniker was in post as Assistant DRT, leading the liaison with the French and Belgian administrations. He was promoted to Lieutenant Colonel in February 1915, and later wrote the official history *Transportation on the Western Front 1914-1918*, published in 1937.

By January 1915, French was able to write to the War Office in London with documents communicating the arrangements with the French and Belgians. These predicted that 'we will eventually be called upon to provide staff for operating railways in the area occupied by the British Armies'. It was not possible to say how many might be needed, but it was suggested that units might be based on the organisation of the French *Sections de Chemins de Fer de Campagne* (SCFC). Each of the 10 *Sections* consisted of nearly 1,500 men recruited in peacetime from French Railway Companies. When mobilised they were clothed and equipped as soldiers. One of them (the *10ème*) was recruited from metre gauge Companies.

The Battle of the Somme was the major stimulus for the appointment in August 1916 of Sir Eric Geddes, who had been General Manager of the North Eastern Railway in Britain, and a Special Commission, to consider the whole question of transportation for the British armies in France. The British Commander-in-Chief, from December 1915 Field-Marshal Sir Douglas Haig, wanted light railways used extensively along the whole British front, to help relieve the deterioration of the roads, and the need for manual labour by the troops. In October 1916 Geddes was appointed Director-General of Transportation (DGT), with a rank of Lieutenant-Colonel and a headquarters near Montreuil.

In November, Twiss was replaced as DR by Colonel W.D. Waghorn QMG CRCE, promoted on appointment to Brigadier General, and on 2 December 1916 his office moved from GHQ in Montreuil to the DGT HQ at the Château de Monthuis nearby. The CRCE now reported directly to the DGT for matters of broad gauge (standard and metre gauge), civil engineering and construction, and the Chief Mechanical Engineer similarly for the provision of engines, rolling stock and workshops. Brigadier General J.W. Stewart (Australian) was appointed Deputy DGT for construction, for supervision and inspection of all works of a civil engineering nature. The DGT also took over the Railway Operating Division (ROD), but they were charged with maintaining liaison about operating, including technical matters, with the French *Nord* Railway. The DGT established a separate Directorate of Light Railways, with a Director (DLR) reporting directly to him. Later, in March 1918, the title of the Directorate of Transportation was changed to Directorate of Railway Traffic.

Railway Companies

Many Royal Engineers and Dominion Railway Companies took part in work in the Ypres sector during the war. Railway companies often moved with the army to which they were attached, so that many units were in the Ypres sector for only part of the war.

Before 1914 there were only two regular RE Railway Companies, the 8th and the 10th (8RC and 10RC), based at Longmoor Camp in Hampshire. There were also three special reserve Railway Companies, the 2nd and 3rd Royal Monmouth (2 and 3RMRE) and the 3rd Royal Anglesey Railway (3RARE) Companies. 8RC arrived in France on 15 August 1914, and by the end of 1914 all five of these companies were working on the Western Front. More details of the work these companies did in the Ypres Sector in the early part of the war are given in Chapter Five.

Later in the war there were many more construction and operating Companies. The Companies of the Royal Engineers were numbered consecutively regardless of their speciality. Therefore reference, for instance, to the 296th Railway Company does not mean that there were more than 296 railway companies, but that there were more than 296 Companies of which the 296th happened to be a railway company. The exceptions were the Light Railway (LR) companies, which were numbered separately (see below). Canadian, Australian, New Zealand and South African railway troops were also available. The Canadian construction companies were organised into battalions, each consisting of four companies, which were often deployed on differing tasks, although usually in the same area. The skilled workers of the Engineers were supported by much larger numbers of less skilled workers. These included British and Indian Labour Companies, the Belgian, Chinese and Egyptian Labour Forces, and German prisoners of war. They were also often supported by working parties of infantry troops who were out of the front line.

Typically, a Railway company would be responsible for constructing and maintaining railways in a designated area. They were also responsible for operating their lines during construction, but, later in the war, operations were handed over to the ROD when construction was complete. The work of the railway companies was mainly on standard gauge railways, but there was some work on metre gauge lines, which were included together by the British army as 'Broad Gauge' railways. A few Railway Companies were more specialist, for instance the 287th and 297th, which were bridging companies. In 1915 and 1916 there was some work on light railways (mostly 60cm gauge), but from later 1916, with the establishment of a Directorate of Light Railways, this was mostly undertaken by LR companies, and by some of the Battalions of Canadian Railway Troops (CRT). However other Railway Companies were lent to the local Light Railways for short periods at times of particular need.

The first Canadian Railway troops to arrive were the Canadian Overseas Railway Construction Corps (CORCC). There is a mystery about their first arrival. Their War Diary records that they arrived at Calais on 25 August 1915, and moved to Alveringhem, north of Ypres, where they were attached to the Belgian forces, and undertook trench tramway and other, non-railway, work. On 30 September their CO left for London. There is no further information as to what happened to this unit, but it must be presumed that the whole Corps returned to the UK. On 1 November, the Corps left Longmoor for Le Havre, under the same CO. Both Companies started standard gauge work south of Poperinghe. This unit remained in France and Belgium, mainly on standard gauge work, until at least December 1918 (see also Chapter Six).

The thirteen Battalions of CRT were formed between December 1916 and March 1918. One was formed in Halifax, Nova Scotia, seven in England (one at Bordon and six at Purfleet), and five 'in the field' in France. Some had previously been Pioneer or Labour Battalions, and one, 2CRT, an Infantry Battalion. Each Battalion consisted nominally of about 1,000 men, in four companies. Many of these undertook broad gauge and light railway work, often changing from one to the other at different times.

From 26 May 1918, CRT were brought together as the Corps of Canadian Railway Troops. These were placed under HQ CRT for administration, reinforcements, promotions, and other personnel matters. However, they remained under local orders within each army for day to day work. Units affected were the Canadian Overseas Railway Construction Corps (CORCC, two companies), the 13 Battalions of CRT, the 1st and 2nd Tramway Companies, Canadian Engineers (TC CE), the 13th (Can) Light Railway Operating Company (LROC), the 58th (Canadian) Broad Gauge Operating Company (BGROC), the 69th Wagon Erecting Company, and the 85th Engine Crew Company.

In all there were, by the end of the war, 32 RE Railway Companies on the Western Front. In addition, there were the 13 Battalions of CRT, a total of 52 Companies. There were also occasional railway units which were not officially part of the RE; in the Ypres sector an example was the 17th Battalion Northumberland Fusiliers (Pioneers).

Railway Construction Engineers and Groups of Railway Construction Companies

By April 1915, there were three Headquarters for groups of railway companies, each under the command of a Railway Construction Engineer (RCE) appointed by and reporting to the CRCE at GHQ of the BEF. Although each RCE was an individual officer, we have referred in this book to each RCE as being plural, meaning the Railway Construction Engineer and his headquarters staff.

Each RCE had command of a group of Railway Construction Companies (RCCs). An RCE and its associated Group of RCCs were usually but not always given roman numerals, which have been used throughout this book. In practice in British Army War Diaries and other documents the terms RCE and RCC were not always used consistently.

In the end there were at various times seven such RCEs and groups, RCE I-VI and an RCE for the Lines of Communication in the back areas (RCE Comms). However, they did not relate consistently to the five British armies.

The RCEs and RCC groups were responsible for construction, repair and maintenance work on standard and metre gauge railways undertaken by railway companies working in their army area. In 1915 and early 1916, RCEs had some part in the development of policy for light railways and trench tramways. After the establishment of the Directorate of Light Railways in late 1916, they were not greatly involved with LR matters, but did sometimes lend Railway Companies for LR construction and maintenance. They maintained liaison with the ADLR for the army area in which they worked. Because the movement of these groups, and their relationship with the different armies, is complex, it is summarised for the groups involved in the Ypres sector in Table 4.1.

The four RCEs and groups in existence by the end of April 1915 were RCE I, RCE II, RCE III, and RCE IV. RCEs I, III and IV were formed at the large British Army railway depot at Audruicq, near Calais. We know from other records that RCE I was formed on 15 December 1914 when the first CO was appointed and was probably based at Audruicq. In May 1915 they moved to Abeele and then Hazebrouck to work with RCE III on the doubling of the Hazebrouck to Ypres standard gauge line. There are no surviving unit records until October 1917, and from then, and probably long before, RCE I served the First Army and was based at Houdain.

We do not know where RCE II was formed, but it may have been at Abbeville (Somme Sector), which is where they were based at least from April to July 1915. RCE III was formed on 8 February 1915 and based at Audruicq from 15 February. They moved to Abeele on 1 May 1915 to work with RCE I on the doubling of the Hazebrouck to Ypres standard gauge line. They continued to serve in the Second Army (Ypres) area,

Table 4.1 Ypres Sector 1915-1919
Allocation of Railway Construction Engineers (RCEs) to British Armies, each with a group of Railway Construction Companies (RCCs)
Note - Fourth Army renamed Second Army from 20 December 1917 to 17 March 1918 while Second Army in Italy (Second Army left from 13 November 1917)

	HQ based at (main locations)	Army	from	to
RCE I (1st Group RCCs)	Audruicq, Abeele, Hazebrouck	Second (? and First)	December 1914	May 1915 or later
RCE II (2nd Group RCCs)	Poperinghe, Crombeke Road, Dosinghem, West Cappel			
Reformed by renaming RCC III 1 June 1916		Second	1 June 1916	28 November 1917
(Previously at Abbeville (Somme Sector) in 1915)		Fifth	1 August 1917	18 January 1918
		Fourth (renamed Second 20 December 1917)		
			28 November 1917	17 March 1918
		Second	17 March 1918	23 February 1919
RCE III (3rd Group RCCs)				
formed Audruicq 8 February 1915	Audruicq, Abeele, Poperinghe, Zwynland			
renamed RCC II 1 June 1916 (see above)		Second	8 February 1915	1 June 1916
(re)formed 20 February 1917 at Doullens (Arras Sector)				
	Strazeele	Fourth (renamed Second)*	17 January 1918	16 March 1918
	Strazeele, Renescure, Lille	Second south*	16 March 1918	prob Nov 1918
	Lille, Ascq	Fifth	prob Dec 1918	19 March 1919**
RCE IV (4th Group RCCs)				
formed by April 1915	Audruicq, Bouloogne	Second	June 1915 or before	
	Strazeele	Second	16 July 1917	November 1917
	Strazeele	Fourth (renamed Second 20 December 1917)		
			November 1917	16 January 1918
RCE VI (6th Group RCCs)				
	Malo-les-Bains (Dunkerque)	Fourth	by 23 June 1917	28 November 1917
RCE for lines of Communication (RCE Comms)				
	St-Omer, Abbeville, Candas	Lines of communication		31 August 1919
			November 1915 or before	

* From January to April 1918 also sometimes attended First Army transport conferences
** disbanded at Montreuil 19 April 1919

and in June 1916 were renamed RCE II; we do not know what had happened to the original RCE II and group of RCCs. The 'new' RCE II continued to work in the Ypres area for the rest of the war but were re-allocated to the Fifth Army during the Third Battle of Ypres in 1917. RCE III was re-formed at Doullens (Arras Sector) on 20 February 1917, but in January 1918 moved north and served armies in the Ypres or Lille areas for the rest of the war.

There are no unit records for RCE IV until September 1917, but we know from the RCE III War Diary that they were at Audruicq from 15 April 1915, and were probably formed then. They remained at Audruicq at least until October 1915. From July 1917 to January 1918 RCE IV were in the Ypres Sector, serving the Second Army while RCE II were re-allocated to the Fifth Army during part of 1917. RCE VI existed only from late June or early July 1917 until 28 November 1917, to serve the Fourth Army while it was on the Belgian Coast (see Chapter Seven). The RCE for the lines of communication (RCE Comms) served the back areas throughout.

The only RCEs that never served in the Ypres Sector were RCE V, who were mostly in the south in the Somme Sector, and the Assistant RCE at Péronne. This latter covered Third Army South in the Péronne area (Somme Sector) from July 1917 to January 1918, while the Fourth Army were on the Belgian Coast and RCE IV were with the Second Army at Ypres.

The Railway Operating Division (ROD)

The ROD was an amorphous group of maintenance and operating railway staff. Officially part of the British army, they were in fact under rather looser discipline, it being accepted that the main task was to operate railway services. In November 1915, they took over operating the line between Hazebrouck and Poperinghe, in the Ypres Sector (see Chapter Six). By January 1917, they had taken over all standard gauge and metre gauge locomotive operating on British lines and some on French lines, except advanced lines still under construction. They also operated some light railways, pending the formation of the LROCs from late 1916.

French Railway Engineers

The French had an equivalent organisation of railway engineers. From 1889, the 5th Génie of French Military Engineers was entirely devoted to railway work. Pre-war there were three Battalions each of four Companies. During the war this increased to 85 Companies, with 450 officers and 21,500 men, and up to 100,000 men as supporting labour, who included Chinese, Indochinese, and Malagasy personnel.

Metre gauge railways

The role of the metre gauge railways in the First World War has been estimated to be small. Col. Henniker in *Transportation on the Western Front* (1937) considered their important use to be confined to a few months in spring 1918. Certainly, in 1916 the decision was taken to develop 60cm gauge (light) railways, and to develop metre gauge railways only in relation to existing lines, and to link them. However, as described in the following chapters, there were new metre gauge developments in the Ypres Sector, and the metre gauge lines were operated throughout the war.

Metre gauge locomotives

In September 1915, the War Department ordered ten 18 ton 0-6-0T metre gauge locomotives of *bicabine* (*Vicinaux Belge*) type, known in British railway parlance as 'tram engines'. In March 1916, a further forty were ordered. A further twenty similar but heavier locomotives of 26.5 tons were supplied to the British War Department by the American Locomotive Company (Alco) and were used with the Belgians north of Ypres (see Chapter Six).

Light (60cm gauge) railways

This gauge was made popular by the French Decauville Company in the nineteenth century. They developed a system of prefabricated light track with steel sleepers. It was widely used in France for shorter industrial lines, and for some public passenger and goods lines. Because of this origin, 60cm gauge lines and track are often known as 'Decauville' lines and track, even if not manufactured by them, or not prefabricated. Although sometimes loosely equated with the imperial gauge of 2ft (610mm), 60cm (600mm) is in fact 1ft 11⅝in. This is close to another gauge of some British lines, including the Ffestiniog and Welsh Highland railways, which are 1ft 11½in (597mm) gauge.

The French and the Germans quickly realised the potential usefulness of light railways for military purposes, and both adopted the 60cm gauge well before the war. The British also recognised this, although rather later, and developed a system based on the 2ft 6in gauge. However it became clear that the use of a different gauge would be unwise, and the 2ft 6in equipment was sent to the Suez Canal Zone. Much has already been written about light railways in the First World War, and the reader is referred to the bibliography. Here we have included a general summary to provide background. Light railways of 60cm gauge were also built and operated by the British in Egypt, around Salonika (Northern Greece), and to a limited extent in Italy.

Trench Tramways

From October 1915, the British army on the Western Front sanctioned short narrow gauge lines known as trench tramways. These were mostly 60cm gauge but occasionally 40cm or 50cm gauge. These lines were laid to bring goods into the trench areas using trolleys, usually manually propelled. There were no locomotives, and the rails were very light, 9lb/yd (4.5kg/m), or sometimes made of wood. These tramways were also sometimes called 'push car lines', especially by the Canadians. Supplies were brought to standard gauge railheads as close to the front as practicable, and the goods moved from there to the trenches or trench tramway ends by horse and cart or motor lorry. British trench tramways were initially built and operated on an *ad hoc* basis by individual units.

At the beginning of major light railway development in 1916, it was decided that although the light railways should extend to the end of the trench tramway systems, they should not be joined. The very light rails made them unsuitable for the light railway rolling stock. However, this policy led to the need to have transhipment points, some under enemy observation. Ammunition for heavy guns was delivered by light railway to group stations from which it was distributed to batteries by hand trolley along numerous spurs. Other goods were delivered to bulk delivery points, with tramways beyond.

By autumn 1917, ten Army Tramway Companies (RE) had been formed on a divisional basis. The main work was laying tramlines from light railway group stations to heavy batteries as they moved forward, and salvaging the tramlines from battery positions as they were vacated. In later 1917, some tramways were allowed a light railway connection, but only four wheel wagons were allowed to travel through from one to the other. A large number of 1 ton, four wheel box wagons were provided to the tramways, and some low power petrol tractors, mainly the 10hp tractors built by McEwen and Pratt. Selected tramways also received the heavier rails, allowing use by bogie wagons, and some were effectively absorbed into the light railway systems. Towards the end of 1917, some men from the Tramway Companies were sent to the Light Railway Workshops at La Lacque, near Isbergues, for training on petrol tractors.

After the difficulties experienced with this model of trench tramways during the Third Battle of Ypres in 1917, it was decided in early 1918 to reform the Tramway Companies. The new companies were initially known as Forward Transportation Companies, but by February or March 1918 they had been named Foreways Companies (FWC). A depot was set up at Savy-Berlette, later renamed the Forward Light Railway Training School.

War Department Light Railways

By February 1916, there were increasing problems with roads breaking up and road transport getting bogged down in mud. Feeder lines of 60cm gauge between standard gauge railheads and the front were agreed. In March 1916, it was also agreed to use heavier rails, 20lb/yd (about 10kg/m), to allow mechanical traction. Some track, locomotives, and other rolling stock were ordered. Also in March 1916, the British Third and Fourth Armies took over from the French some existing 60cm gauge lines in the Somme and Arras sectors. The (First) Battle of the Somme, which began on 1 July 1916, increased the pressure on transport and the difficulties on the roads, especially when the autumn rains came. By October 1916, the British were operating 130km (81 miles) of light railways, 49km (30 miles) taken over from the French. Motive power was mainly mules and men, and there were only about 20 locomotives and petrol tractors, and 200 wagons. French and German light railways were much more advanced.

When Sir Eric Geddes arrived as DGT in October 1916, he appointed a DLR. It was estimated that 200,000 tons of goods per week needed moving from standard gauge railheads, and during intensive fighting this could reach 2,000 tons per mile of front per day. An ADLR was appointed in each army area, supported by a Light Railway Construction Engineer (LRCE), and a Superintendent of Light Railways for operations. Sometimes larger areas were split into two ADLR areas.

The DGT decided to establish a complete system of light railways behind the whole length of the front. The first order was for 1,000 miles of track, 700 steam locomotives, 100 petrol tractors and 2,800 wagons, and 25,000 men to construct and operate these. The track estimate was based on 10km per km of front for the 62km (39 miles) where intensive fighting was expected, and half that for the remaining 83km (52 miles) of front, plus 25 per cent for contingencies and 200 miles (320km) in reserve. Later with greater line lengths behind the front, this was increased. 'Feeder' lines extended forward, linking the standard gauge railheads with the front. These would be in loops, linked by lateral lines at the medium and heavy artillery positions, so that if one line was cut by shell fire there would always be another route. Standard gauge railheads were at least 7 miles (11km), and often 10 miles (16km) or more, behind the front. At the other end, the light railways would extend to the beginning of the trench tramway systems (see above), about 3,000yd (2.5–3km) from the front line. Geddes remained in France as DGT until May 1917.

Construction of light railways was mostly undertaken by specialist railway troops supported by Labour Companies. However, many Operating Companies were called upon to construct part or all of the lines they were to operate, and most were expected to carry out maintenance and repairs. By the end of 1916, 95 miles (153km) were being operated, carrying 7,500 tons per week. Construction was slowed in early 1917 by very bad weather, with continuous frost for five weeks from the beginning of February. Track materials were not delivered from the UK in sufficient quantities until March 1917. From April to October 1917 about 110 miles (177km) of line were constructed each month. After the German advances of spring 1918, the average from early August to November 1918 was 61 miles (98km) constructed, and 163 miles (262km) reconstructed, per month. The reconstruction was mainly of former British lines recaptured, or of German lines captured. During this period in 1918 an average of 4,330 skilled and 4,610 unskilled men were employed on light railway construction. All of these figures relate to the whole British force in France and Belgium, not just to the Ypres sector.

By September 1917, it had been agreed that there should be a north-south 'main' line, called a lateral route, behind the whole length of the British front line, and linking all the armies. This would allow transfer of goods and men without using the standard gauge lines and without transhipment. The first 'main line' was planned approximately 6,000yd (5.5km) behind the front line and was complete by March 1918. By this time, it had been decided to construct another about 12,000yd (11km) behind the front line, with further lines running back from this. This would allow light railways to bring goods from dumps and railheads further back, and to evacuate stock in the case of a German advance, but this second lateral line was never completed.

Light Railway Companies

For operations, LROCs were recruited in England, and in the Dominions, or by taking men with the right background from other units in France. Recruitment began in England in January 1917, with the formation of the 1st LROC at Longmoor on 20 January. Eight further British LROCs (numbered 2-4, 6, and 9-12) were formed at Longmoor or the nearby Bordon Camp from 4 February to 17 May 1917. The 5th (New Zealand) LROC was formed at Codford Camp on 4 February 1917. We are not sure where the 7th and 8th (South African) LROCs were formed but both were in France by early summer 1917. The 13th (Canadian) LROC was formed at Aldershot on 9 June 1917.

Numbering of the Companies was often confusing, especially for the three Australian LROCs. These were originally variously named and numbered, but by the time they reached France they were numbered in sequence with the RE LROCs as the 15th, 16th and 17th (Aus) LROCs respectively. The 15th was formed from professional railwaymen in Victoria, Australia in November 1916 and originally called the 'Victorian Railway Unit' and was probably later called the 1st Australian LROC before arriving in Europe. They arrived in France on 29 May 1917, moving to Belgium (Ypres sector) on 3 June. The 16th was formed in Australia on 21 March 1917, and originally called the 2nd Australian LROC. They arrived at Bordon Camp in England on 21 July 1917, and were renamed the 16th (Aus) LROC on 28 July. The 16th were in France from 6 September 1917, moving on to Belgium (Ypres Sector) on 26 September. The 17th was formed in France, at Fricourt Brigade Camp on 11 June 1917. They were formed from the 1st (Anzac) LROC which was disbanded at Fricourt on that date. The 1st (Anzac) was probably formed in France in December 1916, as the operating company for Anzac Light Railways. The 17th was initially known as the 17th (Anzac) LROC, but by early 1918 were the 17th (Aus). Re-designation was ordered on 28 February 1918, and from 5 March the 15th, 16th and 17th became the 1st, 2nd and 3rd Australian LROC (Aus LROC) respectively. We have called the 15th Company 15(1)Aus LROC up to 4 March 1918, and 1(15)Aus LROC after that, with similar changes for the others, and for the Australian Broad (standard) Gauge Operating Companies, also renumbered from 5 March 1918. On 7 September 1918 the 3rd officially became the 3rd Aus. Light Railway Forward Company (3Aus LRFC).

Companies with higher numbers (such as 17Aus LROC), were generally formed in France by 'combing out' suitable personnel from other support units and from the infantry. The 29th to 31st and the 33rd were formed at Boulogne in February and March 1917. The formation of the 31st is described by T.R. Heritage in *The Light Track from Arras* (2nd Ed 1999). The 32nd was formed 'in the field'. The 34th was raised from men of the XV Corps on 28 February 1917, in the Fourth Army area, as the XV Corps Light Railway Troops, and renamed 34th LROC on 25 May 1917. The 35th were probably the first proper Light Railway Company, having been formed as the XIV Corps Light Railway Company at Trones Wood on the Somme battlefield on 7 December 1916, and renamed the 35th LROC on 10 May 1917. Finally, the 54th LROC was formed 'in the field', almost certainly in the Somme sector, in May 1917.

In addition to the 13th (Can) LROC, the Canadians had two 'Tramway Companies'. The Canadian LROC was formed from the No. 1 Section Canadian Corps Tramway Company on 14 November 1917, at Lens Junction (Arras Sector). No. 2 Section became the Canadian Light Railway Construction Company.

In early 1918, they became the 1st and 2nd Tramway Companies, Canadian Engineers respectively. They mostly supported the Canadian Corps in the First Army. They were not Tramway Companies in the sense of Trench Tramways, and did operate more forward sections of light railways, with petrol tractors but not steam locomotives as motive power. This served to a large extent as a model for the LRFCs formed later in 1918 (see Chapter Eight).

The LROCs were supported by five Light Railway Train Crew companies (TCCs), numbered the 18th to the 22nd. At least three were formed in England, the 19th and 20th at Bordon Camp in February and March 1917 respectively, and the 22nd at Longmoor Camp in May 1917. In some cases, these were closely associated with particular LROCs. By summer 1917 the 19th TCC had been split up, with men supporting the 4th, 6th and 9th LROCs, but in later 1918, the 19th was reformed as an LROC, operating independently until early 1919. The 18th TCC also became a 'proper' LROC. Eventually there were 32 LROCs and equivalents on the British Sectors of the Western Front. There were also two companies of miscellaneous trades, the 23rd and 24th, three Workshop companies, the 25th to 27th, and one tractor repair company, the 28th. These were based mostly at the Central LR Workshops, initially at La Lacque and later at Beaurainville.

A typical LROC consisted of 3 to 4 officers and between 220 and 260 men, most commonly about 250. Of these about 85 would be train crews, 40 telephone and control post operators, 30 on station yard and traffic duties, 50 to 60 on shed duties (repairs and maintenance), and the rest in other support roles.

In addition there were four United States Regiments for the construction and operation of light railways, the 12th, 14th, 21st, and 22nd US Engineers. Of these the 12th and the 14th worked with the British south of the Ypres Sector until July 1918, after which they joined the main force of US infantry.

Telephones and other signals services were provided by the Signals Corps. More detail of rolling stock is given in the next section, but in general steam locomotives were used further away from the front line, and petrol or petrol electric locomotives were used nearer the front where the greater visibility of steam locomotives, from the smoke and steam during the day or visible fire at night, would be a hazard.

The LR Companies which were not formed in France and Belgium usually arrived in France through the port of Le Havre, before deployment to an army and sector in France or Belgium. Near Le Havre and Rouen there were major RE railway works and facilities, and an Australian Base Depot. After the war, most were demobilised through Le Havre.

Track

French track of Decauville type was supplied in lengths already riveted to steel sleepers. British track of light railway weight was supplied loose, mostly in 5m lengths, but some were 2.5m and 7.5m. This could be clipped onto steel sleepers before being laid but could also be spiked down onto wooden sleepers. Because of the tolerance of the gauge to sharp curves, heavy earthworks were usually avoided. On average it took 2,000 man-days, 75 per cent unskilled, to construct and ballast 1 mile (1.6km) of track. Maintenance could require up to twenty men per mile in forward areas vulnerable to damage.

Ballast for the track was provided from various sources. Some was obtained from the rubble of ruined buildings, and some, called mine earth, from the spoil heaps of coal mines. Sand was less effective, but was obtained from dune areas on the coast, especially at Calais and at Ghyvelde.

Rolling Stock

In 1916 and early 1917 some mules were used. The British also used some French locomotives, taken over with the early French lines in the Arras sector in 1916, and later some captured German locomotives. These were mostly 0-8-0T locomotives of *Feldbahn* type.

Steam and petrol locomotives were ordered by the British army in small numbers in early 1916 and in increasingly large numbers after that. Petrol mechanical (PM) and petrol electric (PE) locomotives were known by the British army as tractors, and we have used this term for them. We have given technical details of the principal types of steam locomotive used in Table 4.2, and of petrol tractors in Table 4.3. In these tables we have used imperial units, except where otherwise indicated, because these are the units in which most of these locomotives were designed and built. The numbers given are those put into service in France. More were produced. Some of these went to other fronts, and some had not been delivered when the war ended. We have included in Table 4.2 details of the Péchot-Bourdon locomotive, and the 8 tonne 0-6-0 Decauville locomotive, the main locomotives used by French army light railways.

The light locomotives, with well tanks located between the wheel frames and with a low centre of gravity, were used mainly for yard work. Of the heavier locomotives, those from Hunslet were ordered first (see pictures pp 80 and 97), but when it became apparent that they could not meet the demand, a larger order was placed with Baldwin in the USA (picture on p 120). Steam locomotives on light railways required watering points every

Table 4.2 Technical details of British and French Army Light Railway Steam Locomotives 1914-1918

Manufacturer	Hunslet	Baldwin	Alco-Cooke	Hudson	Barclays	Péchot	Decauville
Country	UK	USA	USA	UK	UK	France	France
Dates(s) of manufacture from	08.1916	10.1916	02.1917	06.1916	early 1917		1914
to	09.1917	04.1917	05.1917	08.1917			1918
Wheel configuration	4-6-0T	4-6-0T	2-6-2T	0-6-0WT	0-6-0WT	0-4-4-0T	0-6-0T
Number put in service in France	75	495	100	32	25		390
Length inc. buffers (ft-ins)	19-10¾	19-6⅛	22-1½	15-5¼	14-8⅜	19-8	15-6⅛
Height (ft-ins)	8-11½	9-3¼	8-10½	8-6	8-4⅝	8-5½	5-3¾
Width (ft-ins)	6-3½	6-11	6-9	5-8	5-3	6-8	8-6¼
Weight empty (tons)	10.90	11.04	13.39	5.76	5.13	12.59 (12.79 tonnes)	8.2
Weight loaded (tons)	14.05	14.50	17.19	6.85	6.38	10.4	
Wheelbase (ft-ins)	13-0	12-2	16-6	4-2	4-4	12-7 (2.3 m)	4-7⅛
Driving wheelbase (ft-ins)	5-6	5-10	5-6	4-2	4-4	(2 of 2-11½) (0.9m)	4-7⅛
Diam. of driving wheels (ft-ins)	2-0	1-11½	2-3	1-11	1-10	2-1½ (0.65 m)	1-11⅝
Boiler pressure (lb per sq inch)	160	178	175	180	160	12 Kg/cm²	178
Heating surface (sq ft)	205	254.5	262	126	131	26.99 m²	188.8
Water capacity (gallons)	375	396	395	110	110	1,514 L	264
Coal capacity (cwt)	15	15.7	15	3.5	3.5	400 Kg	10
Diam. of cylinders (inches)	9½	9	9	6½	6¾	6⅞ (175 mm)	8½
Piston travel (inches)	12	12	14	12	10¾	9½ (240 mm)	11
Centre of gravity (ft-ins) Above track level - loaded	2-10½	3-0	2-10	2-7½	2-9½		

T	side tanks
WT	well tanks
Hunslet	Hunslet Engine Company Ltd, Leeds, England
Baldwin	Baldwin Locomotive Company, Philadelphia, USA
Alco-Cooke	American Locomotive Company, USA (constructed at Cooke Locomotive works)
Hudson	R. Hudson Ltd, Leeds, England (construction subcontracted to Hudswell Clarke)
Barclay	Andrew Barclay Company Ltd, Kilmarnock, Scotland
Péchot	Péchot-Bourdon, plus 280 manufactured for the French Army by Baldwin, USA
Decauville	Decauville, including 70 manufactured for the French Army by Kerr, Stuart, Stoke-on-Trent, England (the 'Joffre' class)

5 to 8 miles (8 to 13km), and coaling every 20 to 30 miles (32 to 48km). The 4-6-0 locomotives were liable to derail when running tender first, but less so when the quality of the track improved later in the war. However, the 2-6-2 Alco locomotives (see picture p 238) were still found best when running tender first. Later in the war, Baldwin also produced 2-6-2 locomotives, but only for the American army.

The Péchot-Bourdon was the main locomotive used by French army light railways. A double ended 0-4-4-0 side tank locomotive of Fairlie type, it was developed by Captain Péchot of the French artillery, and they had sixty-two by 1914. During the war, 280 were built for the French army by the Baldwin Locomotive Company in the USA. The French also used 8 tonne 0-6-0 tank locomotives of Decauville type. Between 1914 and 1918, 320 of these were delivered to the French Army by Decauville. A further seventy were supplied to the French by Kerr, Stuart of Stoke-on-Trent, England (the 'Joffre' class).

Petrol tractors (see Table 4.3) were needed for the forward area lines, for their greater flexibility on track which might be of poorer quality, and their ability to avoid enemy observation. Of these, both the Simplex type and the petrol electrics were excellent, but the latter had longer wheelbases, and they were said to be slow. There were only minor differences between the Dick, Kerr and the British Westinghouse types. The former had fixed side openings ('windows') on the cab, and louvers on the sides of the engine compartment, which the latter did not have. On both, the cab entrance was at the back.

The McEwan and Pratt 10hp tractors were intended mainly for trench tramways, and very forward spurs. In practice they were found to be under-powered. In the end they were mainly used in yards, and in forestry areas. The Crewe tractor was designed around a Model T Ford engine and chassis, with interchangeable road and railway wheels. It did not perform adequately on light railways.

The 20hp Simplex tractors were all open to the elements (see picture p 97). The 40hp Simplex tractors were supplied in three types. The 'open' type had some protection at the front and back and on top. Eighty-four were supplied to France but some were subsequently converted. Twenty were built as 'armoured', and a few more, perhaps seven, were converted from the 'open' type. When the side doors

Railways and light railways (60cm gauge) 1914 to 1918

Table 4.3 Technical details of British Army Light Railway Petrol Tractors 1916-1918

Manufacturer		Simplex PM	Simplex PM (1)	Dick, Kerr PE (BW PE)	McEwan & Pratt	Crewe Tractor (Ford)
Horse power		20	40	45	10	20
Dates (s) of	from	2.1916	5.1917	2.1917	6.1917	1916
manufacture	to	11.1918	late 1918		6.1918	1917
Wheel configuration		4 wh (2 axle)	4 wh (2 axle)	0-4-0	0-4-0	4 wh (2 axle)
Number put in service		749	292 (2)	DK 100 BW 100	42	132
Length over buffers (ft-ins)		8-11	11-1½	15-1	9-0¼	11-0
Height (ft-ins)		4-4¾ (or 4-6)	7-8	8-8	8-3¾	5-0⅛
Width (ft-ins)		4-10	6-6	5-6	3-6	4-10½
Weight empty (tons)		1.68	5.73	7.50	0.95	
Weight loaded (tons)		1.93 (3)	6.00 (3)	8.00	1.89	1.07
Wheelbase (ft-ins)		3-6½	4-0	5-6	3-0	4-5
Diam. of wheels (ft-ins)		1-5¾	1-6	2-8	1-6	
Engine		Dorman 2JO	Dorman 4JO	Dorman 4JO	Baguley	Ford model T
Cylinders Number		2	4	4	2	4
Diameter		4⁵⁄₁₆ (ins)	120 (mm)	120 (mm)	4 (ins)	3¾ (ins)
Stroke		5½ (ins)	140 (mm)	140 (mm)	5 (ins)	4 (ins)
Cooling		Water	Water	Water	Water	Water
Petrol capacity (gallons)			26	40	24	

(1) details are for the 'open' type (see text)
(2) all types (see text)
(3) includes 12 stone driver

Simplex	Motor Rail & Tramcar Company, Simplex Works, Bedford, England
Dick, Kerr	Dick, Kerr and Company, Preston.
BW	Those made by British Westinghouse were very similar (see text)
McEwan & Pratt	McEwan, Pratt & Company (taken over in 1913 by Baguley Cars), London and Baguley Works, Burton-on-Trent, England
Crewe Tractor	London & North Western Railway Works, Crewe

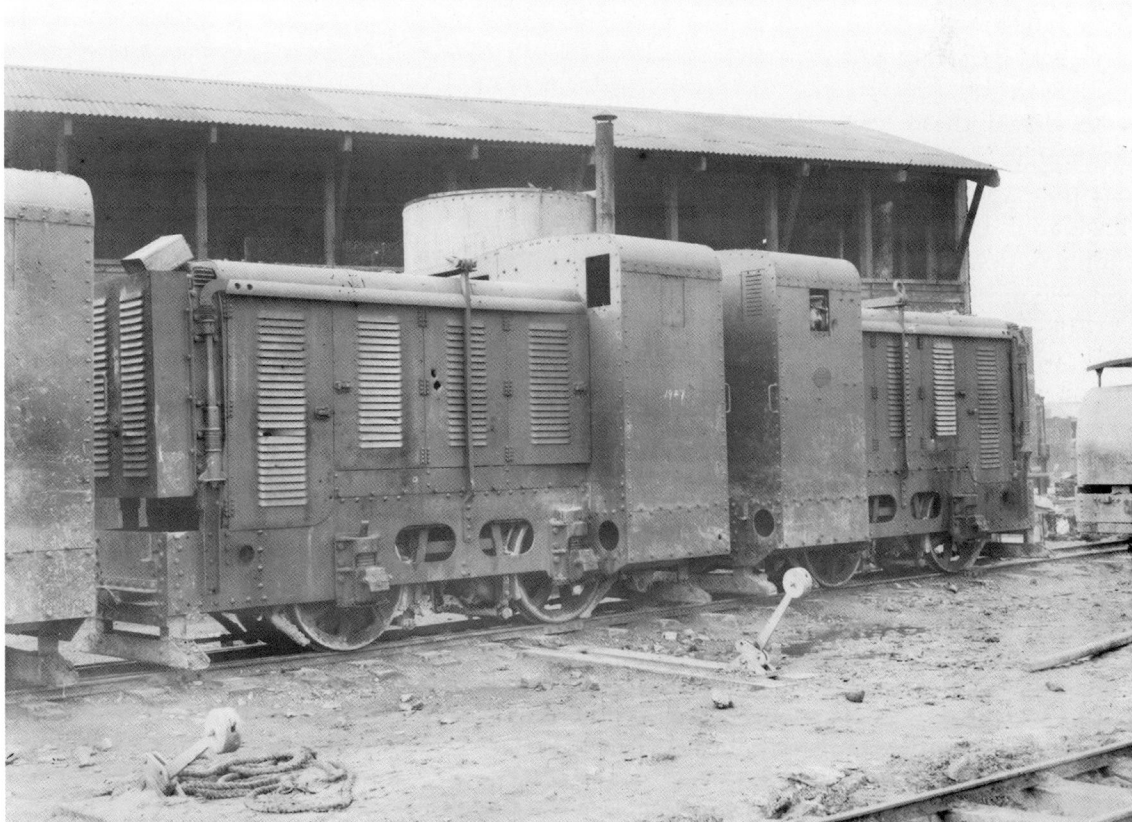

4.1 Two 0-4-0 Dick, Kerr petrol-electric tractors at Mimico (location 94, Figure 7.6), headquarters of 17(3)Aus LROC, in October 1917. (*Australian War Memorial C01361*)

4.2 A Crewe tractor at St-Julien on 13 March 1918, hauling a wagon full of ammunition out of a siding at a dump. These tractors did not perform adequately in field conditions, and sometimes (as in this picture) needed pushing. (*Collection Sandra Gittins*)

were closed these were said to be unbearably hot inside. The compromise was the 'protected' type, with side doors but more opening around the top than the 'armoured' type. One hundred and eighty-eight were supplied to France, and some more produced by conversion.

Wagons were supplied by British manufacturers in a variety of types, listed in table 4.4. The open bogie wagons with sides were able to carry heavy loads of shells, one of their main uses. Special wagons were used to carry some types of artillery. The light P class wagons with slatted sides were for rations and could be moved on trench tramways, either by hand or by light tractor. As a rough guide, it was reckoned that bogie wagons would carry thirty men, or ten tons of ammunition, eight tons of ballast, or five or six tons of most other materials. Box wagons with four wheels would carry ten men, four tons of ammunition, three tons of ballast, or two tons of most other materials.

In spring 1918 it was estimated that 19 per cent of steam locomotives, and 25 per cent of petrol tractors were in maintenance or under repair at any one time. However only 4 per cent of wagons were unavailable for these reasons.

Depots and Workshops

The central light railway depot and workshops were a resource for all the British armies. The initial depot and workshops were at La Lacque, near the standard gauge line between Isbergues and Aire-sur-la-Lys, and repair and maintenance work began there in March 1917. Following the German offensive on the plain of the Lys beginning on 9 April 1918, the depot and workshops were moved to Beaurainville, between Montreuil and Hesdin on the standard gauge line from Étaples to Arras. The new facilities were repairing rolling stock by July 1918. The workshops at Beaurainville are fully described in *Narrow Gauge in the Arras Sector* (2015).

In addition to the fixed central light railway depot and workshops, one and later two standard gauge trains were fitted out as repair shops for light railway

Railways and light railways (60cm gauge) 1914 to 1918

Table 4.4 Technical details of British Army Light Railway Wagons 1916-1918

Class	Type	Wheels	Length Over buffers (ft-in)	Centre of wheels or bogies (1) (ft-in)	Tare (tons)	Maximum load (tons)	Cubic capacity (cu ft)
A	6ft open box, fixed sides and ends	4, 2 axle	8-11½	3-0	0.834	3.666	60
A	6ft open box, loose sides and ends	4, 2 axle	8-8	3-0	0.864	3.636	60
A	6ft open box, folding sides and ends	4, 2 axle	8-8	3-0	0.900	3.600	43
B	8ft open box, loose sides and ends	4, 2 axle	10-8	3-0	0.975	3.525	80
C	12ft open bogie, fixed sides and ends	8, 2 bogie	16-5½	8-0	1.975	7.025	122
C	12ft open bogie, loose sides and ends	8, 2 bogie	16-5½	8-0	1.960	7.040	122
D	17ft open bogie, falling side doors	8, 2 bogie	20-6½	13-9	2.250	9.750	175
E	17ft well bogie, centre falling doors	8, 2 bogie	20-6½	13-9	2.600	9.400	225 (2)
F	17ft well bogie, detachable stanchions	8, 2 bogie	20-6½	13-9	2.100	9.900	323 (3)
H	17ft bogie tank	8, 2 bogie	20-6½	13-9	3.938	8.700	1,500 (4)
K	double sided tipper wagon	4, 2 axle	5-6	1-10			18
L	American side dump car 6ft in inside	4, 2 axle		2-10			40
N	hopper wagon, 6ft 9in inside	4, 2 axle		3-6			
P	light wagon, slat sides and ends (push use only)	4, 2 axle	6-6 (inside)	2-6	0.425		36, 50 if piled
R	ration wagon, push and tractor use	4, 2 axle				1 (notional)	
	bogie workshop wagon (5)	8, 2 bogie	20-6½	13-9			
	covered goods, ± ambulance fittings	8, 2 bogie	23-3½	16-6	4.500	7.500	605

(1) The internal wheelbase of all bogies was 3ft 0in
(2) 167 cu ft if not using well
(3) assumed, 265 cu ft if not using well
(4) gallons
(5) 3 falling side doors, or double swing doors, or workshop office wagon with windows

4.3 A light railway repair train in use, one of the five available to British Army Light Railways, at Angus Locomotive Yard at Ellarsyde (location 26, Figure 7.6) in the summer or autumn of 1917. (*National Library of Scotland 74411179 License CC BY 4.0*)

stock. Also, a 60cm gauge repair train was provided for each army. Each train consisted of a generating car, two machinery cars, a tools car, a stores car, and an office car.

Operations

The development of light railways under the British army on the whole Western Front is outlined in Table 4.5. This shows the rapid build-up during 1917, with maximum support for the battles of that autumn. After a lull in the winter, the maximum mileage and usage was reached just before the German attacks and advances of March 1918 (on the Somme), and April 1918 (the Lys pocket, including the Ypres Sector). As the ground was regained, lines were built, rebuilt or repaired, but by the autumn the advances to the east were leaving the light railways behind.

Apart from ballast and other railway materials, half of the goods carried on light railways was delivered by them to the final destination. This included all heavy artillery ammunition, which was generally given first priority, and a little field artillery ammunition and RE stores, which were second and third priority. Of the other half, roughly one quarter was for the field artillery, not delivered directly, and one quarter was

Table 4.5 British Army Light Railway route miles and tonnage 1917 and 1918, Western Front (Belgium and France)

Month	miles operated	tonnage carried per week
1917		
January	100	10,000
June	360	95,000
September	600	208,000
December	700	160,000
1918		
March	920	250,000
May	350	100,000
August	500	160,000
November	650	50,000

4.4 The central control for the Second Army (North) light railways at Ellarsyde (location 24, Figure 7.6). The District Controls, situated at main depots, are shown on the prominences (probably building supports) with the places served listed underneath. The names of the Mimico, Bedford and Vauxhall controls can be seen. The extent of the control indicates that this was after the Third Battle of Ypres. (*National Library of Scotland 74411178 License CC BY 4.0*)

for the most forward trench areas, including water and rations. Considerable numbers of personnel were also conveyed.

With very few exceptions the light railways were single track. Traffic control was initially based on the methods used on the French lines taken over in March 1916, using verbal or written permission to proceed. Gradually the British light railways established greater local and central control. For local control, manned control posts with or without signals were established at passing loops and junctions, often in dugouts if far forward. These were linked by telephone, and trains could only proceed with permission to use the next section, which might be given verbally or, at some busy places, by semaphore or colour light signals rigged up locally. Each LROC was responsible for operating 20 to 30 miles of track, which they did by telephone from a district control post, with a board showing the position of all trains. Very close to the front line, train control was the responsibility of a named officer or NCO.

Central control was exercised from a control post for each army area. These were able to receive requests from each Corps through their Light Railway Officers, and allocate motive power and wagons as needed. However, most train ordering and scheduling was done within each district control. Initially, there was no traffic between armies on the light railways. With the completion of a north-south lateral line in early 1918, GHQ was able to coordinate activity. One notable use of the light railways to transfer men from north to south is described by Col. Henniker in *Transportation on the Western Front* (1937). Following the German advance towards Amiens on the Somme front from 21 March 1918, General Byng moved the British Third Army Headquarters to Bernaville, west of Doullens, on 25 March. The Deputy DGT (Construction) also opened a headquarters there, with orders to acquire all possible railway construction troops to construct a second defensive line. The 60cm lines and the Lens-Frévent metre gauge line were used to bring men over the north-south lateral line from the Second and First Army areas, using a more direct route than the standard gauge lines and easing the burden on them.

Following the German advance on the Somme front beginning 21 March 1918, the north-south line was lost from near Bapaume to the south. After the further German attack north and south of the Lys river and canal, which began on 9 April 1918, it was also lost for a considerable distance on the plain of the Lys, and on the Messines front south from Mont Kemmel.

There were some well recognised causes of delay and disruption on the light railways. The obvious one was damage from shellfire or aerial bombardment. Mostly this could be quickly repaired, but much manpower was needed to ensure this. Track was also damaged at times by lorries or tanks not using authorised crossing points, and by men using the track as a footpath. Delays in loading wagons at standard gauge railheads, and in unloading at the destination, were more tightly controlled as experience increased. Between November 1917 and August 1918, the productivity of motive power units trebled and that of other rolling stock doubled by reducing these delays.

Reorganisation from June 1918

From the end of June 1918, transportation came under the control of the Quartermaster General (QMG) on the Western Front. One consequence was that construction of all railways, and ports, came under a Director of Construction. Probably the QMG was regaining 'territory' lost to Sir Eric Geddes as DGT in 1916. One might ask why it was thought worth bothering with such a reorganisation at that time, but in spring 1918 it was widely expected that the war would continue into 1919 or 1920.

LRFCs would be formed, for construction, maintenance and operation in the forward areas, still under the DLRs. Forward Companies were seen as operating from the point of hand over from steam traction, at about 3½ miles (5.6km) behind the front line, to the limit of mechanical traction at about 1 mile (1.6km) from the front. Steam traction further back would continue to be the responsibility of LROCs. In this respect the LRFCs resembled the existing Canadian Tramway Companies. There were to be complex regulations for inter-running, if this became necessary. Operations would be entirely with petrol tractors, entirely 20hp Simplexes in the Second Army area, which would be fitted with a small cab. Wagons would be of four wheel (two axle) type only. Class A 3.5 ton box wagons, and class R ration wagons, which could be used as 'push cars' or with tractors, but class P wagons must only be propelled manually (see Table 4.4).

Ten LRFCs were designated, the 231st, 232nd, and 234th to 240th (RE), and the 3rd Aus. LRFC, the latter formed from 3(17) Aus LROC. In practice these were not formed until October 1918, when the events of that autumn overtook them anyway. Some functioned as tractor based LROCs after the Armistice, and some were used to support standard gauge railway units.

Light Railways in the Ypres Sector

The detailed history of light railways in the Ypres Sector during the First World War is given, with the associated stories of the metre gauge railways, and limited necessary information on standard gauge railways, in Chapters Five to Eight, and that of 60cm gauge railways (with some standard gauge lines) in the area after the war in Chapter Nine.

Chapter Five

Railways of the Ypres battlefields 3 August 1914 to 25 May 1915 (end of the Second Battle of Ypres)

The causes of the First World War have been extensively covered elsewhere. On 1 August 1914 Germany declared war on Russia, and on 3 August on France. On 2 August they demanded of the neutral Belgians free passage for their armies, which was part of the Schlieffen Plan of the early twentieth century. When the Belgians refused, Germany invaded anyway. Britain, as one of the guarantors of Belgian neutrality, declared war on Germany on 4 August.

Belgium put up a spirited defence, especially at Liège. They also slowed the German advance, planned on the assumption of easily available railway transport, by systematically blocking and destroying the standard gauge railway network. Despite this, Germany swept through southern Belgium and Luxembourg, and turned south towards Paris.

Regular units of the British army, the BEF, began to arrive in France and Belgium from 15 August. The initial BEF had only 80,000 men, in two Corps, but all were regulars or regular reservists, and highly trained. Their first contact with the Germans was at the Battle of Mons in southern Belgium on 23 August. Here they were at the northernmost part of the German advance, on the left flank of the French Armies. After the Battle of Mons, the BEF retreated 250 miles over two weeks, with the French on their right, almost to the outskirts of Paris. The German advance was finally halted in early September in the First Battle of the Marne. The front line stabilised roughly along the Aisne River.

Finding the way to Paris blocked, the Germans wanted to take the Channel Ports. The French and British moved rapidly north to prevent this (the so-called 'Race to the Sea'), and to support Belgium. After the fall of Antwerp on 9 October, and then Ostend, the Belgians held the Germans along the Yser River from north of Ypres to the North Sea at Nieuport. The BEF was moved north and, reinforced by a third corps arriving from England, occupied most of the front line from Boesinghe, just north of Ypres, to La Bassée, east of Béthune.

The battles for the northern part of the front to the coast (the Battle of the Yser), the defence of Ypres (the First Battle of Ypres) and the Battle for Armentières overlapped from mid-October 1914 into early November.

The Battle of the Yser

On 14 October 1914, the Belgians stood along the west side of the canalised Yser river from Dixmude north to Nieuport, where the Yser estuary enters the North Sea. They were attacked by the German Fourth Army. They were reinforced by a French division, but by 24 October they were nearly exhausted, and ammunition was running low. The Belgian engineers sealed twenty-two culverts in the railway embankment between Dixmude and Nieuport, and opened the sluices at Nieuport to let in the sea, flooding the polders between the canalised Yser river to the east and the railway to the west. This halted the German advance by the end of October. The Belgians also held some ground east of the Yser estuary between Nieuport and the sea. The southern part of the flooded area, and the front line around Ypres at the end of this Battle, are shown in Figure 5.1.

The First Battle of Ypres and the Battle for Armentières

On 14 October, the German Fourth and Sixth Armies also attacked the British positions from Boesinghe (north of Ypres) south. By 16 October the German Sixth Army had taken Armentières, but it was retaken by 18 October. As well as reinforcing the Belgians on the Yser river, General Foch rushed large reinforcements to support the BEF. The bulge east of Ypres, which became famous as the Ypres Salient, was held until the end of the war, and Ypres was never taken. By 18 October, the British were established on the ridge at Passchendaele, and the French north of them were in Roulers. From 22 October, the French held the north part of the salient from north of Boesinghe to Broodseinde, but by the 25th, the north part of the eastern ridge including Passchendaele had been lost. By early November the French also held the line from the Comines railway south-east of Ypres to the Douve valley, with the BEF responsible from Broodseinde to the Comines railway, and from the Douve valley south around Armentières. For the BEF, fighting was very heavy on the Gheluvelt plateau. The battle ended on 11 November in heavy rain and snow, and both sides dug in. Thus began here, as elsewhere on the Western Front, the more static trench warfare of attrition.

Railways of the Ypres battlefields 3 August 1914 to 25 May 1915 (end of the Second Battle of Ypres)

However, the BEF with the French held the salient, bulging about 6 miles into the German front, and the Germans were denied any advance to the channel ports. For the Belgians holding West Flanders (*West-Vlaanderen*) west of the Yser river and canal, the area was important as the only unoccupied part of their country. The front line around Ypres at the end of the First Battle of Ypres is shown in Figure 5.1.

The British army rapidly expanded by the end of 1914. On 26 December the BEF divided into two; the Second Army, under General Smith-Dorrien, was given responsibility for the Ypres salient and south to Armentières, and the First Army was deployed south from there, with GHQ at St-Omer.

Railways

The period covered by this chapter was one of taking over the existing lines, organising their increased use to supply the front line, and making plans. Only at the end of this period did serious railway development begin, with the doubling of the line from Hazebrouck to Poperinghe. However, from the beginning of the conflict all sides recognised the key role which railways would play in the maintenance of supplies, and to the outcome of the war. Standard gauge and metre gauge railways from August 1914 to May 1915 are shown in Figure 5.1.

Standard Gauge Railways

With the outbreak of the war, standard gauge railways in the areas not occupied by the Germans came under the military control of the Belgian and French Armies. With the arrival of the BEF, specialist railway troops were brought in and a command structure was developed for them.

In August 1914 there were only three standard gauge railways into the Ypres sector, into what by November had become the front line area. From north to south these were:

The line from Dunkerque to Dixmude via Furnes (single track)
The line from Hazebrouck to Poperinghe and Ypres (single track)
The main line from Calais to Armentières via St-Omer (double track)

British Army railway troops organisation
The line of railway command has already been described in Chapter Four. During November 1914, the DRT discussed with 8RC what permanent way and bridge materiel should be sent from England for use in the event of an advance. More pessimistically, earlier in November, during the First Battle of Ypres, the DRT ordered that in the event of railway destruction becoming necessary, there must be communication with French GHQ. Because of the constant and mixed up troop movements at that time, one could not say which lines were 'ours', even if British troops were repairing or maintaining them. Bridges and permanent way might be required by the French, even in Belgium.

British Army railway companies
The first of the two regular army railway companies in existence before the war, 8RC, were mobilised at Longmoor, the main RE railway depot in Hampshire, on 4 August. They arrived in France at Le Havre on 15 August 1914. This RC had been formed before the war and was probably a company of regular skilled railwaymen. The second was the 10RC. Although a party of them had been in Belgium from 6 October, the rest of 10RC was mobilised at Longmoor on 25 October. Leaving 27 men at Longmoor to help train the 'new' railway companies being formed, they arrived in France at Le Havre on 28 October, and arrived in the back area of the Ypres Sector at Arques, near St-Omer, on 14 November.

The pre-war reserve units were 2RMRE, 3RMRE, and 3RARE. These units left Longmoor for Le Havre on 11 November 1914, and arrived in the BEF area on 15 November. All three worked on non-railway work, mainly trenching and associated works, in the Ypres sector and south around Aire and Béthune for several months. In addition to these pre-war railway companies and reserve companies, at least five and probably six new railway companies were formed at Longmoor between October 1914 and April 1915, and numbered 109RC to 114RC. All of these except 110RC worked at some time in the Ypres Sector.

Dunkerque to Furnes and Dixmude and branches
This was the northernmost of the three lines into the Ypres sector, and nearest the North Sea coast. A single track line, it ran east from Dunkerque in France to Furnes in Belgium, and on to Dixmude, on the front line after the Battle of the Yser. During the period of this chapter, it was entirely in the hands of the Belgians and the French. The line north-west from Dixmude to Nieuport formed the western edge of the flooded area south of Nieuport and was closed.

Hazebrouck to Poperinghe and Ypres
The single track line from Hazebrouck in France to Poperinghe in Belgium was built and operated by the Belgian *Flandres Occidentale* Company. It opened in 1870. At Poperinghe it linked with the line to Ypres and Courtrai, which had opened in 1853 and 1854. This was single track from Poperinghe to Ypres, and from Ypres to Comines, but double track east of Comines to Courtrai. All standard gauge lines east of Ypres, those

to Comines and Courtrai, to Roulers, and to Thourout via Staden, were closed in 1914. The parts near to Ypres were destroyed or damaged in the First Battle of Ypres, which ended in November 1914.

In March 1915, RCE I and RCE III took over responsibility for overseeing work on this line. It was at this time the only standard gauge line into the British sector in the Ypres salient, and by April the single line was being worked at full capacity. By 22 April the Second Battle of Ypres had begun, and doubling this line to Ypres was agreed between the DRT, RCE I, RCE III, and M. Yseboodt, the Belgian Railways District Engineer. On 27 April, Belgian locomotive crews joined the 10RC. From the beginning of May 1915 all the Railway Companies in the Ypres Sector (8, 10, half of 109, 111, 112 and half of 113) were working on the doubling. All the skilled railway construction companies involved had labour companies in support. During this period, maintenance and operating remained in the hands of detachments of the 26th French Railway Company.

By the end of May 1915, the doubling of the line from Hazebrouck to Poperinghe was complete except for some finishing off by small parties. Work on doubling east of Poperinghe towards Vlamertinghe started on 14 May, when part of 10RC began work on this section. By 25 May, the last day of the Second Battle of Ypres, a platoon of 109RC had moved to Vlamertinghe to begin work on doubling from there to Ypres.

Armentières area

Armentières had been briefly captured by the Germans in October 1914, but was back in Allied hands from 18 October. However, the main line from Calais to Armentières via St-Omer (double track) had been extensively damaged at Armentières, particularly with the destruction of the bridge over the Lys River just west of the town.

During the period of this chapter the main line from Calais to Armentières, which is wholly in France, was probably open at least for military traffic as far as Steenwerck. It was jointly maintained and operated by the *Compagnie du Nord* and, especially nearer the front line, French military engineers. From January 1915 part of 8RC also worked in the Armentières area.

Re-organisation of British Army standard gauge railways, May 1915

From 6 May 1915 Brigadier-General Twiss, the DRT, divided the railhead (supply) areas into two, Northern (Second Army) and Southern (First Army). These would be controlled by a Deputy Director of Railway Transport (DDRT), based at GHQ at St-Omer.

The Northern group of rail heads (those relevant to this book) would be supplied through the ports of Boulogne and Calais. The *Gares Régulatrices* (regulating stations, yards where the supplies were sorted for trains to particular railheads, armies and units) were placed at Boulogne and would eventually also be placed at Calais, this latter becoming later the vast supply yards at Vendroux. The detrainment regulating station was placed at Hazebrouck. Supply routes were to be from Boulogne and Calais through St-Omer, but also from Boulogne to St-Omer via the single line inland from Hesdigneul (on the coast line from Boulogne to Étaples) to Arques. The railways themselves remained under the control of, and were operated by, the French and Belgian authorities.

Metre gauge railways

Metre gauge railways in the Ypres Sector in Belgium and in France which were in operation in the period between the First and Second Battles of Ypres (12 November 1914 to 21 April 1915) are shown on Figure 5.1.

Belgium

In August 1914, metre gauge railways in the Belgian part of the area covered in this book remained open. The initial thrust of the German invasion of Belgium went through the south of the country to Mons, then southwest towards Paris. However, when the fighting came to the Armentières and Ypres areas, and north to the sea at Nieuport, all lines in this Sector closed to civilian traffic in October or November 1914 (see Table 3.1, Chapter Three). The whole area became a forward zone. The exception was line 85 from Courtrai to Menin via Mouscron. This closed between 27 October and 23 November 1914, after which reduced services were resumed. It remained open until July 1917, probably because it was in the major part of Belgium held by the Germans from a very early stage of the war, and was kept open with their agreement.

After the Battle of the Yser, the First Battle of Ypres, and the Battle for Armentières, in October and November 1914, all railways in the battle zones east of Ypres, the Yser Canal, and 'further north' the Yser river were extensively damaged. Some, notably line 121 from Ypres to Gheluwe (mostly along the Menin Road), did not re-open until after the end of the war.

Baron Empain, the Belgian entrepreneur who owned or managed some Vicinal metre gauge groups in Belgium, and some Companies and lines in France, took action when war became unavoidable. He formed a military style unit called the *Section Vicinale des Chemins de Fer en Campagne* (SVCFC) (Vicinal Field Railways Section). This was initially based at Antwerp, but after its fall on 9 October 1914, they were moved to west of the Yser River. Here they were organised into a Company with two platoons, one for operating and one

Railways of the Ypres battlefields 3 August 1914 to 25 May 1915 (end of the Second Battle of Ypres)

Table 5.1. Operation of metre gauge lines in the Allied part of the Ypres Sector in Belgium, October 1914 to 25 May 1915

Line no.	Section operated from	to	Operated by 1914	1915
2	Furnes	Square Wood (2km west of Nieuport)	BREB	BREB
115	Poperinghe	La Panne, via Furnes	SVCFC	
	Poperinghe	Beveren		French RW Engineers
	Beveren	La Panne, via Furnes		SVCFC
29	Furnes	Elverdinghe	SVCFC	
	Furnes	Oostvleteren		SVCFC
	Oostvleteren	Elverdinghe		French RW Engineers
107	Poperinghe via Oostvleteren	Nordschoote	SVCFC	French RW Engineers
107	Eleverdinghe	Zuydschoote	SVCFC	French RW Engineers

BREB   Belgian Army Railway Engineers Battalion
SVCFC   *Section Vicinale des Chemins de Fer en Campagne* (Vicinal Field Railways Section)
French Railway Engineers - mostly and probably entirely the *10ème Section des Chemins de Fer de Campagne* (10CFC)

for maintenance of the track. Rolling stock repairs were carried out by the Belgian Railway Engineers Battalion (BREB), who operated and maintained the standard gauge railways.

Metre Gauge Lines in Belgium

The lines which were operated in the area held by the Allies between the First and Second Battles of Ypres, and in most cases after the latter, are shown in Table 5.1, as well as in Figure 5.1. Initially SVCFC operated all these lines, except Line 2 between Furnes and Nieuport, which was operated by BREB as far as the wood later known to the British as Square Wood (*Bois Carré*) (see also Figure 7.7). The rest of the line was considered too close to the front line at Nieuport, and the Nieuport depot was closed. There were depots

5.1 The yard of the metre gauge Belgian *Vicinal* station at Rousbrugge-Haringhe (line 115). The station building is at the back. A large number of men in uniform are standing around, probably Belgian troops, and to the right a convoy of horse drawn wagons is headed towards Ypres. Postcard written from Rexpoëde on 22 April 1915. (*Authors' collection*)

at Furnes, which was the traffic centre, at Beveren (on line 105), and at Poperinghe, which was the main repair shop. The lines were used for supplies and for troop movements for the Belgian and French Armies. As far as we know there was no civilian passenger service after Autumn 1914.

The lines south of Beveren (on line 115) and Oostvletern (on lines 29 and 107) were handed over to the *10ème* SCFC (French Railway Engineers) at the end of 1914. SVCFC moved their repair shops from Poperinghe to the depot at Beveren. SVCFC headquarters was at Houthem, also on line 115.

Some lines which were in the area held by the Allies were not operated during the period of this chapter. The line from Coxyde Village to La Panne via Coxyde-Bains (line 193) was probably not operated and was later taken up to provide materials for elsewhere. Line 75 from Ypres to Steenwerck (France) was also probably not used during this period, although the section from Steenwerck towards Kemmel was operated sometimes by the British later in the war (see Chapter Six). The branch from Kemmel to Warneton was largely on the German side of the front line.

Elsewhere, in occupied Belgium, east of the front-line areas, metre gauge SNCV services were re-established by the existing operators, with the approval of the government in exile. They were increasingly important to the civilian population because of the control and priority use by the occupying power of the standard gauge system.

Rolling stock
During this period, the rolling stock was probably entirely of SNCV origin. Baron Empain is reported to have evacuated 80 steam locomotives, 330 passenger carriages, and 400 wagons. These came from the Dixmude and Oudenarde systems, and from the Littoral group, which was his own fiefdom. We presume that these were additional to those already based at Nieuport, Furnes, and Poperinghe, or elsewhere on lines remaining in operation in the Allied area. We do not have any further details of the locomotive types.

Links into France
By the end of the period of this chapter French Engineers had begun the construction of a metre gauge link between Pont-aux-Cerfs, on the line from Hondschoote to Bray-Dunes in France, and SNCV line 115 near Houthem in Belgium. This opened in June 1915. Another link from Herzeele in France to near Watou on line 115 may also have been under construction before the end of the Second Battle of Ypres.

France
Following the declaration of war by Germany on France, all the metre gauge railways in French Flanders came under military control.

Hondschoote to Bray-Dunes
In August 1914, civilian traffic was suspended, and the line was requisitioned for military use. From 1915 and probably before, it was completely taken over and operated by the *4ème* subdivision of the *10ème* SCFC, with the temporary cooperation of the Company responsible for the Anvin-Calais line.

Other lines
The other lines in this area were those from Hazebrouck to Hondschoote and from Rexpoëde to Bergues (CF) and from Bergues to St-Momelin and from Bollezeele to Herzeele (SE). From the winter of 1914-1915, *10ème* SCFC, already in control of the line from Hondschoote to Bray-Dunes, coordinated the activities of these Companies, under military control. This allowed trains to circulate directly across the whole network.

Rolling stock
In these early stages of the war, the rolling stock used was principally, and probably entirely, that already on the lines concerned.

Metre gauge operations
On the line from Herzeele to St-Momelin via Bollezeele there had been three trains each way per day before August 1914. The section from Bergues to Bollezeele was opened on the 3 August 1914, the day war was declared. Because of the requirements of military traffic, the civilian service on the existing line and on the newly opened section became two slow mixed trains per day between Herzeele and Bollezeele and two between Bergues and St-Momelin. A timetable from early in the war, probably August 1914, is shown in Table 5.2. This was labelled 'May 1914' but cannot be from then, because the Bergues to Bollezeele section is shown as open, and the service on all sections is reduced to the wartime two trains per day.

Services from Hazebrouck to Hondschoote and from Rexpoëde to Bergues were similarly reduced, but we do not have any timetables. There were no civilian services between Hondschoote and Bray-Dunes.

Bombardment of Bergues
In May and June 1915 Bergues station came under sustained shelling. On the Bergues–Bollezeele line SE converted a passenger carriage into a mobile ticket office to move out of the station to safety when necessary.

Railways of the Ypres battlefields 3 August 1914 to 25 May 1915 (end of the Second Battle of Ypres)

Table 5.2 Bergues - St-Momelin & Herzeele - Bollezeele
Timetable from August 1914 or soon after
Société générale des Chemins de fer Economiques (SE)

Bergues (CdN)	10.55	19.50	
SG line Hazebrouck - Dunkerque, origin of MG line to Rexpoëde, Hondschoote & Hazebrouck (CF)			
Bierne	11.04	19.59	
Steene	11.17	20.12	
Pitgam	11.32	20.27	
Drincham	11.42	20.47	
Bollezeele	12.00	20.55	
junction with line to Herzeele			
Volkerinckhove	12.13	21.08	
Lederzeele	12.20	21.15	
St-Momelin	12.36	21.21	

Bollezeele	13.05	19.22	
Junction with line Bergues - St-Momelin			
Zeggers-Cappel	13.22	19.39	
Esquelbecq CdN arr	13.30	19.47	
dep	14.11	20.11	
SG line Hazebrouck - Dunkerque			
Wormhoudt	14.28	20.28	
Herzeele	14.58	20.58	
junction with line Hazebrouck - Rexpoëde, Hondschoote & Bergues (CF)			

Herzeele	09.15	15.15	
junction with line Hazebrouck - Rexpoëde, Hondschoote & Bergues (CF)			
Wormhoudt	09.49	15.49	
Esquelbecq CdN arr	10.03	16.03	
dep	10.30	16.30	
SG line Hazebrouck - Dunkerque			
Zeggers-Cappel	10.39	16.39	
Bollezeele	11.00	17.00	
Junction with line Bergues - St-Momelin			

	(1)	(2)	
St-Momelin	06.33	07.30	14.40
Lederzeele	06.49	07.49	14.55
Volkerinckhove	06.57	07.57	15.03
Bollezeele	07.11	08.11	15.15
junction with line to Herzeele			
Drincham	07.28	08.28	15.28
Pitgem	07.39	08.39	15.39
Steene	07.51	08.54	15.54
Bierne	08.09	09.09	16.09
Bergues (CdN)	08.18	09.18	16.16
SG line Hazebrouck - Dunkerque, origin of MG line to Rexpoëde, Hondschoote & Hazebrouck (CF)			

CdN *Compagnie du Nord*
SG Standard Gauge
MG Metre Gauge
CF *Compagnie des Chemins de fer des Flandres*

(1) Mondays only (Monday market day in Bergues)
(2) Except Mondays

5.2 During the bombardment of Bergues, the line from there to St-Momelin made a carriage into a mobile ticket office to move out of the station to safety when necessary. The caption refers to May-June 1915, but the writing on the carriage says *Juillet* (July). The French script written at the bottom says (in part) 'Look over the station which I must use to go on leave to your house'. (*Authors' collection*)

Tramway Armentières–Halluin
From the time of the battles around Armentières in October 1914, most of the line, from east of Houplines, was on the German side of the front line, and the line was extensively damaged. Before that, all civilian use had ceased. The line was not brought back into use by the Allies during the war, but we do not know if the Germans made use of the Halluin end. Later in the war, the British assisted in the recovery of rolling stock from the depot at Houplines (see Chapter Six).

Tramways of Armentières
These tramways ceased operation in August 1914. During the battles for possession of Armentières in autumn 1914, the lines were severely damaged, especially the overhead electric lines. The tramways were never reconstructed.

Tramway de Cassel
This short electric tramway, which linked the station at Bavinchove (on the main line from Hazebrouck to Dunkerque) up the hill to Cassel town centre, continued to operate throughout the war. Cassel was an important military base; most notably for a long period it was the Headquarters of the British Second Army.

Light Railways (60cm gauge)
La Panne to Adinkerke
This existing 60cm gauge line was operated by horse tram. The line was taken over by Allies in the war. We have no further information except that relating to the time when the British were on the Belgian Coast in summer 1917 (see Chapter Seven).

La Panne to St-Idesbald
This existing 60cm gauge line ran along the coast from La Panne to the Hôtel des Dunes at St-Idesbald. It had become redundant when the SNCV metre gauge line opened on 1 July 1914, and it closed on 3 August 1914. The track was probably taken up soon after.

Army light railways
At this stage of the war the limited British railway troops were working on standard gauge railways. By May 1915 a few light railways were being developed further south, mainly around Béthune, often with wooden rails. Both the French and the Belgians began light railway development earlier than the British, but we have seen no definite evidence of light railway development in the Ypres Sector by May 1915.

The Second Battle of Ypres
On 22 April 1915, the Germans launched a major offensive against the Ypres Salient. The Ypres Salient was held by the French from near Poelcapelle to the south end of the Belgian position at Steenstraat. The area from Poelcapelle south was held by the British Second Army under General Smith-Dorrien.

At this offensive the Germans used poison gas (chlorine) for the first time. A gap of four miles opened in the British line. The Germans are reported to have been surprised by the success of the gas, which was only an experiment, and failed to fully exploit the situation. Canadian troops rapidly plugged the gap and resisted another gas attack on 24 April. There was some hard fighting especially on the Gheluvelt Plateau, with high casualties. General Smith-Dorrien wanted to withdraw towards Ypres. He was dismissed by Field-Marshal Sir John French, Commander of the BEF, and replaced by General Plumer. In the event, the British and the French did withdraw towards Ypres, but the reduced salient was held and the attacks lost momentum. The Battle was officially declared over on 25 May 1915.

The front lines at the beginning and end of the Second Battle of Ypres are shown on Figure 5.1. Despite the loss of territory, no additional standard or metre gauge railways were lost, because all those east of Ypres had not been reopened after October 1914. What happened after the end of the Second Battle of Ypres is described in Chapter Six.

Chapter Six

Railways of the Ypres battlefields 26 May 1915 to 6 June 1917

The general situation
This chapter covers the long period from the end of the Second Battle of Ypres (25 May 1915) to the beginning of the Battle of Messines, on 7 June 1917. It was a period of small battles and skirmishes with no major offensives on either side. Belgian and French troops held the coast around Nieuport while British troops of the Second Army were the main force in the area of the Ypres salient, Messines Ridge, the Douve Valley and south to Armentières. The Allies used the period to develop a support network for the armies in this sector.

Field Marshal Haig had wanted to attack in Flanders in summer 1916 but the French had suffered badly at Verdun in the spring and so the British agreed to a plan by General Joseph Joffre for a combined attack on the Somme (1 July to 18 November 1916). By March 1917, the British and French had agreed that the British should attack around Arras in the spring and then in the summer in Flanders. The spring 1917 British campaign at Arras had gone quite well. Meanwhile the USA declared war on Germany on 6 April 1917, following unrestricted German U-Boat attacks on shipping in the Atlantic.

Weather
The winter of 1916-17 was a severe one. There had been deep frosts from the middle of January 1917 into February with temperatures down to -8F (-22C) at night. Ten inches (25cm) of ice was recorded in the moat at Bergues. The freezing conditions interfered with construction work and some earthwork needed dynamite. There were snow storms as late as the end of March.

Railways
This was a period of enormous railway development. By end of it, the Ypres salient and back areas west of the Ypres Canal had a dense network of lines. The area was also full of roads, duckboard tracks, supply dumps, and camps. In 1915 and early 1916, the main focus was on standard gauge railway development, although there was some metre gauge work. From late 1916 there was intense development of light railways. The state of play in early June 1917, just before the Battle of Messines, is shown in Figure 6.1. Railways lost, destroyed or otherwise out of use are not shown on this map, unlike Figure 5.1.

British Army Railway organisation
A DRT had been appointed in September 1914. The DRT was renamed Director of Railways (DR) in October 1915. Sir Eric Geddes was appointed DGT in October 1916 based at GHQ in Montreuil. He separated light railway construction, maintenance and operating from other work, under a Director of Light Railways (DLR). It was also decided on 11 December 1916 that RCE II should take over the whole of the Second Army area.

Railway Companies (RCs)
By the end of May 1915 there were eleven RCs of the Royal Engineers in France and Belgium and of these eight had done railway work in the Ypres Sector, and seven of them were still there. 1916 and 1917 saw a rapid expansion. By May 1917 there were twenty-seven on the Western Front of which fifteen served in the Ypres Sector during all or part of the period of this chapter. Only one of the twenty-seven worked on light railways. The 17th Battalion Northumberland Fusiliers, and the two Companies of CORCC, who arrived in autumn 1915, also worked on railways during this period. Later the 13 Battalions of Canadian Railway Troops (CRT) were formed, some in France or Belgium. One was formed in 1916, eleven in 1917 and one in 1918. Four of these, arriving in early 1917, served in the Ypres Sector during this period.

Railway Operating Division (ROD)
The first sections of this Division (later called Companies) were raised in April 1915, and they expanded rapidly during 1915 to 1916. Their main base was at Audruicq on the main line from Calais to St-Omer. They operated standard gauge and a few metre gauge lines throughout the war. They also operated a limited number of light railways until late 1916 when the separate light railway structure took over. They were supplemented by three Australian and two Canadian broad gauge operating companies (BGROCs). On 1 November 1915, the ROD took over operating the line from Hazebrouck to Poperinghe. This was the first ROD operated line.

Construction and maintenance of standard gauge railways
Lines up to early June 1917 are shown in Figure 6.1. A detailed description of all the standard gauge work

is beyond the scope of this book, but all the works are summarised in Table 6.1. However, some additional information is provided on those lines and depots which were most critical for the supply to the light railways. If construction projects began in the period of this Chapter and were completed in the period of Chapter Seven, they appear in both Table 6.1 and in Table 7.4.

Table 6.1 Standard gauge railway works in the Ypres Sector June 1915 - May 1917

Area / Line or other facility	Construction work start	finish	undertaken by	major additional works & notes
Calais area				
Les Attaques RE yard	01.1916	06.02.1916	114RC	
Fontinette petrol depot sidings	26.01.1916	31.01.1916	114RC	
Calais sand sidings	31.01.1917	pr 02.1917	2RMRE	
Vendroux supply depots	14.04.1917	(1918)	298RC	
Rivière Neuve ROD locomotive depot	11.05.1917	06.1917	298RC	
Audruicq				
Extension to RE materials yard & ROD yard	08.11.1915		2CORCC	
Advanced ammunition (ordnance) yard	26.01.1916		114RC	
Ordnance yard extended	05.06.1916		119RC	
Locomotive shed sidings (additional)	12.06.1916		119RC	
Building new & rebuilding old ordnance yard	01.08.1916		114RC	
Construction No 1 railway depot	01.10.1916		114RC/part of 113RC	
New triage (goods shunting) lines	04.1917	16.04.1917	298RC	
Dunkerque				
New yard	05.1916		CORCC	
New docks triage, DT from docks to triage	18.03.1917	29.04.1917	298RC	ST timber pile and trestle bridge (297RC)
Ghyvelde/Bray-Dunes				
Lines to sand quarry	12.1915 or before			prob CORCC
multiple changes & extensions	1916/17		CORCC & 112RC	
St-Pierre-Brouck/Zeneghem				
Supply yards (on Watten to Bourbourg line)	23.08.1916	26.12.1916	CORCC	
Watten chord	25.09.1916	pr 12.1916	CORCC	from Watten-Calais line to Watten-Bourbourg line
Wagon repair siding	02.1917		298RC	
Bergues - Proven line (new British line)				
Single track	21.03.1916	30.06.1916	10RC (+½ 113RC) & CORCC	
Yser River bridge	18.05.1916	12.06.1916	10RC (+½ 113RC)	
Rousbrugge Yard	13.06.1916		CORCC	
Doubling track	09.01.1917	10.04.1917	10RC to 26 January, then 112RC	
Second Yser River bridge	04.02.1917	07.04.1917	112RC	
Bergues Canal Sidings, & branch to	08.01.1917	10.03.1917	112RC	
Bandaghem sidings, including for CCS	19.03.1917	02.05.1917	112RC	
Bergues Exchange extensions & alterations	14.03.1917	31.03.1917	112RC	Included doubling connection north to Bergues French station - difficulties with *Compagnie du Nord* over this work
additional 6 sidings	01.06.1917			Company and completion not known
Rexpoëde detraining platform and siding	02.04.1917	28.05.1917	112RC	
Rousbrugge Station, platform and additional siding		06.06.1917	112RC	
Dosinghem CCS siding	04.06.1917		264RC	
Adinkerke - Proven (new Belgian line)				
Adinkerke - Waayenburg (single track)		by 03.1917	BREB	including Yser River bridge
Waayenburg - Proven (Junction with Bergues - Proven line)		27.04.1917	BREB	junction at Mendinghem just west of Proven Station
Poperinghe Avoiding line (new British line)				
Remy North - Crombeke Road	16.11.1915	20.03.1916	CORCC, 10RC	junction at Remy north (Hazebrouck - Poperinghe line)
Railhoek store park/permanent way depot	26.01.1917	31.05.1917	10RC, 262RC	additional loop (262RC) June 1917

Railways of the Ypres battlefields 26 May 1915 to 6 June 1917

Area / Line or other facility	Construction work start	Construction work finish	undertaken by	major additional works & notes
Crombeke Road - Proven (new British line)				
Single track line	05.04.1916		10RC	
Crombeke chord	19.03.1917	12.04.1917	112RC	
Lovie Loop (passing loop)	01.04.1917	26.04.1917	112RC	
Poperinghe - Ypres (existing Belgian line)				
Double track Poperinghe - Vlamertinghe	01.06.1915	15.11.1915	112RC	Hazebrouck - Poperinghe already double tracked
Double track Vlamertinghe - Ypres	20.06.1916	22.08.1916	112RC, 10RC	
Ypres station works	05.06.1915		French 26RC, 109RC, 111RC	
Pacific RE Park sidings/ railhead	04.09.1915	13.09.1915	10RC	Re-modelled May 1917 (262RC)
Chord to Ypres - Boesinghe - Staden line	23.10.1915	05.11.1915	112RC	Improved and re-aligned March - April 1916 (10RC)
Ypres station re-modelling	25.09.1916	08.01.1917	10RC	
Ypres - Boesinghe - Staden (existing Belgian line)				
Brought back into use	1915			at least to Reigersburg
Ypres - Roulers (existing Belgian line)				
Repaired for trolley traffic to Menin Road	25.06.1915	29.06.1915	10RC	Repaired again August 1915 (10RC) to Cambridge Road
Repaired again to Cambridge Road	09.1916	09.10.1916	10RC	
Ypres - Comines (existing Belgian line)				
Repaired for trollies	08.1915		137th Brigade	Not known for how far from Ypres
Edwaarthoek, Westenhoek, Oakhanger, Peselhoek & Shellhoek complex				
Edwaarthoek line and rail head	04.10.1915	06.12.1915	119RC	converted to loop May 1917
Peselhoek line and sidings	10.1915	08.12.1915	112RC	Forth Bridge over Poperinghe Canal modified August 1916. Later junction with and part of Midland Railway
Shellhoek siding	21.05.1916	06.1916	112RC	
Oakhanger RE Park sidings	11.09.1916	30.09.1916	10RC	
Westenhoek rail head and TS Yard	09.05.1917	10.06.1917	262RC	
Elverdinghe Loop				
Elverdinghe loop	21.10.1915	15.02.1916	10RC, 115RC	Relaid with English track July 1916 (10RC)
Elverdinghe Loop extension		20.03.1916		
Further extension from Elverdinghe Château	03.06.1916		112RC, 10RC fr 22 June	
Part removal of track north part of loop	24.04.1917	07.06.1917	112RC	Probably at Elverdinghe end of extension
Chord and loop to Midland Railway	04.06.1917	16.06.1917	114RC	
Proven - Elverdinghe - Boesinghe line (new British line - aka the Northern Line)				
Proven - International Corner Station	08.06.1916	31.07.1916	112RC	
International Corner - Elverdinghe	27.06.1916		CORCC	
Elverdinghe - Boesinghe	05.08.1916		CORCC	not consistently open before Third Ypres
DT Proven - International Corner	09.01.1917	21.06.1917	112RC	Paused 12 March to 17 April
International Corner station extensions	14.04.1917	31.05.1917	114RC	
DT International Corner - Elverdinghe	25.05.1917	30.07.1917	112RC	
Swiss Cottage Rail Head & dumps				
Construction with MG connections	16.04.1917		112RC	
MG connections removed in favour of LRs	04.06.1917	15.06.1917	112RC	
Crombeke Road - Reigersburg (new British line, aka Great Midland Railway)				
Construction (single track only)	01.01.1917	06.1917	10RC, 17NF (from April 1917), 114RC (from 24 May 1917)	
Peselhoek depot (Midland Railway)		02.05.1917		
Peselhoek Ammunition RH extension	18.05.1917	03.06.1917		
Peselhoek RE Park sidings	03.06.1917		262RC	
Pottenhoek branch and sidings	21.05.1917	15.06.1917	4CRT 262RC	
Steentje Mill and Oosthoek sidings	24.05.1917	10.06.1917	10RC	
Poperinghe Canal Bridge		01.06.1917	17NF	
Chord and loop to Elverdinghe loop	04.06.1917	16.06.1917	114RC	

Narrow Gauge in the Ypres Sector

Area	Line or other facility	Construction work start	finish	undertaken by	major additional works & notes
Reigersburg - St-Jean - Wieltje (new British line, Midland Railway extension)					
	Yser Canal bridge	in place 04.1917			
Hazebrouck - Poperinghe (existing Belgian line)					
	Boeschepe quarry line	24.08.1915	29.10.1915	later CORCC	included 18" gauge gravity funicular
	Hopoutre detraining siding	27.01.1916	10.03.1916	10RC	
	Borre yards	07.03.1916	21.07.1916	120RC, CORCC	also ambulance sidings July 1916
	Borre ROD locomotive depot	02.09.1916	03.10.1916	112RC, 10RC	further work early 1917
	Janghlati (Boeschepe) HQ siding	16.05.1917	03.06.1917	112RC	prob for RCE II HQ train
	Borre workshop sidings	30.05 1917		10RC	
Hazebrouck avoiding line (new British line)					
	Construction	22.02.1917	02.06.1917	296RC, taken over by 268RC March 1917	
Abeele - Ouderdom - Dickebusch (new British line, Ouderdom system)					
	Abeele - Wippenhoek spur and rail head	12.10.1915	05.11.1915	115RC	
	Wippenhoek - Ouderdom	11.11.1915	31.01.1916	CORCC	including branch to Reninghelst/Zevecoten
	Ouderdom RE Park siding & rail head	06.01.1916		CORCC	
	Ouderdom - Dickebusch	25.01.1916	18.04.1916	CORCC	
	Reninghelst rail head		22.03.1916	CORCC	
	Busseboom RE Park	06.08.1916	07.09.1916	112RC	
	Ouderdom RE park sidings	18.09.1916		112RC	
	Zevecoten sidings (Reninghelst branch)	19.09.1916	28.10.1916	112RC	
	Ellarsyde (Busseboom) sidings for LR TS	07.01.1917	05.02.1917		
	Sidings for Fuzeville and Heksken LR dumps	22.03.1917	17.05.1917	10RC	
	Abeele - Ouderdom double track	30.04.1917		10RC	
Kemmel line (new British line, single track)					
	Ouderdom - La Clytte - Mont Kemmel	13.12.1915	03.04.1916	CORCC	
	La Clytte RE Park sidings	18.05.1916		112RC	
	Clapham junction (Douve Valley) to Locre/Brulooze	04.03.1917	04.05.1917	262RC	
Hazebrouck - Armentières - Houplines (existing French lines)					
	Houplines railway bridge reconstruction	06.1915		8RC	
	Armentières Annexe station	20.08.1915	30.08.1915	pr 8RC	
	Houplines line rail replacements	04.1916		8RC	
	Rabot line	23.09.1916	03.1917	279RC, taken over by 109RC 11 October 1916	
	Merris yard (Strazeele)	04.03.1917		109RC	
	Jesus Farm water supply and pipeline	02.06.1917	25.07.1917	109RC 10RC Steenwerck - Armentières section of line	
Douve Valley line (new British line)					
	Haagedoorne branch (Bailleul)	07.06.1916	31.07.1916	279RC, 112RC	
	Douve Valley line (to Clapham Junction)	05.10.1916	26.04.1917	279RC, taken over by 113RC LH (later 296RC) 28 Dec 1916	
	extension to De Kennebak	26.02.1917	14.05.1917	296RC	
	Duke of York sidings, ammunition RH		23.05.1917		Haagedoorne branch
	doubling Bailleul - Duke of York	24.05.1917	30.05.1917	296RC	
	Lindenhoek branch	26.03.1917	06.05.1917	296RC	
Steenwerck - Neuve Église (MG line 75 conversion)					
	Converted to SG	08.1915	pr 11.1915	10RC, poss 109RC	
Steenwerck - Petit Pont (new British line)					
	Steenwerck - Romarin	23.06.1916	23.07.1916	8RC, 109RC	
	Romarin extension to Petit-Pont	11.12.1916		109RC	sometimes called the 'Neuve Église branch'
	Duke of Connaught & La Crèche TS sidings	20.03.1917	31.05.1917	109RC	

aka	also known as
DT	double track
BREB	Belgian Army Railway Engineers Battalion
CORCC	Canadian Overseas Railway Construction Corps
NF	Northumberland Fusiliers (Pioneer Battalion)
RC	Railway Company (RE)
RE	Royal Engineers
RH	Rail Head

Vendroux and other supply depots

Much of the work in 1917 was part of the establishment of huge supply dumps and workshops between Pont-de-Coulogne and Audruicq, along the axis of the main railway from Calais to St-Omer. A major ammunition depot was placed at Audruicq. However, the largest depot, a general *camp de ravitaillement* (supply camp), was established at Vendroux. This was between the canal from Calais to St-Omer and the canal to Guînes, which branched off just south-east of Pont-de-Coulogne. The railway from Calais to St-Omer ran just south of the canal to St-Omer, and the Anvin to Calais metre gauge line left this just east of the bridge over the origin of the Guînes canal and followed this to Guînes. The Vendroux depot was therefore beside both railways and both canals, and was close to the port of Calais and the main lines to Boulogne and Amiens, to St-Omer and Hazebrouck, and to Dunkerque. Another supply depot was developed at Zeneghem, just north of Watten on the line to Bourbourg.

Ghyvelde to Bray-Dunes

This line in the dunes on the north side of the Dunkerque to Furnes line served a sand quarry and was an important source of sand ballast for the tracks. The quarry was operated by CORCC from December 1915 to March 1917 and by 112RC from the end of March 1917 to June 1917. Huge amounts of sand were loaded, for example, on 23 December 1916, CORCC with labour and civilians loaded more than 1,000 tons of sand.

Bergues to Proven line

This was the main new British line into the Ypres area from France. It supplemented the Hazebrouck to Poperinghe line. This line branched off the existing French Hazebrouck to Dunkerque line just south of Bergues station. It was built by CORCC from the Bergues end and 10RC from the Proven end. Construction began on 21 March 1916. The line was completed and fit for traffic on 30 June. Double tracking began in January 1917 and was completed by 10 April 1917.

Edwaarthoek, Westonhoek, Oakhanger, Peselhoek & Shellhoek complex

This eventually became a vast complex of yards, including the Westonhoek standard gauge and light railway transhipment yards (see Figure 7.5). It began with the Edwaarthoek Rail Head in autumn 1915. The last stage was the Westonhoek transhipment yard which was started in May 1917. This reflected the development of light railways in 1917. The Peselhoek sidings merged with those of the Midland Railway in 1917. A proposed metre gauge link from the Cheesemarket at Poperinghe to Oakhanger was never built.

Proven to Elverdinghe line

This was a new British line, also known as the Northern Line. It joined Proven to Elverdinghe in 1916, with double tracking started on 9 January 1917. Doubling work was stopped in March 1917 to concentrate on the Bergues to Proven line, and restarted in April, but was not completed until July. The main stations and yards were at Dosinghem, St-Sixte, International Corner, Ondank, and Elverdinghe. Although some work was undertaken to extend the line in 1916, the line was completed to Boesinghe at the beginning of the Third Battle of Ypres (see Chapter Seven).

Crombeke Road to Reigersburg (Great Midland Railway)

This was the last of the three west to east lines towards the Ypres salient. It lay between the two others, Proven to Boesinghe to the north, and Poperinghe to Ypres to the south (see Figure 6.1). It was constructed as single track and never doubled. Work began in January 1917 but was still continuing in June when the Poperinghe Canal Bridge was completed. The main stations and yards were at Peselhoek, Trois Tours and Reigersburg where there was a triangular junction with the Ypres to Boesinghe line.

Abeele to Ouderdom and Dickebusch (Ouderdom system)

This new British line was surveyed in July 1915, but most was held up from August to November by General Allenby (CO V Corps) who feared it would draw fire into his camps. It was built as far as Dickebusch by June 1917, with multiple depots and sidings, and doubled to Ouderdom in Spring 1917. Major light railway transhipment yards and dumps were established early in 1917 at Ellarsyde (Busseboom), and, on the Reninghelst branch, at Fuzeville and Heksken.

Kemmel line

This new British line linked the Abeele to Dickebusch line, from a junction just east of Ouderdom, with the Douve Valley line. It was always single track and ran north to south to the west side of Mont Kemmel. The first section to La Clytte and Mont Kemmel was surveyed from August 1915 and built from December 1915 to April 1916. It was linked through to the Douve Valley in spring 1917.

Douve Valley line and branches

This new British line was always single track, except from Bailleul to Duke of York on the Haagedoorne branch. The Haagedoorne branch was built in summer 1916, and the rest from October 1916 to April 1917. It was extended to De Kennebak Yard (on the Neuve-Église to Kemmel road) by May 1917, and probably extended towards Wulverghem by June 1917.

Steenwerck to Neuve-Église
Line 75 from Steenwerck to Ypres was converted from metre gauge to standard gauge, from Steenwerck to Neuve-Église, in 1915.

Metre gauge railways
During this period metre gauge lines continued operation, much as described in Chapter Five, and there were some new developments in 1915. By early 1917 most developments were in light railways. Some metre gauge proposals were cancelled and at least one, Swiss Cottage transhipment sidings, dismantled. Metre gauge lines in operation in early June 1917 are shown on Figure 6.1.

Metre gauge developments
The main developments from June 1915 to early June 1917 are shown in Table 6.2. Some existing Belgian lines were dismantled to provide *materiel*, particularly line 193 Coxyde–Coxyde-Bains–La Panne. In August 1915 the Chief Engineers of the British Armies conferred with the DRT to consider their policy for the extension of standard and metre gauge railways. Diamond crossings were to be put in, and signalling arrangements made, with the *10ème* SCFC, where new standard gauge lines crossed metre gauge lines. At that time light railways were only considered suitable as 'trench tramways' which ran up to 3,000yds from the trenches and were manually operated.

Vendroux supply depot
This large supply depot was commenced in spring 1917 by the main Calais to St-Omer line about 5km (3 miles) from Calais centre. Part of the Anvin to Calais metre gauge line ran through the site of the depot. In early 1917, French railway troops installed sidings and

Table 6.2 Metre gauge railway works in the Ypres Sector June 1915 - 6 June 1917

Line or other facility	Construction work start	finish	undertaken by	notes
Vendroux supply depot - sidings and loop line at Écluse-Carrée,		early 1917	French Engineers	
Dual gauge (4 rail) Rivière Neuve (Calais) to Vendroux supply depot		11.06.1917	298RC	provided 3 SG and 1 MG line in this 1.7km (1 mile) section
Connection Pont-aux-Cerfs to Houthem		14.06.1915	pr *10ème* SCFC	linked Bray-Dunes - Hondschoote line (NF) to SNCV line 115
Connection Herzeele to Watou		10.1915	5SCFC	linked Hazebrouck - Hondschoote line (CF) to SNCV line 115
Connection Proven to Crombeke		pr Autumn 1915		linked SNCV line 115 (Proven) to line 107 (Crombeke)
Furnes avoiding line		1915 or 16	Belgian Engineers	linked lines 115 and 29 S of Furnes with line 115 NW of Furnes
Extension to TS facilities Esquelbecq		pr Autumn 1915		MG line Bollezeele - Herzeele SG Hazebrouck - Dunkerque
Ghyvelde depot and yard	11.08.1916	20.09.1916	CORCC, 10RC	At Ghyvelde village, on Bray-Dunes - Hondschoote line (NF)
Stone sidings at Proven	19.03.1917	26.04.1917	112RC	line 115, at Proven station on Bergues - Proven SG line
MG extensions at Railhoek	02.06.1917			
Swiss Cottage Rail Head & dumps				
Construction with MG connections	16.04.1917	24.05.1917	112RC	Branch from line 107 just south of Westvleteren
MG connections removed (for LRs)	04.06.1917	15.06.1917	112RC	
St-Momelin wharf - new sidings	09.01.1917	07.02.1917	119RC	End of line from Bergues & Bollezeele (SE)
Steenwerck - Neuve Église (SG conversion)	21.08.1915	28.09.1915	10RC, 112RC	Southern part of line 75 Ypres - Steenwerck
New sidings at De Seule	12.06.1916	28.06.1916	109RC	
New TS yard Neuve Église (SG/MG)	08.08.1916	10.09.1916	109RC	
Neuve Église - Lindenhoek & Kemmel (line 75)				
Repair & reconstruction	26.03.1916	05.1916	109RC	Maintenance at least to Lindenhoek at least to May 1917
Armentières - Houplines, repairs	22.05.1916	03.06.1916	109RC	Armentières - Halluin line, maintenance at least to April 1917

pr	probably		CF	*Compagnie des Chemins de fer des Flandres*
LR	light railway (60cm gauge)		NF	*Compagnie des Chemins de Fer d'intérêt local du Nord de la France*
MG	metre gauge			
SG	standard gauge		SCFC	*Section de Chemins de Fer de Campagne* (French railway engineers)
TS	transhipment			
RC	Railway Company (Royal Engineers)		SE	*Société Générale des Chemins de Fer Économiques*
CORCC	Canadian Overseas Railway Construction Corps		SNCV	*Société Nationale des Chemins de Fer Vicinaux (Belge)*

a loop line at Écluse-Carrée, including a siding 190 metres long alongside the canal to Guînes. The halt at Écluse-Carrée was at the south-western corner of the depot. To improve standard gauge access, 298RC converted the line to 4-rail dual gauge from the Rivière Neuve junction, near Calais-Fontinettes station, to Vendroux supply camp. This section was 1.7km (1 mile) long and was completed on 11 June 1917. The depot was extended in 1918 (see Chapter Eight).

Pont-aux-Cerfs to Houthem connection
This was a line about 4km (2.5 miles) long linking Pont-aux-Cerfs on the Hazebrouck to Bray-Dunes line with Houthem on the Poperinghe–Furnes line (115) across the border. It was put into service on 14 June 1915. The line was probably constructed by the *10ème* SCFC.

Herzeele to Watou connection
This linked the French line from Hazebrouck to Hondschoote and Bergues (CF) with the Belgian line 115 (Poperinghe to Furnes). It had been surveyed in 1913 but we do not know if this was for military or civil purposes. It was reconnoitred by RCE III in June 1915. The line was 7km (4.5 miles) long. There were triangular junctions at both ends. It followed the road except for a deviation south of Houtkerque village. It was built by the *5ème* SCFC and opened in October 1915.

Proven to Crombeke connection
This linked the SNCV line 115 at Proven north-east to line 107 at Crombeke. It was probably built in Autumn 1915 and certainly finished by the end of 1916. We think it was built by the French *10ème* SCFC. It was about 5km (3 miles) long and it completed a west to east link from Bollezeele to the Belgian line 29 at Oostvleteren. Transhipment facilities at Esquelbecq on the Hazebrouck to Dunkerque standard gauge line were extended in autumn 1915 by French engineers.

Elverdinghe to Zuydschoote
Line 29 was already in operation from Furnes to Elverdinghe. In autumn 1915, line 107 was reconstructed and re-opened north-east from Elverdinghe to Zuydschoote, west of the Yser canal. Some reports say it was opened to Steenstraat, immediately east of the Yser Canal, but the front line at this time was on the Yser Canal and Steenstraat was on the German side.

Furnes avoiding line
This line was built in 1915 or 1916 around the west side of Furnes by Belgian Engineers. It linked lines 115 and 29 south of Furnes with line 115 north-west of Furnes.

Other possible new lines in Belgium
It is reported that there were three new links in Belgium, in the Belgian operated area. We found no definite evidence from primary sources and none are shown on maps of summer 1917 or 1918. If built, it was almost certainly in 1915 or 1916 with dismantling by 1917. These links were south of Houthem (line 115), to just north of Nieuwe-Herberg, where line 29 rejoins from a detour east through Pollinchove and Alveringhem; a line straight up the Ypres–Furnes main road from Linde to Nieuwe-Herberg, cutting out the east detour of line 29; and a link between Kleine-Leysele (line 115) and Linde (line 29). Because of doubt concerning their existence they are not shown on Figure 6.1 or Table 6.2.

Ghyvelde depot and yard
This extensive yard was certainly built by the British in August-September 1916. Construction seems to have been started in a big hurry. On 8 August 1916, RCE II met a French Colonel at Ghyvelde-Ville in connection with 'a scheme for stabling sidings for metre gauge rolling stock there'. Clearly it was linked to the British orders for metre gauge locomotives and wagons. Locomotives began to arrive in December 1916 (see rolling stock below). The next day, CORCC received orders to build the yard to contain 5 miles (8km) of track. The CO visited on 10 August and the party were moved in a CORCC train to a siding at Bray-Dunes standard gauge station. On the next day construction started. The site was very close to Ghyvelde-Ville station. A metre gauge train was provided to take troops to work and track laying began on 16 August. On 23 August CORCC were replaced by half of 10RC. Work was completed by Sept 1916. It was described by RCE II as 'Ghyvelde Garage'. It was dismantled in June and July 1918.

Other works on the Belgian (SNCV) line 115 La Panne to Poperinghe via Furnes
Metre gauge stone sidings at Proven, on the standard gauge Bergues to Proven line, were constructed from March to April 1917 by 112RC. On 2 June 1917 extensions to metre gauge track at Railhoek yard began. This was where line 115 crossed the standard gauge Poperinghe Avoiding line. The wording of the order suggests some metre gauge facilities were there already. Neither are shown on 1918 railway maps.

Swiss Cottage metre gauge branch and sidings
The branch to Swiss Cottage yards was constructed in April 1917 by 112RC. It ran from Line 107 (Poperinghe to Dixmude) just south of Westvleteren into the Swiss

Cottage yards. Work began on 16 April 1917 but was stopped on 24 May. On 31 May, RCE II attended a meeting at Swiss Cottage on remodelling for a light railway connection. From 4 June 112RC were picking up and stacking 3 miles (5km) of metre gauge track that had been laid. This work was completed by June. A railway map of June 1918 shows about 1 mile (1.5km) of branch remaining from Line 107 to the edge of the Swiss Cottage yards.

St-Momelin Wharf
From early January to early February 1917, three platoons from 119RC constructed new sidings on the Aa canal. These included a new wharf at the end of the line from Bergues and Bollezeele, and some construction for the British Army Inland Waterways Transportation.

Steenwerck to Neuve Église (line 75)
Conversion from metre to standard gauge from Steenwerck to Neuve-Église was started on 21 August 1915 by detachments of 10RC and 112RC. The purpose of the conversion was kept secret, but this was probably to provide heavy artillery sidings at Neuve-Église. Work was completed on 8 September 1915 and work on new sidings by October. From November work was taken over by 109RC who laid gun sidings at Neuve-Église and did extensive repairs and maintenance to the track. 109RC ran at least two standard gauge locomotives on the line. The line was also used to carry ammunition. New sidings were built at Le Seau in June 1916 and a new or extended transhipment at Neuve-Église in August. 109RC remained at work on the converted line at least until May 1917, but the line probably fell out of use some time during 1917.

Neuve Église to Lindenhoek and Kemmel (line 75)
From 26 March to 17 April 1916 parties from 109RC were working on track between Neuve-Église and Kemmel. This was night work because of heavy shelling. On 12 April a metre gauge locomotive was unloaded at Neuve-Église, delivered by standard gauge from Steenwerck. Work was suspended on 17 April but then continued through May. Work included repairing breaks from shell damage as far as Kemmel. 109RC remained at work, at least from Neuve-Église to Lindenhoek, until at least May 1917.

Armentière to Halluin
From 22 May to 3 June 1916, a party from 109RC prepared the tramway line to evacuate rolling stock from the Houplines depot which was near the front line. Early in June a metre gauge locomotive and six coaches were delivered to Armentières station from Houplines by 109RC and 8RC. Between 1 and 8 July, a party was preparing tramway vehicles at Houplines for running out to Armentières, and nineteen vehicles were delivered to Armentières station ready to be loaded by the French. From 28 August to 10 September 1916 a party from 109RC were loading a metre gauge crane at Houplines depot. In October and November, 109RC also loaded tramcars, for evacuation from the Armentières electric tramways. In December 1916 the 3rd Australian Pioneer Battalion (3APB) were also dismantling tram cars, and repairing tram lines in the streets of the town. Small parties from 109RC remained at Houplines at least until April 1917 and at Armentières at least until June.

Lines not constructed, or completed
(not shown in Table 6.2 or Figure 6.1) Lines 115 and 107 ended together at Poperinghe Cheesemarket (near the station). In October 1916 there was reconnaissance for an extension about 1 mile (1.5km) north-east to the Oakhanger/Westonhoek yards. Work never started. Light railways to Westonhoek were imminent by this time. In the same month there was an RCE II conference with the CO VIII Corps regarding use of the metre gauge line south-east of Elverdinghe towards Ypres. It was probably not pursued. In May 1917 a branch of line 29 from Elverdinghe to the Waanebeke standard gauge yards was started but work was abandoned within a few weeks.

Metre gauge operations
Allied occupied France and Belgium
At the beginning of this chapter, the end of the Second Battle of Ypres, the SVCFC operated the metre gauge lines north of Beveren (line 115) and Oostvletern (line 29). The lines south of this had been handed over to French Engineers, mostly from the *10ème* SCFC, at the end of 1914. With this handover, the SVCFC had moved its repair shops from Poperinghe to Beveren and its section HQ to Houthem, both on line 115. The main SVCFC traffic control centre moved from Furnes to Klein Leysele in 1916 since Furnes was now very near the front line. Subsidiary depots were at Elsentap near Houthem and Klein Leysele. It is probable that the metre gauge depot at Ghyvelde also supported this area through the Pont-aux-Cerfs to Houthem connection. Transhipment facilities were at Furnes, on the line from Dunkerque, and later at Klein Leysele, on the Adinkerke to Proven standard gauge line, completed in spring 1917. Bollezeele became the main locomotive depot in France, and operations

extended into the Belgian area (see Picture 0.2, page ii). Transhipment facilities were at Esquelbecq, and at the smaller Watou depot on line 115.

By spring 1917 the ROD were also operating the following Belgian lines:

115 Poperinghe to Beveren
107 Poperinghe to Oostvleteren and Noordschoote
29 Oostvleteren to Elverdinghe and
107 Elverdinghe to Zuydschoote
The Proven to Crombeke connection
Part or all of 75 from Neuve-Église to Kemmel

The *10ème* SCFC were operating the east-west lines from Bollezeele, and through the Herzeele to Watou link, possibly as far as Poperinghe on line 115. In March 1917, RCE II conferred at Bergues with the French regarding their taking over a metre gauge line north of Poperinghe. This was almost certainly line 107 from Poperinghe to Oostvleteren and Noordschoote, which led into the part of the front line north of Boesinghe which was taken over at the beginning of July by the French First Army (see Chapter Seven).

German occupied Belgium
Civilian services
East of the front-line areas, metre gauge SNCV services continued to be provided. These became more important as the war progressed, because the Germans increasingly took over use of the standard gauge lines. Powered baggage cars had been constructed by *Ateliers Franco-Belge* in 1909, for the coastal lines and were used from 1915 to haul goods trains on the suburban lines of the main population centres.

Military use
We have not made a detailed study of the German military use of metre gauge lines during this period. However, the Germans probably used line 132 from Ostende as far as Beerst, and line 137 from Bruges to Leke. We think they used spurs from line 137 which had been built around Vladsloo, probably along the path planned for the branch of line 137 to Dixmude, and line 150 from Roulers towards Langemarck, although this had been damaged at the Langemarck end. They also took over the path for an extension to Bixschoote for a light railway. The Germans also sent some SNCV locomotives to France, notably to the *Réseau de la Woeuvre* from Montmédy southwards, where the French had evacuated all the stock before withdrawal.

Rolling Stock
The record of rolling stock, especially locomotives, during this period is very complex. There was a general inter-running between lines, however it was found that French stock frequently derailed on the Belgian lines, particularly at points and crossings and on the sections of tram-style line set in the roads. These often had tight curves.

Locomotives
We have summarised available motive power for this period in Table 6.3.

Existing French stock
For locomotives already operating on the French Flanders lines see Table 6.3.

Existing Belgian stock
As already noted, eighty SNCV locomotives are said to have escaped into Allied territory in 1914. It is likely that thirty-seven were available on the French Flanders lines during this period. All or most were of SNCV Type 18. Some were probably sent to other parts of France, but the balance was running on the lines operated by SVCFC. The ROD also used SNCV stock until the locomotives ordered in 1916 from Britain arrived later that year and in 1917. The French *10ème* SCFC also used some SNCV locomotives. At least three were in French Flanders by May 1916, and six were at Herzeele by September 1916 and were still present in February 1917.

Imported from other French lines
On 8 January 1916, four Corpet-Louvet 0-6-0T locomotives arrived at Ghyvelde from the *Chemins de Fer Économiques des Charentes*. It is reported that the Charente locomotives caused less trouble on the Belgian tramway lines than other conventional French locomotives. This may be because they had no leading or trailing wheels. On 18 February 1916, a Pinguely 0-6-0T, CFV No 1 from *Chemins de Fer Vicinaux de la Haute-Saône*, was sent to Hazebrouck.

In 1915 and 1916, the *Tramways de l'Eure-et-Loir* sent four 0-6-0T Corpet-Louvets, nos 4, 11, 12 and 18, to Flanders. On 6 April 1916 an 0-6-0T Corpet-Louvet No 11 of *Tramways of Eure-et-Loir* was expected but it did not arrive in Flanders until 29 May 1916 after spending some time on *le Meusien* line near Verdun. It left again in July for Montdidier on the Somme. *Voies Férrées Économiques du Poitou* had sent 14 0-6-0T Corpet-Louvets to the Western Front by 18 February 1917, of which 10 were with the *10ème* SCFC in Flanders.

Ordered by British and French Armies
Some technical details of these are shown in Table 6.4. The French ordered fifty 20 tonne 0-6-2T locomotives from Baldwin of Philadelphia in 1916. They were delivered between August and October 1916. These

Table 6.3 Metre gauge locomotives in the Ypres Sector June 1915 - May 1917

Line or other facility	Company (operating)	Manufacturer & weight (tonnes)	configuration empty	arrived	No of locos	left	works numbers	company numbers
from existing French lines								
Hondschoote - Bray-Dunes-Plage	NF	Corpet-Louvet	0-6-0T	resident	1		808	1
		Corpet 16	(0-6-0T)	resident	1		(346)	(2)
Hazebrouck - Hondschoote & Rexpoëde - Bergues	CF	Corpet 16	0-6-2T	resident	6		509, 523-525, 530	1, 31-33, 36
Herzeele - St-Momelin & Bergues - Bollezeele	SE	Blanc-Misseron-Tubize 23	2-6-0T	resident	6		3671-3676	3.661-3.666
from existing Belgian lines								
	SVCFC	SNCV Type 18	0-6-0T	1914	not known			
	ROD	"		1914	not known			
	10ème SCFC	pr SNCV Type 18		bef 09.1916	6			
imported from other French lines								
from *Chemins de Fer Économiques des Charentes*		Corpet-Louvet	0-6-0T	08.1.1916	4			
		Corpet-Louvet	0-6-0T	by 18.2.1917	12*			
from *Chemins de Fer Vicinaux de la Haute-Saône*		Pinguely	0-6-0T	18.2.1916	1			1
from *Tramways de l'Eure-et-Loir*		Corpet-Louvet	0-6-0T	23.1.1915	1	18.11.1916	719	4
		Corpet-Louvet	0-6-0T	13.4.1916	1		1077	12
		Corpet-Louvet	0-6-0T	25.5.1916	1		1085	18
		Corpet-Louvet	0-6-0T	29.5.1916	1	7.1916	856	11
from *Voies Ferrées Économiques du Poitou*		Corpet-Louvet	0-6-0T	by 18.2.1917	10			
ordered by the French Army	10ème SCFC	Baldwin 20	0-6-2T	Aug-Oct 1916	50**		(see Table 6.4)	1-50
ordered by the British Army	pr SVCFC	Alco 26.5	0-6-0T	1915	20		55227-246 (see Table 6.4)	
	ROD	Robert Stephenson 21	0-6-0T	1916-1917	30		(see Table 6.4)	201-230
	ROD	Hawthorn Leslie 21	0-6-0T	1917	20		(see Table 6.4)	231-250

Information in brackets - possible or probable only

MG	metre gauge
SG	standard gauge
TS	transhipment
RC	Railway Company (Royal Engineers)
pr	probably
CF	Compagnie des Chemins de fer des Flandres
NF	Compagnie des Chemins de Fer d'intérêt local du Nord de la France
ROD	Railway Operating Division (Royal Engineers)
SCFC	Section de Chemins de Fer de Campagne (French railway engineers)
SVCFC	Section Vicinale de Chemins de Fer de Campagne (Belgian MG railway engineers)
SE	Société Générale des Chemins de Fer Economiques
SNCV	Société Nationale des Chemins de Fer Vicinaux (Belge)
Alco	American Locomotive Company, USA (constructed at Cooke Locomotive works)
Baldwin	Baldwin Locomotive Company, Philadelphia, USA
*	including the 4 arrived 8.1.1916
**	21 in the Ypres Sector in February 1917

Railways of the Ypres battlefields 26 May 1915 to 6 June 1917

6.1 Corpet 0-6-0T No. 25 (metre gauge) brought from the *Chemins de Fer Économique des Charentes*, and possibly used on the Flanders lines. The Charente locomotives were said to cause less trouble on the Belgian tramway lines than other conventional French locomotives. (*Photo Georges Mangin, collection Didier Oberlin*)

Table 6.4 Technical details of Metre Gauge locomotives ordered for the British & French Armies 1916-1917

Ordered by	French	British			
Manufacturer	Baldwin	Alco	Robert Stephenson	Robert Stephenson	Hawthorn Leslie
Country of manufacture	USA	USA	UK	UK	UK
Year(s) of manufacture	1916	1915	1916	1917	1917
No ordered (1)	50	20	10	20	20
Delivered	Aug-Oct 1916	Oct-Nov 1915	1916	1917	1917
Manufacturers nos.	43854 - 43864　44066 - 44090　44168 - 44177　44219 - 44222	55227-246	3663-3672	3675-3694	3215-3234
Service nos.	1-50 (2)		201-210 (3)	211-230 (3)	231-250 (3)
Wheel configuration	0-6-2T	0-6-0T (bicabine)	0-6-0T (bicabine)	0-6-0T (bicabine)	0-6-0T (bicabine)
Length (m)	7.89	7.29	6.48	6.48	6.48
Height (m)	3.10	3.20	3.88	3.88	3.88
Width (m)	2.20	2.66	2.32	2.32	2.32
Weight empty/loaded	20/ (tonnes)	23/26.5 (tons)	18/21.5 (tons)	18/21.5 (tons)	18/21.5 (tons)
Diam. of driving wheels (m)	0.87		0.865	0.865	0.865
Boiler pressure (atmos)			12.3	12.3	12.3
Heating surface (m^2)					
Diam. of pistons (m)	0.33		0.28	0.28	0.28
Piston travel (m)	0.41		0.40	0.40	0.40

(1) not all used in Ypres Sector (see text)
(2) *10ème Section, Chemins de Fer de Campagne* (*10ème* SCFC)
(3) Railway Operating Division (Royal Engineers) (ROD)

Alco	American Locomotive Company, Schenectady, New York, USA
Baldwin	Baldwin Locomotive Company, Philadelphia, USA
Hawthorn Leslie	R & W Hawthorn Leslie, Newcastle-on-Tyne, England
Robert Stephenson	Robert Stephenson & Co, Darlington, England

were of the conventional French metre gauge type. In 1915 the British ordered 20 0-6-0T locomotives of the *bicabine*, 'tram engine' type from Alco. They were delivered through La Rochelle in Autumn 1915. They were operated by SVCFC on Belgian lines, even though they were rather wide and heavy for them. In 1916 the British ordered fifty lighter 0-6-0T *bicabine* locomotives. These were closely modelled on the SNCV class 18 locos but manufactured in Imperial units. At first ten, and then twenty more, were delivered in 1916 and 1917 from Robert Stephenson of Darlington. Then twenty were ordered from Hawthorn Leslie of Newcastle upon Tyne. They were delivered in 1917. All these locomotives were operated by ROD mostly in the British area of the Ypres Sector. However, then or later a few were moved further south, at least one to the Béthune to Estaires line.

Availability in early 1917

On the 18 February 1917, the *10ème* SCFC reported the presence of the following locomotives in the Flanders (Ypres) Sector. We assume therefore that these locomotives were for their use:

Twenty-one 0-6-2T Baldwins imported from USA in 1916
Twelve 0-6-0T Corpet-Louvets from *Chemins de Fer Économiques des Charentes*
Ten 0-6-0T Corpet-Louvets from *Voies Férrées Économiques du Poitou*
One 0-6-0T Corpet-Louvet from *Tramways de l'Eure-et-Loir*
Three probably 2-6-0T Blanc-Misseron-Tubize from the *Société Générale des Chemins de Fer Économiques* (for the Herzeele–St-Momelin & Bergues–Bollezeele line)
Six 0-6-0T probably Type 18 from Belgium (SNCV)

Other rolling stock

Passenger carriages and wagons were mostly those already on the systems in Belgium and France. One thousand two hundred additional wagons of SNCV type were ordered for the British army in 1916.

Light railways

Boescheppe quarry gravity funicular

In autumn 1915, a standard gauge branch was built to Boescheppe quarry, half a mile south of Abeele, from the Hazebrouck to Poperinghe line. In the quarry a gravity operated 18in gauge funicular was built to extract ballast. On 10 October the 18in gauge tip tubs were found to be top heavy and too wide and it was deemed necessary to spread the track at the crossing place. However, this led to difficulty in keeping the wire rope on the runners. Despite this, trial runs were made on 20 October and the funicular started working regularly on 26 October 1915. On 29 October, the first loaded standard gauge train of ballast left the quarry.

Light Railways in 1915

As in the previous period these were regarded as 'Trench Tramways' with manual motive power, called 'push-car lines'. On the 6 August 1915 the DR met

6.2 Baldwin 0-6-2T No.50 (metre gauge), one of the 50 of this type supplied in 1916 to the *10ème Section des Chemins de Fer de Campagne* (French railway engineers). Twenty-one of this type were in use on the Flanders lines in February 1917. (*Collection Bernard Rozé*)

the Chief Engineers of the British Armies to formulate a policy of extending railways, and discussed the employment of trench tramways. Discussion ended with the recognition that they were suitable only for running back from the trenches about 3,000 yds. Beyond this some form of locomotive or horse transport 'would have to be introduced'. This implies that trench tramways were regarded as 'push-car' only. It was agreed that RCCs would assist locating and putting down these lines, and providing material and lorries.

The strange first deployment of the Canadian Overseas Railway Construction Corps (CORCC)

The first volume of their war diaries covers 24 August to 30 September 1915. The Corps of 17 officers and 441 men arrived in Calais from England on 25 August and travelled that day to Alveringhem in Belgium, on SNCV Line 29. They reported to General Bridges, head of the British Mission to Belgium, and a major of the RE was attached to them. They were attached to the Belgian 2nd and 6th Divisions. On 27 August they looked over work with Belgian officers and on the following day 2nd Company (2CORCC) started work in a materials yard. The 1st Company (1CORCC) arranged duties with officers of the Belgian 2nd Division *Génie*, and on 30 August they started laying 60cm track in front line trenches about 1.5 miles south of Dixmude. 2CORCC moved to Forthem on metre gauge Line 29 to work unloading materials. From 31 August to 30 September both Companies worked laying track and undertaking other work. In early September, their CO was sent to London by the British Mission and returned four days later, and on 30 September he left for London under orders of the QMG. The War Diary ends here. We can only assume that the whole unit must have returned to England, since the next War Diary for the CORCC begins on 31 October at Longmoor (Hampshire), with the same CO. On 1 November 1915, 15 officers and 446 men, in two Companies, left Longmoor, and sailed from Southampton to Le Havre. They were deployed on standard gauge work, one company at Audruicq (between St-Omer and Calais), and the other at Abeele (on the line from Hazebrouck to Poperinghe).

Light Railways in 1916

Some light railway construction was being planned and carried out from early 1916 but it was patchy. In January 1916 RCE III met the Chief Engineers of V and VI Corps (Second Army) about the feasibility of 'mechanically operated trench tramways'. By August and September there was much discussion of 60cm gauge light railway schemes for the Second Army.

Light Railways in later 1916 and in 1917

By autumn 1916 all had changed. This followed the Battle of the Somme and its supply problems, and the report of Sir Eric Geddes. On 30 October 1916, the DR met Geddes, the incoming DGT, with the CRCE and the CO CORCC Colonel Ramsey, to discuss a range of lines including 60cm gauge. By the end of 1916, implementation of the light railway programme was in full swing with metre gauge developments relatively in decline. On 4 December an ADLR for the Second Army arrived and on 21 December he and RCE II examined the whole area from Elverdinghe to Steenwerck looking at schemes for improving transport. There was a marked acceleration in 1917, with the arrival of more construction companies, especially the Battalions of Canadian Railway Troops (CRT), and of dedicated operating companies. In the Ypres Sector light railways were being used in preparation for the Battle of Messines and Third Ypres (see Chapter Seven).

Light railway systems 1916 and 1917

Light railways up to 6 June 1917 are described by line systems from north to south. These works are summarised in Table 6.5. The light railways are shown on the overview map (Figure 6.1) without detailed labelling. Details are shown for each area in Figures 6.2, 6.3 and 6.4.

6.3 German soldiers and a light railway in the ruined village of Wytschaete in 1916.

Table 6.5 Light railway works in the Ypres Sector 1916 - May 1917

System	line	Section	Construction work start	finish	undertaken by	notes
Proven-Boesinghe system	A1	St-Sixte - Elverdinghe	pr 02.1917			
	A1	Elverdinghe - Euston (jct)		bef 01.06.1917		
A lines	A2 & A3	NW of Elverdinghe		pr bef 06.1917		
	A4	Woking jct - Dawson's Corner (Reading) Swiss Cottage LR remodelling		pr bef 06.1917 06.1917	pr 7CRT	
	A5	Woking jct to Euston	bef 06.1917	25.06.1917	4CRT 2CRT	additional works by 4CRT end 05.1917
Westonhoek system	B1	Hagle jct (Triangle) - Mission jct		by 08.04.1917		new ammunition spur (Muskoka Spur) 25 June - 1 July 1917 2CRT
		Brandhoek branch		by 06.1917		
B lines	B1	Mission jct - White Pole Corner (Ypres)		by 03.06.1917	7CRT	
	B1	link to D system Triangle (B1) - Toronto (D1)		05.1917	7CRT	to the south
	B8	Atherley jct (B1) to Machine Gun Farm		by 08.04.1917	7CRT	
	B9	Mission jct (B1) - NE of Brielen		ongoing 06.06.1917	7CRT 2CRT	
	B4	White Pole Corner - St-Jean		06.1917	7CRT	
		Yser Canal crossing	03.05.1917	04.06.1917	7CRT 2CRT	
	B5	White Pole Corner - St-Jean via Potizje	by 06.1917		7CRT	
	B6	B5 nr White Pole Corner - EYpres ramparts via Ypres Station Square	by 03.06.1917	by 03.06.1917	7CRT 2CRT	
	B11	B5 NYpres - Ypres Station Square (B6)	by 03.06.1917	by 03.06.1917	7CRT 2CRT	
	B12	Westonhoek - Hagle jct (Triangle)	01.05.1917	ongoing 06.06.1917	7CRT 2CRT	DT began last week 05.1917
		Westonhoek TS yards	29.05.1917	01.07.1917	7CRT 2CRT	
Busseboom system	(D1)	Pacific - Brandhoek	30.05.1916	12.07.1916	CORCC	became part of D1 line
	(D5)	Pacific - Busseboom	09.1916	24.10.1916	4APB	incl Busseboom dump & TS sidings
D & most	(D1)	Brandhoek - nr Vlamertinghe	27.09.1916	pr 10.1916	4APB	
F lines	(D5)	Busseboom - Ellarsyde		01.1917		re-ballasted 04.1917, including Ellarsyde dump & TS sidings
	(D1)	Pacific - Quintin		by 02.1917		
	(D1)	nr Vlamertinghe - Pioneer jct		by 02.1917		
	(D2)	Pioneer jct - Kruistraat		by 02.1917		
	(D3)	Pioneer jct - Dickebusch		by 02.1917		
	F1	Zealand - Napier jct	03.1917	04.1917	5(NZ) LROC, 7CRT	
	F3	Napier jct - Dickebusch	03.1917	04.1917	5(NZ) LROC, 7CRT	
	D4	Hull (D3) - Brisbane Dump (D2)		04.1917	7CRT	also Brisbane dump built at D2/D4 jct
	D6	Ottawa (D3) - Rupert (D2)		04.1917	7CRT	
	D10	Dunedin (F3) - nr Yale (D1)		04.1917	7CRT	

Railways of the Ypres battlefields 26 May 1915 to 6 June 1917

	D11	Frankton (D2) - Ypres Asylum		04.1917	7CRT
	D12	Dawson (D6) to White House (D4)		04.1917	7CRT
		Winnipeg Marshalling Yard (D1 line)		04.1917	7CRT
		Link to Ouderdom system fr Wellington (F3)		04.1917	7CRT
	D8	nr Yale (D1) - Pocklington		05.1917	7CRT
	D11	Ypres Asylum - Ypres (later B6 jct)		05.1917	7CRT
		Pacific yard	04.1917	05.1917	7CRT, 262RC	're-modelling' in 05.1917. North Atlantic loop
	F2	Montreal (D1) - Wellington (F3)		05.1917	7CRT
		link to B system Toronto (D1) - Triangle (B1)		05.1917	7CRT	to the north
Ouderdom	K1	Ouderdom - Willesden (jct)	06.10.1916	12.1916	112RC
system	K2	Barbed Wire jct - Willesden (jct)		pr by 04.1917	poss 112RC
		Ouderdom - Fuzeville RH & Heksken TS yard	03.1917	by 04.1917	10RC	small extension Heksken 05.1917
	K3	Milky Way jct (K2) - Rossignol		04.1917	7CRT	pr not extended to Kemmel until June 1917
	K4	Wallebeek jct (K3) - Kemmel		04.1917	7CRT
	K5	Willesden jct (K1) - Vijverhoek (D4)		04.1917	7CRT
	K8	Rossignol (K3) - Parrett Farm		04.1917	7CRT
	K10	Milky Way jct (K2) - Kim Camp		04.1917	7CRT
	F4	Napier jct (F1) - Robson		05.1917	7CRT	second link with Busseboom system
Douve Valley		25th Divisional tramway (II Anzac Corps)	NK		
		De Kennebak - front line DT	22.05.1917	ongoing 06.06.1917	4ATC
Steenwerck		Steenwerck - Romarin	28.06.1916	by 30.11.1916	112RC, 109RC
- Romarin -		La Crèche TS yard (LR sidings)	02.04.1917	07.04.1917	109RC
Ploegsteert		Romarin south & east loop	28.04.1917	ongoing 06.06.1917	8CRT	R1 and at least parts of R2 & R6

( )	line later given this title
MG	metre gauge
SG	standard gauge
TS	transhipment
DT	double track(ing)
RC	Railway Company (Royal Engineers)

pr	probably
bef	before
jct	junction
APB	Australian Pioneer Battalion
ATC	Army Tramway Company (Royal Engineers)
CORCC	Canadian Overseas Railway Construction Corps

The Proven to Boesinghe system (A lines)

The railways in this area are shown (with the B and B9 systems) in Figure 6.2. The lines were centred around the Proven to Boesinghe standard gauge line (the Northern Line). Between August and October 1916 there was extensive discussion with VIII Corps who were responsible for this area at the time. A final scheme was surveyed by RCE II beginning on 25 November. This consisted of the A lines. The AB and Z lines were constructed later. We have little information about construction until the arrival of 4CRT on 27 May and 277RC on 28 May. Construction began in early 1917 with the A1 line from St-Sixte/International Corner to Elverdinghe and extended to Euston (between Elverdinghe and Boesinghe) by early June at the latest. It is probable that lines north-west of Elverdinghe, at least A2 and A3, were constructed in spring 1917. A4 and A5 south and north from Euston respectively were being worked on by 4CRT in early June. It is possible that the A6, A7, and A10 lines had also been built, or at least started, by this time.

The A lines were served with goods to transport from transhipment sidings at St-Sixte and the nearby International Corner on the Proven to Elverdinghe line. Initially the dumps and transhipment at Swiss Cottage were to be served by metre gauge lines, but this was changed to light railway. On 31 May 1917, RCE II and others met at Swiss Cottage to agree the remodelling, and the work was done in June. Thereafter it is likely that much of the goods transported on the A system came from the Swiss Cottage yards.

Westonhoek system (B lines and B9 system)

The railways in this area are shown (with the A system) in Figure 6.2. This was initially known as the Vlamertinghe–Ypres or B lines system but finally known as the Westonhoek (or Westenhoek) system. We have identified the B9 system separately because this became a Corps system during the Third Battle of Ypres (Chapter Seven).

7CRT was formed at Purfleet, which was the Canadian railway base, on 8 March 1917, and HQ and all four companies arrived in the Poperinghe area in early April 1917. They were employed on the construction and maintenance of the Westonhoek, Busseboom and Ouderdom systems (see also the next two sections). Between 8 and 30 April they graded 9.67 miles (15.6km) and laid 1.23 miles (2km) of track. The

6.4 A light railway train hauled by Hunslet 4-6-0T No. 309 carries a working party of British troops near Elverdinghe in the snow in February 1917. (*National Library of Scotland 74300808 License CC BY 4.0*)

works map indicates that they ballasted the B1 line from The Triangle through Mission Junction and well on towards Ypres. The Triangle junction linked the B1 line, which included the link line to D1 (Busseboom system) across the main Poperinghe to Ypres line, and the B12 line to the Westonhoek transhipment yard. They also graded the B8 line from Atherley junction to the area of Machine Gun Farm, also well in towards Ypres.

4CRT did some additional work on the B1 line in late May, but the main works were undertaken by 7CRT. D Company started work on the B12 line on 1 May. On 15 May, a location was reconnoitred for the Westonhoek yards, plans were inspected and on 29 May D Company moved to Westonhoek for construction of the yards. By the end of May about 2 miles (3.2km) of track had been laid in the yards and associated spurs and sidings, and ¼ mile (0.4km) of the B12 line had been double tracked, probably at the yard approach. The yards were declared in shape to receive transhipment of ammunition on 2 June. A plan of the Westonhoek and associated yards, as they were in April 1918, is shown in Figure 7.5.

On 1 May, B company took over construction in and near Ypres and began constructing dugouts in a very heavily shelled area. The B4 and B5 lines from White Pole Corner to St-Jean and Potizje had already been started. On 3 May, parties started work on the B4 line Canal crossing ½ mile (0.8km) from the end of the Yser Canal. A site was chosen where the Canal had significant banks on each side. The intention was to link up the B4 line from White Pole Corner to St-Jean. It was done as a 'dirt fill' since the Canal here had no water. The crossing had a drainage culvert and a canal-side passage for men. It was designed for the light railway and tanks to cross and required 12,000 cu yds of dirt. It was declared completed and ready for traffic on 4 June 1917. From 11 May they also worked at night on the B5 line south from St-Jean towards Potizje.

In May 7CRT also worked on the B9 line from Mission junction which they completed to just north-east of Brielen by the end of May, and the B6 line from White Pole Corner to Ypres Station Square. During May they laid almost 14 miles (22.5km) of track, as well as maintaining 29 miles (47km). At this time 7CRT had many casualties especially around Ypres. These included a captain and a lieutenant killed in early May.

By early June, 7CRT had suffered more heavy casualties and it was decided they should be relieved by 2CRT. HQ, A and B Companies arrived on 3 June at Poperinghe from Bapaume, having travelled by train overnight. They arrived at the lower end of Poperinghe yard at 7.22am, but heavy shelling held up their move into the main yard. They had to stay on the train until 9am. That same evening, HQ was visited by LRCE II who outlined the work ahead. They were to take over all B lines east of Mission Junction, except the B9 line, and also take over the D11 link from south Ypres to Ypres Asylum on the Busseboom system. By the next day arrangements had been completed. Major Gibson of 2CRT walked all the lines east of Mission junction with Major Lumsden (ADLR II). All these lines were under constant bombardment with high-explosive and gas shells. Forward of Ypres they were also swept periodically by machine-gun fire. Major Gibson commented that the lines in Ypres would prove particularly difficult owing to flying brick from shattered buildings. As the work was so hazardous it was arranged that they should be relieved every few days.

On 5 June Captain Anderson, two Lieutenants and sixty men of A Company 2CRT left HQ at 8am, by motor lorry to the road crossing of the B1 line at Hagle. They were to entrain at 9am but the train did not come until 11am, then they were delayed by traffic on the line before they detrained at 1pm at the Yser Canal bridge on the B4 line. 7CRT were relieved and returned on the same train. 2CRT moved into the dugouts of corrugated iron and sandbags which were in the east bank of a stream which paralleled the Yser Canal. They were hidden from air observation by trees and foliage. On the immediately opposite bank was a 4.5in Howitzer battery and 100yds behind a 6in Howitzer battery. On 6 June, which was the evening before the Battle of Messines, these fired all night, so they had little sleep. They worked during the day in Ypres but did the more dangerous work further east at night. There were carpenters working on the Canal bridge during the day, and in and around Ypres much brick rubble and wrecked walls needed clearing.

Dangers further east are illustrated by work on the B5 line on the St-Jean to Potizje road on that same night, 6 June. A Captain and fifteen men met seventy-five men from 16 Rifle Brigade at 8.30pm. They formed two parties, one in St-Jean village and one further south. However, the latter found it impossible to start work until 9.30pm because nearby Howitzer batteries were being heavily shelled. Then further shelling set fire to nearby farm buildings, in the light of which the party came under machine gun fire. This was just under a mile from the front line.

Rolling stock (B lines)

We have no information on rolling stock used by the operating company, however when 2CRT arrived for construction in early June 1917, they were given two Simplex 20 petrol tractors. They transferred two of the heavier petrol electric tractors to 7CRT, as these were considered unsuitable for work in the forward areas.

Busseboom System

The Busseboom system to the end of May 1917 is shown in Figure 6.3. This system was started immediately south of the main line from Poperinghe to Ypres, 1 mile (1.5km) east of Poperinghe. When it was begun in spring 1916 it was one of the first proper light railway lines in the Ypres Sector. It began from Pacific Sidings, constructed as a Royal Engineers park in September 1915. At the end of May 1916, a lieutenant and thirty-seven men of CORCC moved to near Vlamertinghe. By 12 July they had constructed the first section from Pacific Sidings to Brandhoek (about 2 miles, 3km). Most of the CORCC detachment left but four men remained, presumably to operate the line. We do not know its motive power, but it may have been mules. It was known as the 'Canadian' or 'Dinkie' line.

In September 1916 the 4th Australian Pioneer Battalion (4APB) moved into the area and from 10 September to 24 October 1916 worked on the 'Decauville tramline' doing maintenance and construction. They worked with prisoners at Busseboom sidings and 'Corps Dump' and almost certainly constructed the line to there from Pacific. This became the D5 line. They began construction at Brandhoek near Vlamertinghe on 27 September and built at least 1 mile (1.6km). They probably operated lines as well as doing construction and maintenance. They suffered casualties, including eight killed. They moved away in October 1916.

By the time 5th New Zealand LROC (5NZ LROC) arrived to operate the system on 27 February 1917, the lines had been extended but we do not know who constructed these lines. A line west from Pacific to Quintin and east from near Vlamertinghe to Pioneer junction had been built. This was later all called the D1 line. Also, there were lines from Pioneer Junction east to Kruistraat and south to Dickebusch, later called the D2 and D3 lines. A loop, later called the D4 line, from just west of Kruistraat to the Dickebusch line at Hull, had also been built. 5NZ LROC initially undertook some maintenance but were hampered by lack of equipment, for example initially they only had two shovels. The track was in poor condition with light (12lb/yd = 6kg/m) rails but some were replaced with heavier (20lb/yd = 10kg/m) rails in April to May 1917.

7CRT arrived in the area on 2 April 1917 and 8CRT on 22 April. 7CRT did the main construction from April as well as working on the B lines from Westonhoek. Before the Canadians arrived, 5NZ LROC had started some construction, including the F1 and F3 lines linking Pacific to Busseboom. All this work was taken over by 7CRT in April 1917. Also, the same month, they built the D6 line from Ottawa on the D3 line to the D2 line at Rupert and the related D11 line from near there on the D2 at Frankton to an old tramway at Ypres Asylum. This line included an 18ft bridge. Standard gauge sidings for the depot at Ellarsyde, extending the D5 line beyond Busseboom, had been built in January 1917. After ballasting, the light railway extension was handed over to 5NZ LROC, probably in April. That same month, 7CRT also built the D4, D6, D10 and D12 lines and the Winnipeg marshalling yard. The latter was a half mile east of Pacific on the D1 line and it became the main marshalling yard for the system. However, when 5NZ LROC began to use it heavily for ammunition, the lines, 'built on a swamp' according to the New Zealanders, had not properly settled and there were sixteen derailments on the first night. Later things settled down with maintenance and there were fewer problems.

Pacific Yard (TS) was worked on in April 1917 by 7CRT and in May re-modelled by 262RC, and the 'North Atlantic' loop into the yard was built. In April 1917, with A and B Companies based at Brandhoek, 7CRT laid nearly 6 miles (10km) of track. In May, with only A Company at Brandhoek, the focus was more on maintenance, with 1.7 miles (2.7km) of new track laid. In May 1917 7CRT also built the D8 line and extended the D11 line from Ypres Asylum into the south of Ypres town. The Busseboom system was linked with the Ouderdom system to the south in April 1917. In May 7CRT also completed the link with the B system north from Toronto on the D1 line.

Operations

5NZ LROC had been formed at Codford Camp, Salisbury Plain, on 5 February 1917 from New Zealand railwaymen who had seen active service in Gallipoli and in France. They arrived at Poperinghe on 25 February. Their main depot was initially at Quintin yard and transhipment ¾ mile (1km) south-west of Poperinghe on the Hazebrouck to Poperinghe line. Their war diaries are lengthy and anecdotal, strong on detail but in some places short on dates. They took over operating the system as it existed at the time, probably at the end of February. They took over operations from 'a few men from ROD RE whose duties were simply to keep the rolling stock in order', although almost certainly they were doing more. This is typical of the New Zealanders' attitude to other peoples' work.

5NZ LROC describe, sometimes quite colourfully, the early operating problems, for example the light rails led to many derailments and crews often re-railed their own locomotives. They carried pieces of wood to help, as otherwise they might have had to walk back to their depot, which could be several miles. In addition, the breakdown gang often worked under shell fire and in bad weather. The section from Brisbane to Café Belge, on the loop (later D4) line, was especially vulnerable to shell fire. Initially they had no water supply after leaving the depot, so crews had to take buckets to shell holes. Insects and even frogs tended to block the steam injectors and tanks. Dead matter caused the tanks to smell badly. Later, two water tanks were provided at Pacific yard with a small Merryweather stationary pump. The supply was drawn from a neighbouring creek. The trains were not always popular with battery commanders since their smoke could not be hidden and there were threats made to blow the trains off the line. In addition, sometimes parts of the line were pulled up by construction units without notification, and there were several accidents.

21TCC was formed at Longmoor and arrived in France on 27 March 1917. They were soon attached to 5NZ LROC, initially assembling track at Quintin Yard, but soon operating trains full time. Just before the Battle of Messines, on 3 June 1917, 15(1)Aus LROC arrived, to help operate the system. They had been formed in Victoria, Australia and arrived in France through Le Havre on 29 May.

Command problems

These arose when non-railway officers tried to give orders to light railwaymen. Since they did not understand railway operations this led to unpleasant arguments, with refusals by train crews to take orders, and accusations of breaches of military discipline. Some train crews were placed under arrest. In March 1917 the High Command ruled that light railway personnel must receive orders only from their own command structure, in the same way as already decreed for broad gauge operating personnel.

Control systems

Initially, the absence of control systems caused trains to meet on the single track with the risk of collisions especially at night. In addition, mule trains from trench tramway connections, operated at Divisional or Corps level, were sometimes operating on the same lines. This led to 'sad results' for the mules. To help matters, guards would carry torches bought in Poperinghe. The only telephones were those at nearby Casualty Clearing Stations or gun batteries, often engaged on other business. However, matters improved by April 1917 with the issue to guards of coloured signal lamps. Telephones were placed at four stations. These provided station to station control and things began to move more easily with a reduction in accidents.

This was followed by the establishment of a central control at Pacific to regulate the supply of wagons, order engine power, and see that crews were relieved at proper times. There was still a lack of adequate telephones. Central control soon moved to Zealand, at the junction of D5 and F1 lines, with a train controller and telephone operator. Also, a liaison officer was established, based at Ellarsyde Yard, to regulate the business of wagon supplies and delivery of orders.

Traffic

In March 1917 only six trains were required to meet all orders from the units. However, with the handover in April of the extension to Ellarsyde (later the D5 line), more trains were run, including those carrying ballast and RE Stores from the Ellarsyde transhipment. 5NZ LROC started ammunition deliveries in a small way from Pacific to Vancouver on the D1 line. This developed until the Company served every battery on the front within the Busseboom system. The D11 line to Ypres Asylum was used mostly to carry ammunition for the 18 pounder batteries. As the system expanded, 5NZ LROC began to run a ration train daily. This was timed so that men isolated in forward areas could have time to cook their meals. With preparations for the Battle of Messines, there was a lot of haulage of RE materiel to the forward area. When they first took over in March 1917, 5NZ LROC carried 600 tons of goods per month, but by early June they had carried 30,000 tons of ammunition alone.

Rolling stock

When 5NZ LROC took over, their motive power was two Hunslet 4-6-0T, two Hudson 0-6-0WT (well tank), and two Decauville 0-6-0T steam locomotives. The Decauville locomotives were probably the two found derelict at Dickebusch Lake in January 1916 by RCE III and recovered. Presumably they had been abandoned by the French when they had served on this front in 1914 and possibly 1915. (For details refer to the tables of British and French locomotives in Chapter Four.) Later these were supplemented by one Simplex 20 petrol tractor for light work.

In April or May 1917 the Angus locomotive yard at Ellarsyde was constructed, probably by 7CRT. This became the main motive power depot for the Busseboom system. With its repair shops it is claimed to have been the largest light railway locomotive depot in the Ypres Sector. One of the light railway travelling workshop trains was based there (see picture 4.3 p. 55). During April or May 1917, new rolling stock arrived; ten 4-6-0T Baldwins and four Barclay 0-6-0WTs. The Baldwins proved a great success and allowed old engines to be overhauled. The Barclays were smaller but 'powerful for their size'. They were intended to replace the Decauville locomotives. These arrivals put 5NZ LROC on a better footing with the units they served. Later 5NZ LROC arranged with the Romarin System, probably 12LROC, to send two Hudson and two Decauville locomotives in exchange for two 4-6-0T Hunslets. 5NZ LROC also received ten 10hp McEwan & Pratt (Baguley) tractors but, as elsewhere, they were found unsuitable for traffic as insufficiently powerful. They were given to trench tramways to replace mules.

One or more PE tractors were tested but found by 5NZ LROC to be too slow. Simplex 40 tractors were ordered instead. They also acquired Crewe tractors. The flexibility to use these with either railway or road wheels was found by 5NZ LROC to improve supervision of ammunition delivery to batteries and helped secure ration deliveries to outlying personnel. Two ordinary motor cars were also made available to enable officers to get in touch personally with the various controlling HQs and to speed things up. To summarise, by early June, just before the Battle of Messines, the Busseboom operating companies had ten Baldwin 4-6-0T, four Hunslet 4-6-0T, and four Barclay 0-6-0WT steam locomotives, supplemented by an unknown number of Simplex 20 and 40 petrol tractors, and some Crewe tractors.

Wagons

By the end of February 1917, thirty-one bogie and sixteen box (two axle four wheel) wagons were on the system. By the end of May, fifty-six additional wagons had arrived, at least sixteen of these were bogie wagons. Thus, they had a total of 103 wagons of which at least 47 were bogie wagons with a 10 ton load. Box wagons had a 2 ton load.

Ouderdom system

The Ouderdom system to end May 1917 is shown in Figure 6.3. A scheme for light railways in the area from Ouderdom south to Kemmel was prepared by RCE II in September 1916. Consultation with the Chief Engineer for IX Corps began in early October. Construction by 112RC assisted by a PoW Company began from Ouderdom to Willesden in early October 1916. 112RC continued to construct and maintain the Ouderdom system during November 1916. After the main Company moved to Bergues on 7 January 1917, one officer and seventy men remained at Ouderdom for work under ADLR II until 16 April 1917.

The timing of the next developments is uncertain but the Fuzeville rail head and Heksken dumps and transhipment were constructed in March 1917 by 10RC, although it is possible that they only did the standard gauge work. On 13 April 1917, C Company of 7CRT moved to Ouderdom and undertook most, and possibly all, the construction in the system, until they left on 28 May. In April they laid over 6 miles (10km) of track, and in May over 1 mile (2km), when there was more attention to maintenance.

Operating

ROD, who had been consulted in September 1916, almost certainly operated the system in its early stages. Fifty men of 33LROC, recently formed in Boulogne, arrived at Ouderdom on 15 March 1917, and a further 100 arrived two days later. They were attached to a detachment of 112RC to 'assist in construction of Ouderdom system'. From later notes it seems that the assistance was mainly in moving materials from Ouderdom Yard to places where it was required by construction companies. On 11 April, a further draft of eighty-four men arrived. By 1 May all controls and stations were manned, for taking ammunition and materials to the forward areas, and these trains began to run on 4 May. From 5 to 31 May the daily average of ammunition and RE materiel hauled was 900 tons. At the end of May, detailed diaries end. 14LROC was formed at Longmoor on 16 May 1917, and was based at Ouderdom from 27 May.

Rolling Stock

In October 1916, 10RC constructed forty 60cm trollies for the Ouderdom line. When 33LROC arrived at Ouderdom on 15 March 1917 they must have taken over existing rolling stock. Further locomotives and other stock were supplied by Second Army in April that year, but we have no details.

Douve Valley

Railways in this area up to 6 June 1917 are shown in Figure 6.4. Although the Douve Valley standard gauge railway, leaving the main St-Omer to Armentières line at Bailleul, had been extended to De Kennebak (just north of Neuve-Église on the Kemmel road) by mid-May 1917, there were no major light railway developments until during and after the Battle of Messines. There were, however, many trench tramways. Although not classed as a light railway, there was already by mid-May a divisional tramway from De Kennebak, which ran east along the valley to the front line in the valley below Messines.

On 20 May 1917, 4th Army Tramway Company (4ATC) RE moved from Arras to the de Kennebak Yard. They were instructed by the Chief Railway Engineer of the 25th Division (II Anzac Corps) to double-track the divisional tramway from De Kennbak to the front line. They had started work the next day and the following day were visited by the Corps' Chief Engineer and were instructed to double-track the 25th Division line to the NZ Division line. This probably means that the line divided, because for the Battle of Messines the NZ Division was south and the 25th Division was north of the main Neuve-Église to Messines road. By 5 June, 4ATC were able to confirm to the Chief Engineer Second Army that they had taken over traffic on the tractor operated section from the 25th Division. This is the section further from the front line, and on the eve of the Battle of Messines they were still working on the line. We have no details of motive power, but it was obviously mechanical.

Steenwerck to Romarin and Ploegsteert

Railways in this area up to 6 June 1917 are shown in Figure 6.4.

Trench Tramways

From October 1915 to May 1916, 109RC worked on tramways near the front line north-east of Armentières. From 27 October to 7 November 1915, they were laying trench tramways at Le Touquet and Mesnil. Parties were billeted at Armentières. On 11 November 1915, 109RC moved to Steenwerck, and the main part remained there until December 1917. From 26 March to 31 May 1916, one NCO, and sometimes a party of men, were based at Ploegsteert or Ploegsteert Woods. This was presumably to supervise trench tramways and probably to make repairs. From 24 to 29 April, they ran ammunition to a gun battery each night at dusk. We have no details of the layout of tramways or of their motive power, but it was probably manual or by mule.

Light railways

From June to July 1916, 112RC were based at Steenwerck. On 28 June they started work on a light railway from Steenwerck to Ploegsteert. When 112RC left, the work was probably taken over by 109RC, who remained at Steenwerck, and were also constructing the standard gauge line to Romarin. We have little further information but on 30 November 1916 109RC were loading a light railway locomotive at Romarin, and on 3 December they were unloading one, either repaired or as a replacement.

By January 1917, 109RC were surveying at Ploegsteert, and in March they were working on light railway track in this area. 109RC constructed the La Crèche standard gauge sidings and dumps from March to May 1917. In early April they also put in the light railway transhipment sidings there which they connected to the Steenwerck to Romarin line. By the end of May these were used to load rails and other equipment. On the 27 and 28 April, HQ and three Companies of 8CRT arrived at Le Seau and started construction on a railway 'east towards Ploegsteert'. This probably means constructing the loop south and east from Romarin and back north to west of Ploegsteert village. Work continued through May but there was often heavy shelling especially in early June around Hyde Park Corner. Work on the section shown on Figure 6.4 as under construction probably finished by the start of the Battle of Messines on 7 June.

Operating

Seven South African LROC (7SA LROC) operated at Romarin until 1 June 1917. They had been formed at Bordon and arrived at Romarin on 18 May 1917. 12LROC took over on 1 June. On 30 May, at Romarin, a Dick, Kerr PE tractor was in collision with a motor lorry, and the tractor driver was pinned under the overturned tractor, suffering a compound fracture of his arm. On 4 June a gun battery was blown up, and four bogie wagons and two box wagons loaded with ammunition were destroyed. We have no other details of operations or rolling stock.

Armentières area

The 3rd Australian Pioneer Battalion (3APB) arrived at Armentières on 28 November 1916. They were billeted in the town and were subject to shelling and gas attacks. They changed billets in December. They undertook various kinds of work between November 1916 and May 1917 when they moved to Nieppe, west of Armentière. Here they did much work on trench tramways close to the front line east and north-east of Armentières and Houplines. They repaired and maintained existing trench tramways and built new ones.

Operating

We have no details of operations or rolling stock available but since we do know that they had engine drivers, they clearly had some form of mechanical power, probably light petrol tractors.

Chapter Seven

Railways of the Ypres battlefields 7 June 1917 to 8 April 1918 Messines and the Third Battle of Ypres, to the final German offensive in the north

This chapter describes railways in the Ypres Sector from the beginning of the Battle of Messines on 7 June 1917 to the beginning of the German offensive in the north on 9 April 1918. The chapter covers the Battle of Messines, the Third Battle of Ypres, the period when the British Fourth Army was based on the Belgian Coast, and the periods after all these up to the German offensive of spring 1918.

Civilian rail services before the Third Battle of Ypres

As described in Chapter Six, although there was heavy shelling and some bombing around Ypres and Messines, there was less shelling in the back areas in France. Also there had been no major offensive on either side from June 1915 to May 1917. Our main source for details of these services is a copy of the *Chaix* (French national timetables) for 1 July 1917.

Standard gauge services

All of the services were reduced compared with those before the war. All of the trains on these lines stopped at all stations and were slow. All were sharing the lines with heavy and at times intense military traffic. Some were mixed passenger and goods trains, with third class passenger accommodation only. With the exception of the line between Hazebrouck and Poperinghe, and possibly one train between Dunkerque and Adinkerke, civilian trains were operated by the *Compagnie du Nord*, the pre-war French operators. A summary of these services is shown in Table 7.1.

Metre gauge services

As for standard gauge services, all were reduced compared with immediately before the war.

Hazebrouck to Hondschoote & Rexpoëde to Bergues, and Hondschoote to Bray-Dunes-Plage

Table 7.2 shows the *Chaix* timetable for civilian services from 1 July 1917. The table in the *Chaix* includes the names of stations between Hondschoote and Ghyvelde (Bray-Dunes), called in this table Ghyvelde-Nord, but there are no posted services. The line between Hondschoote and Ghyvelde (Bray-Dunes) did remain open for military purposes.

The company responsible for the line from Hondschoote to Bray-Dunes, the NF, up to 1914 appears to have left the scene. The line was closed from the standard gauge station of Ghyvelde (Bray-Dunes) to Bray-Dunes-Plage, and army maps of 1918 suggest that it had already been dismantled. Certainly this section was never re-opened after the war.

The timetable posted for civilian services shows one train from Hondschoote to Hazebrouck in the early morning and back in the late afternoon. This connected well with trains from Rexpoëde to Bergues (morning) and back (afternoon). The journey time between Hondschoote and Hazebrouck was slower than before 1914 (about 2½ hours compared with mostly 1½ hours), but the time between Rexpoëde and Bergues had hardly changed.

Bergues to St-Momelin & Herzeele to Bollezeele

Table 7.3 shows the timetable for civilian services from 1 July 1917. There was one train daily each way from St-Momelin to Bergues in the early morning and back in the late afternoon. A second train ran each way on Mondays, market day in Bergues, and starting from Bergues. There was also one train daily each way from Herzeele to Bollezeele in the early morning, and back in the late afternoon, but the connections at Bollezeele were very poor, except on Monday afternoon with the train coming from St-Momelin.

Although the line between Bergues and St-Momelin had not opened until the beginning of the war, journey times in 1914 (best about 1½ hours) were faster than in July 1917 (best about 2 hours). Journey times between Herzeele and Bollezeele were always long, because of long layovers at Esquelbecq, probably for standard gauge connections.

No operating company is posted on the tables for these lines in the 1917 *Chaix*, although before the war this had been the SE. It is likely that the line was operated by the *10ème* SCFC, who handed the Bollezeele base over to the 85th Canadian Engine Crews Company (part of the ROD) in January 1918 (see later in this chapter).

Railways of the Ypres battlefields 7 June 1917 to 8 April 1918

Table 7.1 Ypres Sector Standard Gauge summary timetables from 1 July 1917
Selected stops only shown

Compagnie du Nord (French)

Calais - Lille line (service Calais - Steenwerck only)

		(1)(2)		(1)
Calais-Ville			13.20	
Fontinettes		07.37	13.27	
Pont d'Ardres	arr	07.57	13.44	
	dep	08.05	13.45	
Audruicq		08.21	13.59	
St-Omer	arr		14.35	
	dep		14.50	
Hazebrouck		06.46	15.40	15.56
Steenwerck		07.26		16.48

		(1)		(1)(2)
Steenwerck		07.59		
Hazebrouck		08.31	08.53	
St-Omer	arr		09.38	
	dep		09.50	
Audruicq			10.29	13.47
Pont d'Ardres	arr		10.42	14.05
	dep		10.43	15.01
Fontinettes			11.01	15.21
Calais-Ville			11.08	

Hazebrouck - Dunkerque line

	(1)	
Hazebrouck	07.51	15.51
Esquelbecq	08.36	16.27
Bergues	08.57	16.47
Dunkerque	09.15	17.01

		(1)
Dunkerque	07.04	17.05
Bergues	07.26	17.28
Esquelbecq	07.42	17.45
Hazebrouck	08.22	18.28

Calais - Adinkerke (Belgium) line

	(1)	(1)		
Calais-Ville	07.25		17.33	
Bourbourg	08.23		18.34	
Dunkerque	09.30	09.53	19.12	19.38
Ghyvelde (Bray-Dunes)		10.36		20.21
Adinkerke		10.51		20.36

			(1)	(1)
Adinkerke	08.40		15.20	
Ghyvelde (Bray-Dunes)	09.01		15.39	
Dunkerque	09.38	10.00	16.25	17.30
Bourbourg		10.38		18.18
Calais-Ville		11.36		19.48

Bourbourg - St-Omer line

Bourbourg		18.44
St-Omer		19.35
St-Omer	06.53	
Bourbourg	07.42	

Belgian, ROD operated
Poperinghe (Belgium) - Hazebrouck line

Poperinghe	06.18	13.26	
Abeele	06.44	13.54	
Hazebrouck	07.43	14.48	
Hazebrouck	09.56		16.38
Abeele	11.04		17.43
Poperinghe	11.13		17.58

ROD   Railway Operating Division (RE)

(1) Mixed goods and passengers, 3rd class only
(2) Wednesdays only, market day in Audruicq

The Messines front, during and after the Battle of Messines 7 June 1917 to 8 April 1918

The Messines front at the beginning of the battle can be considered to extend from Hill 60 in the north to the northern edge of Armentières in the south. Hill 60 is immediately on the north side of the Ypres to Comines railway, and Ploegsteert Wood is at the south end. After the 31 July there was no significant change on the Messines front until the German offensive of 9 April 1918, so railway developments on this part of the front are described in this section for the whole period of this Chapter.

The Battle of Messines

The Battle of Messines began on 7 June 1917 and officially ended on 14 June. The Germans held the high ground along the Messines ridge around the villages of Messines and Wytschaete, and then north-east to the area of Hill 60. The high ground commanded much of the Ypres salient, and it was considered essential to dislodge the Germans from the ridge before the main offensive just to the north, the Third Battle of Ypres.

The battle had been planned for a year. Twenty-one tunnels were dug into the ridge to place mines containing more than 1 million tons of high explosive (in total) under key German positions along their front line. The preliminary bombardment began on 3 June. Just after 3am on 7 June, nineteen mines were detonated over a period of 20 seconds, all but the southernmost ones, and one or two which had been abandoned. At 3.12am a party of the 2CRT, working on the light railway on the St-Jean–Potijze road, heard the terrific explosion to the south-east. This was about 3 miles (5km) from the northernmost mine at Hill 60. 'The ground rocked and heaved for fully five minutes', they reported. Immediately every Allied battery in the district opened fire in unison, and this continued steadily for 6-8 hours. In the general

Table 7.2 Hazebrouck - Hondschoote & Rexpoëde - Bergues
Timetable July 1917 (main stops only, excludes *arrêts*)
***Compagnie des Chemins de fer des Flandres* (CF)**

Hondschoote			06.05	Hazebrouck (CdN) (1)		16.08
Killem			06.14	Hondeghem		16.27
Rexpoëde (2)			06.28	St-Sylvestre-Cappel		16.46
Bambecque			06.46	Steenvoorde		17.08
				Winnezeele		17.27
Herzeele		arr	06.55	Herzeele	arr	17.40
		dep	07.05		dep	17.55
junction with line to Esquelbecq & Bollezeele (SE)						
Winnezeele			07.26	Bambecque		18.07
Steenvoorde			07.46	Rexpoëde (2)		18.27
St-Sylvestre-Cappel			08.07	Killem		18.36
Hondeghem			08.25	Hondschoote		18.44
Hazebrouck (CdN) (1)			08.40			
Rexpoëde (2)			06.28	Bergues (CdN) (3)		17.46
Warhem			06.44	Warhem		18.05
Bergues (CdN) (3)			07.00	Rexpoëde (2)		18.19

SG	standard gauge	CdN	*Compagnie du Nord*
MG	metre gauge	SE	*Société générale des Chemins de fer Économiques*

(1) SG lines Calais-Ville - Steenwerck, Hazebrouck - Dunkerque, Hazebrouck - Béthune - Nœux(-les-Mines), Hazebrouck - Merville (Intérêt Local), Hazebrouck - Poperinghe (Belgium)
(2) Junction of lines Hazebrouck - Hondschoote & Rexpoëde - Bergues
(3) SG line Hazebrouck - Dunkerque, origin of MG line to St-Momelin (SE)

Table 7.3 Bergues - St-Momelin & Herzeele - Bollezeele
Timetable from 1 July 1917

	(1)					
Bergues (CdN)	11.18	17.58				
SG line Hazebrouck - Dunkerque, origin of MG line to Rexpoëde, Hondschoote & Hazebrouck (CF)						
Bierne	11.30	18.10				
Steene	11.50	18.30				
Pitgam	12.13	18.53				
Drincham	12.29	19.09				
Bollezeele	12.49	19.29	Bollezeele			15.32
junction with line to Herzeele			*Junction with line Bergues - St-Momelin*			
Volkerinckhove	13.00	19.40	Zeggers-Cappel			15.52
Lederzeele	13.09	19.49	Esquelbecq CdN		arr	16.02
St-Momelin	13.25	20.05			dep	18.09
			SG line Hazebrouck - Dunkerque			
			Wormhoudt			18.28
			Herzeele			18.58
			junction with line Hazebrouck - Rexpoëde, Hondschoote & Bergues (CF)			
			Herzeele			06.35
			junction with line Hazebrouck - Rexpoëde, Hondschoote & Bergues (CF)			
			Wormhoudt			07.09
		(1)	Esquelbecq CdN		arr	07.25
St-Momelin	06.09	14.09			dep	08.52
Lederzeele	06.28	14.28	*SG line Hazebrouck - Dunkerque*			
Volkerinckhove	06.36	14.36	Zeggers-Cappel			09.08
Bollezeele	06.50	14.50	Bollezeele			09.24
junction with line to Herzeele			*Junction with line Bergues - St-Momelin*			
Drincham	07.11	15.11				
Pitgam	07.29	15.29				
Steene	07.49	15.49				
Bierne	08.11	16.11				
Bergues (CdN)	08.20	16.20				
SG line Hazebrouck - Dunkerque, origin of MG line to Rexpoëde, Hondschoote & Hazebrouck (CF)						

CdN	*Compagnie du Nord*	MG	Metre Gauge
SG	Standard Gauge	CF	*Compagnie des Chemins de fer des Flandres*

(1) Mondays only (Monday market day in Bergues)

Railways of the Ypres battlefields 7 June 1917 to 8 April 1918

7.1 Wounded men of the 36th Division being evacuated on a hand propelled trolley on a trench tramway or light railway near Messines on 7 June 1917, the first day of the Battle of Messines. The line is screened from observation on the left, and men with equipment are proceeding past towards the front (*Imperial War Museum ©IWM Q5839*)

German retaliation, with shrapnel and machine gun fire, then a gas wave, further work was prevented. It is said that the explosions were heard in London.

The British Second Army took the ridge and advanced some distance down the far side. Despite a German counterattack, when the battle ended on 14 June the ridge was held. The front lines are shown in Figures 7.1 and 7.2, and the sites of the mines in Figure 7.2. At the north end the British advanced less than half a mile (0.8km) east from Hill 60, and about half a mile (0.8km) south-east down the Ypres to Comines railway.

There was a little further advance by the end of 31 July, the first day of the Third Battle of Ypres, not more than half a mile, but the village of Hollebeke was captured, and the front advanced a few hundred yards further down the Ypres to Comines railway and canal. After that there was no significant change on the Messines front until the German offensive of 9 April 1918.

Railways

The development of railways during and after the Battle of Messines, up to 8 April 1918, is shown in Figures 7.1 and 7.2, with the front lines immediately before the Battle of Messines (6 June 1917), and at the end of the Third Battle of Ypres (10 November 1917).

Standard gauge railways

The Second Army had more than a year to prepare railways for this battle (see Table 6.1, Chapter Six). The front was fed by three standard gauge lines from Hazebrouck, and their branches. The northern part was supplied through the line Hazebrouck to Dickebusch via Abeele and Ouderdom, the southern part through the line to Armentières and branches, and through the Douve Valley line. Standard gauge developments during this period are summarised in table 7.4.

Narrow Gauge in the Ypres Sector

Table 7.4 Standard gauge railway works on the Messines front 7 June 1917 - 8 April 1918

Area / Line or other facility	Construction work start	finish	undertaken by	notes
Hazebrouck - Poperinghe (existing Belgian line, double tracked 1915)				
Borre stores depot and water supply	11.1917	31.01.1918	110RC	
Boeschepe sidings	09.10.1917	21.01.1918	269RC 10RC	
Caestre ammunition dump	03.04.1918		279RC	
Hazebrouck avoiding line (new British line 1917)				
Construction	22.02.1917	21.06.1917	296RC, taken over by 268RC March 1917	
Extensions especially around Carlisle East	01.1918	03.1918	120RC 277RC	linking into Isbergues and Lillers line
Hondeghem detraining halt	22.01.1918	23.03.1918	10RC	with work at nearby Ana Jana sidings
Abeele - Ouderdom - Dickebusch (new British lines, Ouderdom system, begun 1915)				
Water pipeline Yser River to Ouderdom		05.07.1917		
Dickebusch - Diependaal extension	06.07.1917	24.07.1917	10RC	to Elzenwalle, Oaten Wood (Porridge)
Wytschaete branch	30.08.1917	by 30.10.1917	119RC	
English Wood ballast siding	27.07.1917		120RC	for ADLR, probable transhipment
Elzenwalle ramp construction	05.12.1917	11.12.1917	119RC	
Parma Dump and ramp	05.12.1917	18.01.1918	119RC	
Elzenwalle coal siding	26.12.1917		119RC	to supply LR Vauxhall depot
Dickebusch stone & supply sidings	07.07.1917	26.07.1917		
Ouderdom west locomotive yard	12.11.1917	17.12.1917	120RC	
Dickebusch stone siding		10.11.1917	120RC	
Kemmel line (new British line, single track, begun late 1915)				
La Clytte RE Park siding, additional	19.06.1917	28.06.1917	10RC	
Brulooze loop and rail head passing siding	03.01.1918	16.02.1918	119RC	
Paddington chord		01.02.1918	295RC	junction with Douve Valley line
Hazebrouck - Armentières - Houplines (existing French lines)				
Jesus Farm water supply and pipeline	02.06.1917	25.07.1917	109RC 10RC	Steenweck - Armentières section of line
Strazeele (Merris) railway stores yard	13.06.1917	07.07.1917		
Strazeele water supply from Hazebrouck Canal		05.07.1917		pump house Hazebrouck 12-17 July
Bailleul pipe line and ROD reservoir	12.06.1918	12.07.1918	109RC 296RC	
Houplines Bridge - moving	25.07.1917	31.07.1917	109RC	bridge broken by shell fire 23 June
Steenwerck - new RE Park siding	28.11.1917	19.12.1917	109RC 277RC	opened 12 December
Outersteene ambulance train siding	25.03.1918	05.04.1918	277RC	
Douve Valley line (new British line, begun 1916)				
Duke of York sidings, ammunition RH repairs	07.06.1917	22.06.1917	296RC	After explosion 5 June - Haagedoorne branch
extension De Kennebak to La Plus Douve Farm	07.06.1917	10.06.1917	268RC 296RC	begun on day of capture of Messines Ridge
De Kennebak rail head and transhipment	19.06.1917	11.07.1917		standard gauge track completed
extension to Ploegsteert - Messines road	16.06.1917	28.07.1917	295RC 268RC 296RC Portuguese RW Bttn	
further construction	05.08.1917	11.08.1917	109RC	probably gun spurs, loops and sidings
Clapham jct - Paddington DT		27.01.1918	295RC	
Trent ROD water supply		02.03.1918		
Duke of York TS sidings	06.03.1918	27.03.1918		
Steenwerck - Neuve Église (MG line 75 conversion to SG 1915)				
dismantled	04.08.1917	26.09.1917	109RC	25 July 1917 - MG RS removed from Neuve Église
Steenwerck - Petit Pont (new British line, begun 1916)				
Duke of Connaught water supply	11.06.1917	14.07.1917	109RC	from Lys River, for ROD
Duke of Connaught RE yard dismantled	12.12.1917	24.01.1918	277RC	
Romarin loading ramp construction	31.12.1917	04.01.1918	277RC	
Duke of Connaught TS sidings		27.03.1918		

ADLR	Assistant Director of Light Railways		RH	Rail Head
DT	double track		ROD	Railway Operating Division (RE)
LR	light railway (60cm gauge)		RS	rolling stock
MG	metre gauge		SG	standard gauge
RC	Railway Company (RE)		TS	transhipment
RE	Royal Engineers			

For most of the period of this chapter the Messines front was held by the British Second Army under Plumer. On 16 July 1917, RCE IV moved to Strazeele and took over standard gauge railways on the Messines front from RCE II. In November 1917, the Fourth Army took over when the Second moved to Italy. Confusingly, they were renamed Second Army, but the new name was not consistently applied. From 17 January 1918 RCE III replaced RCE IV at Strazeele. When the Second Army returned from Italy on 17 March, RCE III continued to be responsible for Second Army south. From January to April 1918, they also sometimes attended First Army transport conferences. The Fourth Army (now named Fourth Army again) were sent south.

The lines from Abeele to Ouderdom and Dickebusch were operated by the ROD from their base at Borre on the Hazebrouck to Poperinghe line. The 58Can BGROC arrived at Merris Yard on 8 June 1917, and took over from the Royal Engineers, to operate the line to Armentières, and the Douve Valley Line and branches.

There is some confusion about the 'Neuve-Église line' or 'branch'. The metre gauge line between Steenwerck and Neuve-Église (Belgian *Vicinal* line 75) was converted to standard gauge in 1915. This conversion was dismantled in August and September 1917. During 1916, the line from Steenwerck to Romarin was built and extended to Petit Pont by spring 1917. This seems often to have been called the 'Neuve-Église line'. From 27 March to 5 April 1918 the 3rd Survey Section, working for RCE III, prepared land plans for a Neuve-Église branch. This is shown on a British Trench map of July 1918, when this area was in German hands, as a branch off the Petit Pont line between Duke of Connaught and Romarin, and as 'under construction'. It is not shown at all on a September map when the area was back in British hands, so it seems that it was planned but never completed.

Metre gauge railways

The metre gauge line between Steenwerck and Neuve-Église (Belgian *Vicinal* line 75) was converted to standard gauge in 1915. The line north from Neuve-Église was reconstructed and repaired in metre gauge and in use at times at least as far as Lindenhoek.

On 12 April 1916, a metre gauge locomotive was unloaded at Neuve-Église, delivered by standard gauge from Steenwerck. We do not know when operations ceased, but on 25 July 1917 a metre gauge locomotive and seven wagons were loaded onto standard gauge flat trucks at Neuve-Église and forwarded to Hazebrouck. On 4 August 109RC started dismantling the standard gauge track between Duke of Connaught and the standard gauge to metre gauge transhipment at Neuve-Église village. This was all completed by 26 September. There was no further attempt to repair or use any part of *Vicinal* line 75 until after the end of the war.

Light Railways

Light railway developments up to early June 1917 are shown in Table 6.5 for the Ypres Salient front and for the Messines front. Further developments up to 8 April 1918 for the Messines front are shown in Table 7.5.

Ouderdom system of lines

For this system there is some overlap between the Messines front and the Ypres Salient front. Railways in this area relevant to the Messines front are shown in Figure 7.2, and for the whole system in Figure 7.6.

Before and during the Third Battle of Ypres – 7 June 1917 to 10 November 1917

Construction and maintenance

Up to the middle of August, the construction and maintenance of the V system of lines was undertaken by 7CRT, with their HQ near the D1 line just east of Toronto Junction. By 7 June, the V1 line had already been constructed from the K5 junction at English Wood to Voormezeele Junction. This had probably been undertaken by 17th Northumberland Fusiliers in May. On 8 June 7CRT started further construction of the V lines, to follow the advance on Messines Ridge. When 7CRT left the Ouderdom system between 9 and 11 August the V1 and V3-7 lines were completed and in use.

On 10 August, C Company of 9CRT moved to near Dickebusch, to the D6 line yard at Harrow. From here they undertook the general maintenance of the C and V systems, and operated the D6 yard. By the end of August, they had repaired about 120 shell breaks in the lines, and laid more than 1 mile (1.6km) of track. From 25 September, 10CRT were based at La Clytte, and they moved to near Dickebusch on 25 November. They were mainly maintaining and repairing lines in the C system forward areas, often under heavy shell fire, but they also did some construction and maintenance on the V systems.

Operating the Ouderdom system lines

The 5NZ LROC had arrived at the Quintin base at the end of the D1 line (Poperinghe) on 27 February 1917, where they stayed until April 1918. From there they

Narrow Gauge in the Ypres Sector

Table 7.5 Light railway works on the Messines front 7 June 1917 - 8 April 1918

System	line	Construction work start	finish	undertaken by	notes
Ouderdom system					
	K5 DT at English Wood (near V1 jct)	08.11.1917		10CRT	
	V1 English Wood (K5) - Hollebeke Junction (V4)	05.1917	14.07.1917	17NF 7CRT	
	V3 Voormezeele jct (V1) - Bedford jct (C1 C2)	09.06.1917	25.07.1917	7CRT	track laid by 30 June
	V6 D4 jct - Vimy (V3)	06.1917	24.07.1917	7CRT	track laid by 30 June
	V4 Hollebeke jct (V1) - Spoil Bank jct V5	06.1917	18.07.1917	7CRT	
	V5 Langkhot jct (V3) - Bluff jct (C6)	06.1917	31.07.1917	7CRT	
	V7 Swan Château (V6) - Bedford (C1)	26.07.1917	31.07.1917	7CRT	
	V8 Villain (V1) - Iron Bridge (V3)	07.09.1917		8CRT	
	English Wood ballast siding	27.07.1917		120RC	for ADLR, probable transhipment
	V10 branch from V5 to Violet	04.09.1917		8CRT	
	V9 branch from V3 near Iron Bridge	09.09.1917		8CRT	track completed 11 September
	V2 V1 junction (Vesuvius) - Oosttaverne Wood		by 12.1917	8CRT 3APB	
	P lines		by 12.1917	8CRT 3APB	
	W1 Kemmel to Wytschaete		by 17.10.1917		
Ouderdom system related tramways					
	Voormezeele tramway		by 01.1918		
	Battle Wood - Railway Dump tramway		by 01.1918	pr 9NSRP	
	Shrewsbury Forest tramway		21.01.1918	9NSRP	at least in part
	Spoil Bank tramway		31.01.1918	4APB	to Lone Tree Dump - mule track laid February 1918
	extension to Bow dugouts	05.02.1918	ongoing 28.02.1918	4APB	
	Vierstraat tramway (Parma Dump - Ravine Wood)		by 09.11.1917		
Douve Valley & Messines					
	25th Divisional tramway (II Anzac Corps)				
	De Kennebak - front line DT	22.05.1917	by 08.06.1917	4ATC	taken over as light railway 8 June 1917
	Kemmel - De Kennebak		by 12.1917		
	S1 Douve Valley line to Peckham jct (W1)		by 17.10.1917		
	S2 La Petite Douve Farm to S1 (Ontario Farm)				
Douve Valley related tramways					
	Swayne's Farm line				
	Douve Valley tramway				
Steenwerck	Duke of Connaught TS sidings extended		27.03.1918		
- Romarin	X3 Douve Valley line to Dundas (R1 line)		by 17.10.1917		
- Ploegsteert	R6				dismantled from 15 June 1917
	R11 Hyde Park Corner - Dead Horse Corner	01.10.1917	21.10.1917	8CRT	
	R12 Delenelle - Pompadour Farm	17.11.1917	08.12.1917	3APB	
	La Crèche - Merville (First Army area)		02.1918		part of main north - south lateral line
	La Crèche - De Kennebak		03.1918		
	La Crèche - Trent Dump & Duke of York (Bailleul)	03.1918			
	Duke of York TS sidings	06.03.1918	27.03.1918		
Romarin system related tramways					
	Fort Boyd and Prowse Point tramways		by 9 June 1917		repaired by 3APB November 1917
	Warneton line	11.1917	05.01.1918	3APB 2APB	extension of Prowse Point tramway
	Vancouver line	by 11.1917	05.12.1917	3APB	Warneton - Vancouver loop completed 12 Jan 1918
	Caledonian line (Gunners Farm line) & extension	by 12.1917			
	Caledonian loop	19.01.1918	06.03.1918	2APB 3APB	
	Great Northern line (Le Touquet tramway)		by 02.1918		also known as the Californian tramway

( )	line later given this title	ARP	Ammunition Refill(ing) Point
MG	metre gauge	APB	Australian Pioneer Battalion
SG	standard gauge	ATC	Army Tramway Company (Royal Engineers)
TS	transhipment	CORCC	Canadian Overseas Railway Construction Corps
DT	double track(ing)	CRT	(Battalion) Canadian Railway Troops
RC	Railway Company (Royal Engineers)	NF	Northumberland Fusiliers Pioneers
pr	probably	NSRP	North Staffordshire Regiment (Pioneers)
jct	junction	SCFC	*Section de Chemins de Fer de Campagne* (French railway engineers)
ADLR	Assistant Director of Light Railways		

worked partly in the area of the Messines front (K and V lines) and partly to the southern part of the Ypres Salient front (D, F and C lines). At some time during the spring or summer of 1917 they were joined there by the 21TCC.

33LROC was based at Ouderdom by 1 May 1917 (see Chapter Six). They continued there for the Battle of Messines, but we do not know for how long after. There is a photograph of part of this Company in June 1917, with a petrol electric locomotive, and placard saying 'Messines–Wytschaete'. 14LROC was also based at Ouderdom before and during the Battle of Messines and moved a large quantity of heavy ammunition to the front line area. We do not have any details of their later work, but they stayed in the Ypres area with the British Second Army at least until August 1918.

15(1)Aus LROC arrived at the Angus Locomotive Yard (Ellarsyde, D5 line) on 3 June 1917. They had arrived in France only a few days earlier. From this base they operated widely, on both the Ypres Salient and the Messines fronts. They became very fragmented, with a lot of detachments sent to other units, to the annoyance of their CO, and to the detriment of their morale. The Company was re-united when they left on 19 February 1918 for Savy-Berlette, near Arras.

16(2)Aus LROC arrived at Ellarsyde on 12 September 1917, and stayed to 21 October, when they moved to Vauxhall depot at Elzenwalle (V1 line). From here they operated mainly on the V and P lines supplying the northern part of the Messines front around Wytschaete. They remained at Vauxhall until the German offensive of April 1918. In early August 1917, traffic was very heavy, and on 4 August 15(1) Aus LROC opened a new depot at English Wood, on the K5 line near the V1 junction, where new sidings and a transhipment had been constructed. With the growth of the system after the Battle of Messines, 5NZ LROC established another District Control at Hastings (Vijverhoek) on the junction of the D4 and V1 lines, and started to organise a new operating district. Water tanks for locomotives were erected here.

On 20 September the offensive to try to take the Menin Road Ridge (Gheluvelt Plateau) started. From Ellarsyde and the other bases, 16(2)Aus LROC and the other companies operated trains to bring in the wounded and wounded prisoners. With the next offensive on 4 October, to take Broodseinde Ridge, it rained heavily all the previous night, and this continued for several days. All trains on the Ouderdom system were reserved to bring in the wounded. Many 'including German prisoners' were brought in from the forward areas to Casualty Clearing Stations and Forward Dressing Stations. From 5 October, a large amount of ammunition and supplies were moved to the front line areas.

In mid-October, orders had been received that the district up to Vauxhall should be worked by 5NZ LROC, who retained the control at Vijverhoek, Vauxhall District to be operated by 16(2)Aus LROC, and Mimico (D11 line) by 17(3)Aus LROC. Operating from Bedford yard (V8 near junction with V6, C1 and C2) was allocated to 15(1)Aus LROC, which, as their CO complained, meant that they 'must be scattered all over the system'. In this way 17(3)Aus LROC were entirely serving the Ypres Salient front, and 15(1)Aus LROC mostly so. In later October, all the operating companies were fully stretched with all men and motive power working long hours.

By 10 October there were numerous derailments due to the effects of the rain. 16(2)Aus LROC was very split up with detachments stationed all over the system. The reorganisation brought them together and by 14 October most of the unit were at Vauxhall (Elzenwalle, near Voormezeele). Company HQ moved there on 21 October. By 6 November, work on the depot had provided a steam locomotive shed, a petrol tractor shed, and a water system. Previously locomotives had to be attended to without cover, and water drawn from shell holes.

Rolling stock

In June 1917 steam power consisted of American 4-6-0 Baldwins, British 4-6-0 Hunslets, and lighter 0-6-0 Barclays with well tanks. Trucks were mostly bogie. 5NZ LROC received eight additional 4-6-0 Baldwin locomotives at Angus in summer 1917, placed with 15(1)Aus LROC. Soon after 2-6-2 Alco-Cooke locomotives arrived, welcomed because they were less likely to derail in the forward areas.

More bogie wagons arrived during the summer, bringing the total to 650 wagons and 56 locomotives. Even this did not meet needs and loads had to be delivered so that a second trip (which needed to be in darkness near the front line) could be got out of the stock. Two 10 ton cranes were also added, saving time and manual labour when locomotives fell into shell holes.

Accidents and incidents

On 16 August 1917 at 11.30pm, a standard gauge locomotive on the Kemmel line north of La Clytte line ran into a Baldwin 4-6-0T at a crossing, badly damaging the Baldwin and derailing the standard gauge locomotive. The 120RC wrecking crew turned out and attempted to re-rail the locomotive, until the ROD wrecking crew arrived at 4am. The track was cleared at 9.30am. An air raid on 2 September 1917 at Fuzeville station (F4 line) set fire to a train of light railway trucks loaded with 9.2in shells. The yard was cleared by an ROD lieutenant and an artillery lieutenant. Two trucks of shells eventually blew up killing the artillery lieutenant, a light railway lieutenant, and an ROD corporal. The explosion also cut every track in Fuzeville Yard. Repair work in the yard started at 4am on 3 September and was completed by noon. In another air raid on 21 October, 10 bombs were dropped on camps in the Ellarsyde area. One man of 15(1)Aus LROC was wounded, and fourteen Royal Engineers (not railway troops) were killed and twenty-four wounded.

After the Third Battle of Ypres – 11 November 1917 to 8 April 1918
Construction and maintenance
We have no records of any major light railway construction in this area during this period. 8CRT remained in the area until the German offensive of April 1918, but recorded very few details of their work. From December 1917 to February 1918 3APB worked with 8CRT on the maintenance and improvement of the P1 and V2 lines. 2APB took this work over in February 1918.

Operating the Ouderdom system lines
Even after the formal end of the Third Battle of Ypres, shelling of all the back areas continued to be heavy, causing numerous derailments, and aerial bombing was quite frequent. 16(2)Aus LROC based at Vauxhall reported that work was a little slacker, giving them more time to make depot working more comfortable. More troops were being carried towards the front from rest areas, with up to 250 men per train carried in open trucks. December was quieter, with intense cold.

16(2)Aus LROC celebrated Christmas at Vauxhall, but many men were away on detachment and could not join them. Heavy snow on Christmas morning allowed good snowballing. Many of the Australians had not previously seen snow. Dinner was boiled ham and roast beef, potatoes, turnips. plum pudding, and tea, with a liberal issue of rum. Officers and the YMCA gave gifts to the men, and snowballing continued to late at night. There was depression on Boxing Day evening because the post from Australia did not arrive.

Late December and January brought long periods of severe frost, with thaws, when the mud was dreadful. Traffic in materiel and men became heavier again in January and February, mainly personnel trains, but also carrying RE materiel, ammunition, and supplies. Artillery batteries were moved to forward positions. By the end of February the weather was milder, but there were very cold spells again in March.

On 18 February, 15(1)Aus LROC moved to Savy-Berlette near Arras (First Army). The unit was re-united for the move. Most of the Company travelled for 27 hours in standard gauge wagons. 16(2)Aus LROC had 117 South Africans attached for relief, and from 23 February ten men per week were allowed six days' leave in Paris. On 5 March 1918 the Australian 15th, 16th and 17th LROCs were officially renamed the 1st, 2nd and 3rd Australian LROCs, dated back to 15 February, and came for administrative purposes under the CO of Australian Railway Companies. The Broad Gauge ROCs were also renumbered.

From 17 March there was markedly increased shelling, with gas in forward areas, holding up personnel traffic. 2(16)Aus LROC reported only minor damage at the Vauxhall depot, but one case of shell shock. On 22 March all leave was cancelled but the programme of allowing four rest days in rotation continued, and on 24 March there was a practice 'stand to' with all locomotives and tractors manned. 21 March was the beginning of the major German offensive south of Arras. In early April work was very light.

In early April 2(16)Aus LROC at Vauxhall had fifty-four men detached, sent to 33LROC, 14LROC, 3(17)Aus LROC, and Light Railway central workshops. Sixty-one men were attached, mainly South Africans and Royal Engineers. Only two officers were at Vauxhall, the others were acting as Assistant Locomotive Superintendent, Second Army, CO Second Army workshops, and District Locomotive Superintendent at Mimico, the 3(17)Aus LROC base (Busseboom system).

Rolling stock
At the end of 1917 15(1)Aus LROC reported the available rolling stock each week since June. It is clear from later entries that this is stock for the whole operation of the Ouderdom and Busseboom systems. A small sample shows the change over the period:

Week beginning	Steam locomotives	Simplex tractors	PE tractors	Bogie wagons
1 June 17	92	48	10	702
26 Oct 17	159	80	40	1003
28 Dec 17	137	70	42	805
(PE - petrol electric)				

Tramways
Voormezeele, Spoil Bank and Shrewsbury Forest

On 12 January 1918 4APB arrived at Godezonne Farm near Vierstraat, after a tiring move from Péronne. A detachment had already moved on 6 January to Voormezeele to take over the Spoil Bank tunnels. On 14 January D Company took over construction of the Spoil Bank tramway, which had been started but needed 1,000yds (900m) to finish. 4APB took over from the 9th Battalion North Staffordshire Regiment (Pioneers), who had been working on trenches and board walks, but also on the Shrewsbury Forest and Battle Wood tramways. Their handover report to 4APB shows that the Shrewsbury Forest tramway started from Knoll Road near Verbrandenmolen, at the end of the Voormezeele Tramway and the junction with the Battle Wood Tramway. It had been constructed to Halfway House, and was being extended by them to The Glen, very close to the front line. The Battle Wood Tramway had been completed, probably by them, and was being maintained. At that time there was mule traction on this tramway as far as Molen Dump, with manual traction beyond.

The Spoil Bank Tunnels were built in the bank of the Ypres to Comines Canal and were designed to house five battalions of infantry. Spoil Bank was also the junction of the V4 and V5 lines, and D Company 4APB arranged with light railways to take men to work there at 7.30am and back at 3pm. However, on 21 January they moved to a nearer camp site and walked to work. One platoon and 100 infantry as labour took over the Spoil Bank Tramway on 25 January, in bad order and incomplete, and by 31 January the line was running to the end at Lone Tree Dump. The last 400 yards (360m) were very rough, with large shell holes full of water to cross, which were extremely muddy and hard to drain.

During the week ending 28 January, 4APB were also pushing up materiel from Rodney Dump (by the V5 line) and the Voormezeele tramway to Railway Dump at the end of the branch tramway to Battle Wood, by the disused Ypres to Comines railway. This confirms that at that time this was, in the Canadian parlance, a 'push car line'. This is surprising as the North Staffordshire handover in January said that there was mule traction to Molen Dump. However, at the end of February there were tractors running from Rodney Dump to Molen Dump on the Battle Wood line, then 'push tram' to Railway Dump and Olaf Avenue, to which the line now extended. This would make the manual section about half a mile (0.8km). Mule track was laid alongside the Spoil Bank tramway.

In February, 4APB started to extend the Spoil Bank line about half a mile (0.8km) to Bow Dugouts, which required camouflage near the end, because it was under German observation. They were also maintaining the main tramway back to Voormezeele, and the line on to Shrewsbury Forest. At the end of February they handed the lines over and moved to the Douve Valley area. Probably 1APB, who were repairing part of the Spoil Bank line in March 1918, took over. In any case, all these tramways were captured by the Germans in April 1918.

The Vierstraat Tramway

We do not know who built this line, or when. However, in November 1917 5APB surveyed the line in order to make improvements and undertake maintenance. The tramway ran from Parma Dump (P1 and P2 light railways) just east of Vierstraat to Ravine Wood and Rose Wood, near Verhaest Farm. On the way, it crossed the standard gauge line to Wytschaete, and the V2 light railway, near In-de-Sterke Cabaret. It is known that in November 1917 8ATC were also working in this area. The Vierstraat line was still being maintained by 5APB in March 1918, when it was partially converted from 9lb to 20lb rails, suggesting that tractor operation was being planned.

Messines, Douve Valley and the Armentières area

Railways in this area are shown in Figure 7.2.

Before and during the Third Battle of Ypres – 7 June 1917 to 10 November 1917
Construction and maintenance

On 8 June, the ADLR took over the 25th Divisional tramway east from de Kennebak as a light railway. Construction of new lines began on 9 June, by No 4 (D) Company of 8CRT. They found the ground in awful condition owing to heavy shell fire, but work progressed rapidly, in fine weather. Towards the end of June the weather became stormy and the roads were difficult. By 8 August, damage was so bad that the R2 line was partly closed. No maintenance work was done on forward lines, because 3 (C) Company of 8CRT had been temporarily withdrawn from the area. They commenced reconstruction of parts of the R2 line, which had been almost completely destroyed, on 26 August.

There is very little detail in the diaries of 8CRT about construction work. The light railway system was extended east as the R11 line from Hyde Park Corner to Dead Horse Corner in October 1917. Between June and December 1917, nineteen men of 8CRT were killed or died later of wounds.

Operating the Messines system lines

The operating company in the Douve Valley and around Messines was probably one of the South African LROCs (7 or 8SA LROC), based at De Kennebak. On 8 June, what had been the 25th Divisional tramway

(II Anzac Corps), from De Kennebak to the front, operated by 4ATC, was taken over by the ADLR, becoming officially a light railway.

The main operating company in the southern part of the area was 12LROC, based at Romarin. They had taken over the Romarin system from 7th South African LROC on 26 May. On 7 June they conveyed a record tonnage of ammunition, and also ran hospital trains from Hill 63 to Pont d'Achelle Dressing Station. On 11 June they moved 2,250 tons of ammunition, another record for them. Even after the formal end of the Battle of Messines, shelling and bombing continued daily. The camp and yards of 12LROC were hit. On 1 August a Baldwin locomotive was damaged in the yard, and on the next day the R1 line (main line) was impossible to work nearer the front line. The R2 line was cut many times, and on 8 August was partly closed, but there was by this time little traffic here anyway. On 5 August there were two direct hits on the light railway loading platforms at the Duke of Connaught transhipment.

From 10 October, 12LROC ran a daily ration service for the 8th Division (VIII Corps) from De Seule (La Crèche yard) to Hyde Park Corner (R1 line). By the middle of October there was little traffic other than personnel, but personnel increased in early November. There was also a big increase in RE materiel for forward areas from 9 November.

Rolling stock
The locomotive yard for 12LROC was at Organ Farm, on the line from Romarin to Steenwerck station and the La Crèche transhipment yards. On 17 June, the Superintendent of Light Railways (SLR) instructed 12LROC to have ten 4-6-0 Baldwin Locomotives loaded onto standard gauge trucks and shipped to Ellarsyde (Ouderdom system) because of a shortage of motive power there. A further ten Baldwins were despatched from Romarin to Angus (Ellarsyde) on 29 July, but this time they travelled on the light railway system. They left at 10 minute intervals, commencing at 8am. A heavy downpour of rain started later in the morning and continued throughout the day. The S1 and X3 lines were newly constructed, and the ballast had not settled, and it proved impossible to get the locomotives through. Three derailed on the S1 line and one fell on its side at Enderby siding. The remainder put back into the de Kennebak yard. We do not know what happened next but presumably they were eventually delivered to Ellarsyde.

On 31 July (first day of the Third Battle of Ypres) the CO of 12LROC at Romarin received orders from the SLR at 8.30pm to transfer as many bogie wagons as could be put together quickly for ammunition work in the northern area. He and a party left at 9.15pm with eighteen wagons, two Simplex 20 tractors and one Barclay locomotive. A drenching downpour started and continued all night. The party had six derailments *en route*, and saw the track blown up behind and ahead of them but delivered eighteen bogie wagons and two Simplex 20 tractors at Angus (Ellarsyde) at 9.45am on 1 August. The record does not say what happened to the Barclay.

Tramways
We know more about the trench tramways in this area than in some others, because of the presence of all five APBs at various times between June 1917 and early April 1918. Towards the end of this period, they had more involvement with light railways as well. From June to August 1917 3APB and 4APB constructed and repaired tramways, 4APB mainly in the Messines and Douve Valley section, and 3APB mainly in the Romarin section to the south of this. Often only one company of an APB worked on railways, or sometimes only one platoon, while the rest of the Battalion undertook other work, mostly on roads and on trench systems. In June and early July, 4ATC was also working in the Douve Valley.

For details of particular tramways see the section below (after the Third Battle of Ypres). Between June and August 1917 fifty men of 1, 3 and 4APBs were killed or died of wounds, thirty-five in June alone.

After the Third Battle of Ypres – 11 November 1917 to 8 April 1918
Construction and maintenance
8CRT remained in the area as the main light railway construction company until the German offensive of April 1918. In February, the Battalion moved about eighty men with horses, mules and some construction equipment and road wagons, to a site in the rear, west of Bailleul, but the majority stayed at Steenwerck and Romarin.

From late December 1917, a company of 3APB maintained and repaired the Wulverghem system of light railways. These ran from the De Kennebak transhipment yard to La Petite Douve Farm on the Ploegsteert–Messines Road, and north to Peckham junction on the S1 line from Kemmel to the Wytschaete area. The De Kennebak yards are shown in Figure 7.3.

From the same time, another company of 3APB maintained and repaired the Romarin system of light railways. At the beginning of February 1918, 2APB took this over, working with 8CRT. In early March, the APBs concerned went for training and then to the Somme Sector. More light railway lines were constructed and opened in late 1917 and early 1918, mostly in the back areas, and mostly lateral lines linking systems from north to south. The R12 line from

Railways of the Ypres battlefields 7 June 1917 to 8 April 1918

7.2 The De Kennebak yards on the Douve Valley line in winter. It is likely that was the winter of 1917-1918. However, the smart state of the men (without helmets) and the locomotives might suggest 1918–19, but it is probable that these yards were not so fully restored after the German retreat of September 1918. Behind the standard gauge lines at the front are multiple light railway lines, probably in the locomotive yard area (see Figure 7.3). There are multiple steam locomotives on the rear line, including Hunslet 4-6-0T No. 315, Baldwin 4-6-0Ts including No. 862, and probably at least one Alco-Cooke 2-6-2T. At the right hand end is a boxcar converted into a travelling office. In front of the locomotives is a Simplex 20 tractor. The number (18) is probably a local numbering. (*Hastie Collection, Royal Engineers Museum Library 23/552 No.8*).

Delenelle to Pompadour Farm (November and early December 1917) was constructed by 3APB. The line from Kemmel south along the road to de Kennebak had been opened by the end of December 1917. This almost certainly used the formation of the *Vicinal* metre gauge line 75.

The line from La Crèche to the south, to Neuf-Berquin and Merville in the First Army area, was built in February 1918. This was part of the north to south lateral line, also dubbed the 'Cape to Cairo line', which in February was completed all the way to Noyon on the River Oise. The lines from La Crèche north to De Kennebak in the Douve Valley and west to Bailleul (Trent and Duke of York yards, on the standard gauge Douve Valley line) were completed in March 1918. Apart from the R12 line we do not know who constructed these lines, but it is likely to have been 8CRT.

Operating the Messines system lines

The Romarin system continued to be operated by 12LROC. By early December, the daily quantity of goods moved was again nearly 2,000 tons. The R12 line from Delenelle to Pompadour Farm opened for general traffic on 8 December, and on 12 December the Delenelle control was reopened. On 31 December, 12LROC took over the De Kennebak system and were now operating the entire southern system from Peckham and Kemmel to Steenwerck. However, on 12 February 1918 they were moved to Westonhoek to operate the B system lines in the Ypres Salient area. We do not know who took over operating on the Messines front. In any case all of these lines were captured by the Germans in April 1918.

Once the north to south main lateral line was opened in February, it was certainly used. One such use was described in the personal diary of a tractor driver of

1LROC, and the story is told in full in *Narrow Gauge in the Somme Sector* (Pen & Sword, 2019). The journey was made from International Corner to Foreste in the Somme Sector by a detachment of 1LROC, between 2 and 6 March 1918. The first night they stayed at Romarin, and then used the main lateral line to get to Barlin, HQ of First Army Light Railways, south of Béthune, for the second night. They went south in support of the Fifth Army who had transferred there from the Ypres Salient to the Somme area in December 1917 and January 1918. No doubt other units made the same journey for this reason, but there are no records.

Following the German offensive south of Arras, which began on 21 March 1918, the main lateral line was used as part of the effort to concentrate railway troops and labour units north of Amiens, to help with the construction of a defence line. This operation is described by Col. Henniker in *Transportation on the Western Front* (1937) and is fully reported in *Narrow Gauge in the Arras Sector* (Pen & Sword, 2015). Twenty units were transferred from the Second Army area, from Westonhoek and La Crèche, and eight from the First Army, mostly from Merville. The trains consisted of eight trucks each and travelled in convoys of five trains. Each truck could take sixteen men with tents, tools and rations, so that each convoy could accommodate two companies of men. On 27 March, twenty trains left La Crèche in four convoys, carrying 2,404 other ranks, on 28 March nineteen trains left carrying six labour companies, and on 29 March twelve trains left carrying four labour companies. We do not have details of departures from Westonhoek, but 12LROC noted the departure of the first convoys from there on 26 March.

Between 25 and 30 March, 1(15)Aus LROC, based at Savy-Berlette, provided a total of 114 steam locomotives and 45 petrol tractors either at Gouy-Servins (25–27 March) or at Haystack Loop (28–30 March) to meet these trains. Both locations were on the main north–south lateral railway. Presumably this allowed the motive power arriving with the trains to return north, but we do not know what arrangements, if any, were made for the trucks. Even if some trains required two tractors, this means the arrival of well over 100 trains, including the 51 known to have left La Crèche.

The men and materiel were transferred at Villers-au-Bois, Camblain l'Abbé, or Savy-Berlette onto the Lens to Frévent metre gauge line, one suspects mostly at Savy, being the 1(15)Aus LROC base. They travelled to Avesnes-le-Comte, where they were at the disposal of the temporary DGT HQ at Bernaville. After the offensive which began on 9 April, the lateral line was lost, with all the other light railways in this area, and across the First Army area to just north of Béthune.

Tramways

From September 1917 to the end of the Third Battle of Ypres the APBs moved away, and we do not know who constructed and maintained the tramways. 5APB arrived on 9 November to work in the Messines and Douve Valley area. 3APB returned on 14 November to work south of them in the Romarin and Ploegsteert area, later replaced by 2APB.

It is clear that by then the existing lines were in poor condition. The tramways were improved by draining, ballasting, inserting wooden sleepers, and laying duckboards between the rails and covering these with wire mesh netting, to help those pushing trollies. Two hot drinks depots were established, one at Hyde Park Corner. A complete system of telephones was installed in January 1918 by 2APB along the forward tramway lines in the Romarin system, so that patrols could speak to the maintenance depot.

Wulverghem to Swaynes Farm

This tramway ran from Wulverghem up the hill, and the total length was about 3 miles (4.8km). Taken over by 5APB on 9 November 1917, the line was converted to mule haulage on 30 November. The line was extended by 400 yards in March 1918.

Douve tramway

This line left from the same point in Wulverghem and followed the Douve valley for about 3 miles (4.8km). As far as the Ploegsteert to Messines Road it was close to the light railway from De Kennebak, and later maps suggest a junction at the end of the light railway at La Petite Douve Farm. It was converted to mule haulage during December 1917 by 5APB.

Fort Boyd and Prowse Point tramways

These tramways had been in the area held by the British before the Battle of Messines but were very near the front line. They had been built when 3APB arrived during the battle but had been extensively damaged. Later there was a junction with the R2 line near the origin. The Prowse Point tramway branched off north to reach Prowse Point. During June 1917 3APB repaired these lines and tried to extend the Prowse Point line for about 600 yards. On 9 June they were driven back to Canpac Dump at Hyde Park Corner by intensive shelling. Also it was impossible to make much progress because of limited personnel and the very badly broken state of ground. When 3APB returned in November 1917, they repaired the lines, and the Fort Boyd line was open for traffic by 17 November.

Warneton and Vancouver tramways and loop

From December 1917, 3APB and 2APB extended the line from Prowse Point as the Warneton line, which led much of the way towards the ruined town of Warneton on the Lys River. By 12 December it extended about 1½ miles (2.4km) from Prowse Point. This section was also heavily shelled, frequently causing the withdrawal of the men working. The line was fully open by 5 January 1918.

The Vancouver railway was reconstructed and further constructed by 3APB and 2APB from November 1917. Officially this began from Fort Boyd, but in December 1917 3APB regarded it as beginning at the junction of the Prowse Point tramway. By later 1918 the whole Fort Boyd line back to Butler's House was labelled as the 'Vancouver Railway', but by that time it had been in German hands from April to late September 1918. The line was regarded as fit for traffic from 5 December, but not fully ballasted. Because of shelling this could only be undertaken in foggy weather or on moonlit nights.

By 5 January 1918 there were fewer than 400 yards between the ends of the Warneton and Vancouver lines, and by 12 January they had been linked up to form the Warneton-Vancouver loop, although initially this was only fit for urgent traffic. Ballasting, and laying duckwalks with wire netting between the rails to help those pushing wagons, was completed by March. In February, it was considered that most of the loop could be converted to mule traction, but on instructions from the Chief Railway Engineer other work was given priority, and it was intended only to make these forward lines 'good strong tracks for hand traffic'. In January, the Warneton line was extended as a branch from the loop to the Egmont Avenue communications trench, less than half a mile (0.8km). In March 1918, the Vancouver line was in a poor state of repair.

Caledonian line and Caledonian loop

This tramway, also sometimes called the Gunners Farm line, began from Delenelle, at the junction of the R2 and R12 light railways. In December 1917 the line was present to Rabecque Farm. At Pompadour Farm there was a junction with the end of the R12 light railway.

By 19 January 1918, construction of the Caledonian loop had started, linking the Vancouver line to the Caledonian line at Pompadour Farm. The whole link was less than 1½ miles (2.4km) long and construction was undertaken from both ends, partly using rails salvaged from the Le Touquet track. The line included a bridge across the Suffolk Avenue communication trench at Pompadour Farm. The ground was very wet, crossing the Warnave River, and broken by trenches, making construction difficult. The loop was ready for traffic by 6 March. The line beyond the origin of the Caledonian loop was called the Caledonian extension, or the Gunners Farm extension. In December this line was extended by about 400 yards beyond Rabecque Farm.

Le Gheer tramway

This tramway was constructed in March 1918, from the Caledonian loop near Le Gheer towards the front line. Built at least partly with rails and sleepers salvaged from within Ploegsteert Wood, it was at most half a mile (0.8km) long when the line was captured in April 1918.

Great Northern line (Le Touquet tramway)

This tramway ran south from a junction with the Caledonian line, not far south of Delenelle on the R2 light railway, and then turned east at Motor Car Corner to Le Touquet station on the disused standard gauge line from Armentières to Menin. We do not know when it was built or by whom, but by February 1918 the line had been extended east of Le Touquet another few hundred yards towards the front line. It was also known as the Californian Line.

Operations and rolling stock

Although there is some evidence for junctions between tramways and light railways in these areas, there is nothing to suggest that the rules were broken. These said that there was to be no inter-running between the two types of line. When 3APB and 5APB returned to these tramway systems in November 1917, all requests for trucks had to be made to the Trench Tramway Officer in each Brigade Sector. Pushing parties had to be supplied by the company using the truck. From 19 November, 5APB were given the Lindenhoek workshops and 3APB established two small workshops, most likely on the Romarin system. By 30 November, eleven trucks had been repaired, and by 5 December another thirty-nine, and they had started to repair ambulance trucks. Instructions were issued to 5APB that, despite the change to mule haulage on the Douve Valley tramways, the heavier type of truck should be gradually discontinued and the lighter type substituted as soon as possible, but we are not sure which types of truck these were. In February 1918, 3APB had workshops repairing trucks at Racine Dump and Motor Car Corner.

In January 1918 a suggested Divisional Routine Order was issued. This stated that:

All tramline trollies to be under the control of the Divisional Tramway Officer (TO)
    There would be three trolley depots -
    Motor Car Corner (Le Touquet tramway)
    Dellenelle Farm (R2 light railway, origin of Caledonian line and loop)
    Racine Dump (origin of both arms of the Warneton–Vancouver loop)
    Representatives of the TO would be constantly on duty at these

Trollies would only be issued to Officers or NCOs or men bearing an Officer's signed order with unit stamp.

The individual to whom trolley is issued signs a receipt, and is responsible personally for returning the trolley within 12 hours, to redeem the receipt. If not returned there would be disciplinary action.

During March, 2APB located 128 trucks, and numbered, labelled and repaired 94 of these. By the end of March they had truck controls at:

Douve Valley -
Wulverghem (origin of Swayne's Farm (Messines)) and Douve Valley lines
On the Swayne's Farm line near Bell Farm
On the Douve Valley line near Plus Douve Farm

Romarin system -
Origin of the Warneton–Vancouver loop near Racine Dump

Dellenelle Farm (R2 light railway, origin of Caledonian line and loop)
Motor Car Corner (Le Touquet tramway)

The Ypres Salient front 7 June 1917 to 8 April 1918

The Ypres Salient front can be considered to extend from Hill 60 in the south to the edge of the flooded areas between Steenstraat and Dixmude in the north. The British front was from Hill 60 to just north of Boesinghe, then the French, and then at the north end the Belgians. The front lines at the beginning and end of the Third Battle of Ypres are shown on Figures 7.1, 7.4 and 7.6.

Standard gauge railways

Standard gauge railway developments up to early June 1917 are shown in Table 6.1 for the Ypres Salient front and for the Messines front. Further developments up to 8 April 1918 for the Ypres Salient front are shown in Table 7.6, and on Figures 7.1, 7.4 and 7.6.

Table 7.6 Standard gauge railway works in the Ypres Sector 7 June 1917 - 8 April 1918

Area	Line or other facility	Construction work start	finish	undertaken by	notes
Calais area					
	Vendroux supply depots	14.04.1917	(1918)	298RC	24 September works ongoing 298RC
	Rivière Neuve ROD locomotive depot	11.05.1917	06.1917	298RC	extended 28 February to 4 April 1918
	Conversion of MG to dual gauge Rivière Neuve to Vendroux		18.07.1917	298RC	
St-Pierre-Brouck/Zeneghem					
	New empties depot and stores yard sidings	14.06.1917	28.07.1917	113RC	
	Forage yard		16.06.1917	113RC	
	Extension west triage stores yard, and new triage	01.09.1917	02.10.1917	113RC 260RC	
	Ammunition depot extension	24.01.1918	13.03.1918	268RC	
	North ammunition depot	by 24.03.1918		269RC 268RC	work suspended 16 April
	Saw mill sidings	by 24.03.1918	03.04.1918	269RC 268RC	
Bergues - Proven line (new British line), begun 1916					
	Bergues Exchange additional 6 sidings	01.06.1917	15.06.1917		
	Rousbrugge station 2 additional sidings	28.06.1917	19.07.1917	262RC	for French Army traffic
	Bandaghem connections	30.06.1917	02.07.1917	262RC	for French Army traffic
	Mendinghem junction triangle	29.06.1917	12.07.1917	262RC	improved connection to French Army lines
	Jhangilathi siding, Rexpoëde	03.07.1917	10.07.1917	264RC	
	Bergues Exchange new sidings	01.09.1917	13.09.1917	261RC	also French constructing garage sidings for *Nord* trains
	Branch to Hoymille and Les Forts		07.09.1917	261RC	signal cabin at MG crossing staffed from 15 October
					depot and dock lines completed December (262RC)

Railways of the Ypres battlefields 7 June 1917 to 8 April 1918

Area / Line or other facility	Construction work start	finish	undertaken by	notes
Bergues - Proven line (Continued)				
Rousbrugge detraining platform	01.02.1918	12.02.1918	1CRT	
Watou ammunition depot	26.02.1918		1CRT 113RC	
Bambecque ammunition depot	01.03.1918		1CRT 113RC	
Poperinghe Avoiding line (new British line, begun 1915)				
Railhoek detraining siding	24.08.1917	17.09.1917	268RC	
Crombeke Road - Proven (new British line, begun 1916)				
Chord to Proven - Boesinghe (Northern) Line	12.07.1917	22.07.1918	262RC	to make triangular junction
Poperinghe - Ypres (existing Belgian line, double tracked 1915)				
Ypres station - siding and gun spur	05.06.1917	11.06.1917	262RC	
Pacific RE Park - conversion to LR ammunition TS	01.06.1917	16.06.1917	262RC	
Brandhoek CCS sidings	18.06.1917	30.06.1917	262RC	
Vlamertinghe supply, stone and RE sidings	10.07.1917	25.07.1917	262RC	
Machine Gun Farm advanced rail head	11.07.1917	25.07.1917	262RC	
Ypres station stone siding		09.08.1917	262RC	
Vlamertinghe detraining siding	24.09.1917	08.11.1917	262RC	work suspended 4 to 18 October
Ypres station and yard - 5 extra loops		12.10.1917	262RC	all station work finished 31 October
Vlamertinghe unloading ramp	13.10.1917		262RC	
Machine Gun Farm ammunition depot	10.12.1917	23.02.1918		work discontinued
Ouderdom - Brandhoek (new British line)				
Construction	18.10.1917	12.12.1917	4CRT 120RC	linking Poperinghe - Ypres with Ouderdom System
Brandhoek chord	11.01.1918	01.02.1918	120RC	NE leg of junction
DT Ouderdom chord	13.02.1918	03.1918	119RC 109RC	junction with Abeele - Ouderdom line
Ypres - Roulers (existing Belgian line)				
Repairs at Ypres end	29.07.1917		10RC	
Reconstruction Ypres to Wildewood	17.08.1917		10RC 262RC	Suspended 21 August - 14 September
Hell Fire Corner stone siding	04.10.1917	18.10.1917		
No Mans Land station	12.10.1917	15.11.1917	262RC	1 siding 1,000 ft long only because of shelling
Diamond crossing for 60cm line at Wild Wood		21.10.1917	112RC	
Gordon House loop and guns spurs	18.10.1917	27.10.1917	10RC	
Doubling ? Ypres to Hell Fire Corner	30.11.1917		279RC	
Gordon House loop extension	29.01.1918	13.03.1918		extended into Hell Fire Corner cutting
Ypres - Comines (existing Belgian line)				
Diversion for embankment (east from Shrapnel Corner)	03.08.1917	05.09.1917		
Ravine Wood branch and Howitzer spurs	17.08.1917	22.09.1917		with 3 spurs - 4th spur 25 October
Trois Rois stone sidings	20.09.1817		262RC	alongside Ypres - Wytschaete road
Zillebeke Lake siding	28.09.1917	27.10.1917	262RC	cour siding completed 3 December 1917
Shrapnel Corner holding siding and loop	29.10.1917	27.11.1917	262RC 261RC	

Narrow Gauge in the Ypres Sector

Area	Line or other facility	Construction work start	finish	undertaken by	notes
Edwaarthoek, Westonhoek, Oakhanger, Peselhoek & Shellhoek complex (begun 1915)					
	Westonhoek rail head and TS Yard	09.05.1917	10.06.1917	262RC	
	Oakhanger triangle and crossover		27.06.1917	262RC	
	Peselhoek Corps RE Park	01.06.1917	14.07.1917	262RC	
	Peselhoek ammunition siding extension and ramp	13.07.1917	15.07.1917	264RC	
	Oakhanger additional sidings	10.08.1917	04.09.1917	10RC	
	Oakhanger RE Park extension	29.09.1917	15.10.1917	10RC	
	Westonhoek rail head additional siding & loops	30.09.1917	14.10.1917	262RC	
	Peselhoek RE Park extension	01.10.1917	13.10.1917	296RC	
Elverdinghe Loop (Begun 1915)					
	Part removal of track north part of loop	24.04.1917	07.06.1917	112RC	Probably at Elverdinghe end of extension
	Elbe Cottage tank detraining platform and siding		09.06.1917		
	Chord and loop to Midland Railway	04.06.1917	16.06.1917	114RC	
Heidebeek - Steenstraat (previously French line taken over 22 December 1917 - aka the Far North line)					
	Noordhoek loop siding and TS	05.01.1918		1CRT	Z1 LR line and MG dual gauge siding (line 29)
	Connection to Swiss Cottage yards	02.01.1918	12.02.1918	1CRT	
	Heidebeek loading platform	21.02.1918	02.03.1918	1CRT	
	Repair north leg connection to Proven - Waayenberg line	16.01.1918	23.01.1918	1CRT	40 foot bridge
	Reconstruct south leg same junction with bridge	18.02.1918		1CRT	
Proven - Elverdinghe - Boesinghe line (new British line - aka the Northern Line, begun 1916)					
	DT Proven - International Corner	09.01.1917	21.06.1917	112RC	Paused 12 March to 17 April
	DT International Corner - Elverdinghe	25.05.1917	30.07.1917	112RC	handed over to ROD 5 August
	St-Sixte - Siding for LR transhipment	22.06.1917	05.07.1917	112RC	
	Straffehem spur and refilling point	22.06.1917	06.07.1917	264RC	adjacent to MG Proven - Crombeke military link
	Dosinghem - ROD siding	03.07.1917	09.07.1917	112RC	
	Elverdinghe - Boesinghe ST	28.07.1917	08.08.1917	1CRT	
	Noppe Farm 60cm gauge diamond crossing		26.07.1917	112RC	for 'old' A1 line across new DT tracks
	Elverdinghe supply sidings	05.07.1917	12.08.1917	112RC 1CRT	
	Waanebeke branch and sidings	23.07.1917	10.08.1917	112RC	
	Boesinghe triangle south limb	01.08.1917	19.08.1917	1CRT	
	Elverdinghe station re-modelling	12.08.1917	22.08.1917	112RC	
	Elverdinghe - Boesinghe station DT	10.08.1917	30.08.1917	1CRT 112RC	
	Boesinghe station tracks	10.08.1917	11.10.1917	112RC	ROD dugouts 19 October by 1CRT
	International Corner loop and platform	24.08.1917	05.09.1917	112RC	
	Waanebeke rail head	12.11.1917	28.11.1917	1CRT	
Ypres - Boesinghe - Staden (existing Belgian line)					
	ST repaired Ypres to south chord to Northern Line	02.08.1917		262RC	
	DT Ypres to Boesinghe chord (Northern line)	20.08.1917		262RC	
	Line linked from chord to Boesinghe station		27.08.1917	1CRT	
	Reconstruction Boesinghe station to Pilckem	01.10.1917	01.11.1917	1CRT	
	Canon Park stone siding	23.10.1917	01.11.1917	1CRT	
	Pilckem loop	30.10.1917	16.11.1917	1CRT	
	Artillery Wood Howitzer & ammunition spurs	05.01.1918	26.02.1918	1CRT	

Railways of the Ypres battlefields 7 June 1917 to 8 April 1918

Area / Line or other facility	Construction work start	finish	undertaken by	notes
Swiss Cottage Rail Head & dumps (begun April 1917)				
Standard gauge construction	16.04.1917	16.06.1917	112RC	
Swiss Cottage dump extension		12.02.1918	1CRT	
Crombeke Road - Reigersburg (new British line, aka Great Midland Railway, begun January 1917)				
Steentje Mill and Oosthoek sidings	24.05.1917	10.06.1917	10RC	
Arrival Farm Siding, Brielen		12.06.1917	10RC	
Arrival Farm refilling point (ammunition)	08.07.1917	19.07.1917	264RC	light railway loading point
Kortebeek and Kemmelbeek branches		14.06.1917	4CRT 262RC	
Pottenhoek branch and sidings	21.05.1917	15.06.1917	4CRT 262RC	
Chord and loop to Elverdinghe loop	04.06.1917	16.06.1917	114RC	
Dirty Bucket Corner refill sidings		16.06.1917		temporary dump for 264RC from 16 July
Reigersburg RE Park and supply sidings	12.07.1917	25.07.1917	262RC	unloading ramp added 24-28 August 1917
Trois Tours 2 loops and TS	07.08.1917	19.08.1917	296RC 262RC	B9 LR line - remodelled 7-12 September
Reigersburg loop supply siding	20.10.1917	03.11.1917	296RC	
Reigersburg Château sidings	05.11.1917		296RC	
Noordhofwijk salvage dump sidings	18.12.1917	01.01.1918	279RC	TS branch from B14 LR
Trois Tours 3 additional spurs		19.02.1918	113RC	
Reigersburg - St-Jean - Wieltje - Spree Farm (new British line, Midland Railway extension, begun 1917)				
Yser Canal bridge embankment	in place 04.1917			
widened for tanks and LR B9 line	14.07.1917	15.07.1917	264RC	track laid 31 July 264RC
Construction Yser Canal to English Farm	31.07.1917	08.09.1917	264RC 268RC	
St-Jean cour, stone and other sidings	18.08.1917	09.11.1917	264RC 268RC	
Zouave stone siding and TS	24.08.1917	28.08.1917	268RC	B15 LR line - siding extended November 1917
St-Jean and Zouave 60cm diamond crossings		21.09.1917	264RC	B15 LR line
Construction English Farm to Spree Farm	01.10.1917	27.11.1917	264RC 268RC 113RC	
Wieltje station cour siding	11.10.1917	17.10.1917	264RC	
Buff's Road refilling point siding TS	20.10.1917	25.10.1917	264RC	B16 LR line
Zouave LR ballast siding	10.10.1917	05.11.1917	264RC 268RC	B15 LR line
Yser canal bridge reconstruction	01.11.1917	22.11.1917	1CRT	plus new bridge for B9 LR line
St-Jean detraining siding and loading ramps	11.11.1917	14.11.1917	264RC	
Wieltje station shunting neck and other works	12.11.1917	24.11.1917	268RC	
Pittsburgh TS siding	28.11.1917	20.02.1918	113RC	LR branch off B4 line
St-Jean station extension	11.01.1918		113RC	
Wieltje extensions	11.03.1918	25.03.1918	113RC	
Hazebrouck - Poperinghe (existing Belgian line, double tracked 1915)				
Boeschepe sidings	09.10.1917	21.01.1918	269RC 10RC	
Caestre ammunition dump	03.04.1918		279RC	
Borre stores depot and water supply	11.1917	31.01.1918	110RC	

DT	double track	LR	light railway (60cm gauge)	RH	Rail Head
ST	single track	CRT	Canadian Railway Troops	ROD	Railway Operating Division (RE)
TS	transhipment	RC	Railway Company (RE)		
MG	metre gauge	RE	Royal Engineers		

Light Railways

Light railway developments up to early June 1917 are shown in Table 6.5 for the Ypres Salient front and for the Messines front. Further developments up to 8 April 1918 for the Ypres Salient front are shown in Table 7.7, and on Figures 7.4 and 7.6.

The Ypres Salient front, from the Battle of Messines to the Third Battle of Ypres 7 June 1917 to 30 July 1917

The British Fifth Army moved to the Ypres Salient on 10 June 1917, in preparation for the Third Battle of Ypres. They took over the front from Second Army from just north of Boesinghe to south of Ypres. The Second Army continued to be responsible for a small part of the Salient front at the south end around the Comines Canal. At the beginning of July, the French First Army took over responsibility from the Belgians for the front from just north of Boesinghe to Noordschote.

Railways

By the build up to the Third Battle of Ypres, the British Fifth and Second Army areas, east of the Hazebrouck to Bergues railway, and west of the Yser Canal and Messines Ridge, contained an extremely dense network of railways, of standard gauge, 60cm gauge, and to a lesser extent metre gauge. Light railways delivered to the front goods arriving at the rail heads and transhipment points. As had been shown at the Battle of the Somme in 1916, manual, horse, mule, and road transport could not deliver adequately in static battlefield conditions.

Standard gauge railways

As the start of the Third Battle of Ypres approached, final standard gauge preparations were made. On 28 July, 1CRT started extending the Northern Line in single track from Elverdinghe to Boesinghe, but on the first day were driven off by shell fire. However, by 31 July, the first day of the battle, the line had reached to near Joyeuse Farm, two thirds of the way from Elverdinghe to Boesinghe Station. From 28 July RCE II reconnoitred Boesinghe bridge on the disused Ypres-Staden line, and beyond on the east side of the Yser Canal. They found no mines at the bridge. The Midland Railway extension was set out for 600 yards east of the Yser Canal bridge on 28 July, with some work done on the formation. On 30 July, 264RC were preparing tools and loading track material immediately on the west side of the bridge. The Ypres to Roulers line on 31 July was only fit for traffic from about 500 yards from Ypres Station. The Comines line was also only fit for a similar distance.

Light Railways Construction and maintenance

The A, B9 and B system lines up to April 1918 are shown in Figure 7.4. The Busseboom (C and D systems) and related lines are shown in Figure 7.6.

A system lines (including LX, AB, B9 and Z lines)

The main units responsible for construction and maintenance on the A system and related lines from the beginning of the Battle of Messines were part of 4CRT, and 277RC. Both units had been undertaking this work since late May. On 15 June, 4CRT moved to Ghyvelde, on the coast east of Dunkerque. Later, 9CRT, who arrived in early July from south of Arras, joined 277RC in this work. The 17NF were also in this area undertaking light railway work from 26 June.

During July there was much shell fire, with many breaks to be repaired, especially on the A7 line and the northern part of the A5 line. 9CRT also worked on the maintenance of the Z1 and Z2 lines and the construction and maintenance of the Z3 line. In late July work often had to be stopped on the Z3 line because of gas.

On 18 July, 9CRT started the construction of the AB line, linking the A1 line and Swiss Cottage yards to the B12 line and the Westonhoek system, via Peselhoek. They were also constructing the B9 line east of Brielen, handed over to 17NF on 20 July. This line is included with the A system lines for this period because, although it began at Mission Junction on the B1 line, it was largely constructed with the A lines, and finished by November 1917 near Langemarck, connecting with the A system extensions. As well as being the major unit for the B system lines, 2CRT started work on some A system lines on 22 June.

B system lines

For the B system lines, construction and maintenance were undertaken by 2CRT, who had moved their HQ to Windmill Camp, Steentje Mill from Bapaume (Somme Sector) on 3 June 1917. A and B Companies were placed near to the B1 line well west of the Yser Canal, with working parties at Ypres based near the B4 line bridge just west of the Ypres Canal, and shortly afterwards another at St-Jean. These parties were rotated frequently because of the difficult conditions, with daily heavy shelling, and sometimes gas attacks. C and D Companies of 2CRT remained in the Bapaume (Somme) area until they were moved to the Belgian Coast on 20 June. From 6 July 2CRT were joined by parts of 9CRT for B system work.

A dump for track assembly was established by 2CRT at The Triangle (B1 - B12 junction), with material transhipped in Westonhoek Yard. In June and July 1917 2CRT laid 32 miles (52km) of light railway track, mostly on the B system but some on the A system, and maintained up to 43 miles (65km). There were endless problems with the light railway operating units. The delivery of ballast for construction was often delayed. Transport for men to and from work was difficult to arrange, and unreliable. Locomotives and tractors often broke down. On 9 and 11 June, the CO of 2CRT visited ADLR V, and arranged that trains would run from Triangle to White Pole Corner and return, leaving Triangle at 6am, 2pm, and lastly at 9pm to run up any night parties and return with the afternoon shift.

As one example of these problems, on 16 June at 8am, track was loaded at Westonhoek. Four trucks were ready to pull out at 9.30am, but three steam locomotives developed problems and the train did not leave until 1.30pm. Leaky injectors, broken drawbars, or lack of water or coal were cited. Before reaching the assembly dump at Triangle (only about 1½ miles – 2.4km) the fourth locomotive ran out of water, and a petrol tractor completed the journey, at 4pm. This was all reported to ADLR V. Average haul distance was 3½ miles (5.6km), with an average of twenty light railway trucks delivered per day.

Water supply was a problem on the A and B systems. Seven water tanks were planned, with power pumps where necessary. The first additional well and 1,000 gallon tank was at White Pole Corner, Ypres (B1, B4 and B5 lines), completed 4 July, when the second tank was started at Triangle (B12 and B1). Double tracking from Triangle to Atherley Junction was started on 8 June.

In Ypres town, 2CRT tried to continue the work on the B6 and B11 lines already begun by 7CRT. Shelling was intense, and the working parties had to constantly move around. There was similar difficulty laying and repairing the track for B5 through the northern edge of the town. By 8 June, track had been laid for B11 across the open space by the prison, and for B6 to the Station Square. This was difficult work because the streets were paved with stone blocks with a high crown, and the rails of the metre gauge lines were still intact. On the night of 10 June, 200 feet (60m) of track on B11 was destroyed by shelling, and on the night of 12 June a further several hundred feet on both B5 and B11. There were similar problems working on the B5 line near Potizje and St-Jean villages. On 14 June, ADLR V forbade further track laying in Ypres town, but work continued on the B5 line on the northern edge.

Night maintenance parties were provided on forward lines to keep traffic moving. By the end of July, the main lines were in very good repair. Traffic on the doubled B1 line was very heavy with ammunition and ballast for the forward lines. By mid-July, the intensity of traffic was causing delays at the control boxes. There was still a shortage of locomotives. Tractors were used nearer the front, east of the Yser Canal.

C and D system lines (Busseboom system)

In addition to working on the B system in June 1917, and on the Ouderdom system in June and July (see Messines front), 7CRT undertook the main maintenance and construction on the C and D lines in this period.

Operating the Ypres Salient lines – 7 June to 30 July 1917

There is a marked lack of information about the light railway operating companies in the Ypres area during this period. 29LROC are the most likely to have been operating the A system lines, probably replaced by 10LROC, who arrived at International Corner on 10 July. 2LROC and 33LROC are also known to have been in the Ypres area at this time.

One or more of the above may have been operating the B system lines, but by July the operating company were the 7SA LROC, who had probably arrived after they were replaced on 26 May on the Messines front at Romarin by 12LROC. The 22nd Light Railways Train Crews Company (22TCC) moved from the Somme to the Ypres area on 4 June 1917, and remained until December 1917. The five burials for men of this Company between July and November 1917 indicate that they were probably assisting in operating the B System.

For the Busseboom system, the main operating companies were 5NZ LROC, based at Quintin (Poperinghe), and 15(1)Aus LROC based at Ellarsyde. Both these units were partly operating also on the Messines front (Ouderdom System). 21TCC were with 5NZ LROC in June 1917 and probably stayed with them at least until November.

Accidents and incidents

On 22 June at 2.30am there was a collision between a standard gauge train, consisting of a locomotive and forty-one empty wagons, and a light railway train, consisting of a locomotive and four wagons, at a diamond crossing. The accident occurred where the AB line crossed the double track Northern (Proven to Elverdinghe) line. Eight standard gauge wagons were badly piled up. The Craven crane from RCE II workshops at Railhoek was fetched by 112RC, and the standard gauge line was open again within 8 hours.

Preparations for the Third Battle of Ypres

From this period onwards we have separated out the B9 system of lines from the main A and B systems because this very long line was mainly in support of a different Corps.

A system lines –

A group of units were brought together under 277RC for advanced construction of the A system during the

Battle. 277RC had been formed in March 1916 with railwaymen selected from the infantry, 'mostly not young'. Their effective strength was 180 men, but the officers were all skilled railway construction engineers. The CO for the group was Lt. Col. S.H. Hancox, an experienced Australian light railway officer. The lines were supporting XIV Corps (Fifth Army) and their front extended from just north of Boesinghe to Fusilier Farm, by Brixham Junction on the Hindenurgh (or Fusilier) loop. Other units attached included the 7th Battalion of the York and Lancaster Pioneers, 580 men, mostly miners with some light railway experience. Each of the four Divisions of the Corps contributed a Pioneer Company (average 150 men), and four Labour Companies were attached (each 300-400 men). The plan was to use the 16 hours of daylight for two construction shifts, and 8 hours at night for the carriage of ammunition, rations, water, and other essentials. Stations about 1,000 yards apart, to suit battery positions, would hold three trains about 160ft long. Guns would be served by tram lines. Traffic would be opened as far forward as possible each day. There would be no transport for non-railway construction materiel.

A new light railway camp at Watou had been decided on 29 June in conference with ADLR V. This group of units trained at Watou for three weeks in July, and parties were organised for the various construction tasks. For laying, track was brought up assembled on flat trucks, loaded as would be laid, pushed by a steam locomotive or tractor, with two piles of ten lengths on a truck. These were offloaded to six hand trolleys, four lengths per trolley. At the rail head each length was lifted over the trolley and dropped into position. The trolley was pushed forward and fishplate bolting and lining was done behind the trolley. The six trolleys were light, so that they could be lifted off to allow loaded ones to pass. The best training ground rate for construction for the 277RC Group was 1 mile in 4 hours.

Track was assembled at St-Sixte, and 6 miles was stored at Green Dump, with another 4 miles at Elverdinghe dump. The points provided were very unsatisfactory and tended to break up with heavy use, but the design and manufacture later improved. Sea sand was brought as ballast from the transhipments at International Corner and at Woesten. From the middle of July, the existing A1 line was taken over from St-Sixte to Euston Junction, and was put into good order to carry heavy construction traffic. East from Green Dump this line was not used in day time, because it was under observation. Towards the end of July, the line was so frequently broken by shell fire at night that in order to get ammunition through it was necessary to detail a repair party with every train, with a truck of repair materiel put in front of the train. Otherwise, there was great delay, with locomotives and tractors falling into shell holes.

B9 system lines –
These lines were supporting XVIII Corps. Other units were attached to the 17NF on 27 July in preparation for advanced construction during the Battle; four companies of Pioneers, and the 18th NF (Pioneers). Half of this force moved out to positions in readiness on 30 July.

B system lines
These lines were supporting XIX Corps. During July, the Companies of 2CRT and 9CRT rotated through the new training camp at Watou. The King's Own Yorkshire Light Infantry (Pioneers) also arrived there on 3 July and were attached to 2CRT for use as labour in the foreseen advance. Training consisted of laying track, with culverts and other features, as quickly as possible. By 25 July, A and B Companies of 2CRT had completed training. That day, twenty-five men of B Company and the attached Pioneers graded 660 yards, built a bridge in 9 minutes, and laid 360 yards of track per hour, equivalent to 3.2 miles (5.1km) per working day. Special tongs had been developed so that while the bolting party were attaching the track section just laid, the next section could be lifted over them from the truck behind and laid in front. Note that this was different from the trolley method used on the A system lines.

On 4 July, 2CRT started assembling track at Westonhoek Yard for the advance. The assembled sections were collected together in quarter mile (400m) units and placed in forward dumps. On 7 July, 9CRT took over maintenance of all lines from 2CRT, except the B1 line from The Triangle to White Pole Corner and the B4 line on from there to St-Jean, so that 2CRT could have a 'free hand' for the preparations for the offensive. Officers of 2CRT walked the B4 line to the ruins of St-Jean on 9 July and proceeded as near to the front line as possible, to reconnoitre line building during the advance. By 20 July, dumps of track material were being placed at quarter mile intervals from Mission Junction to White Pole Corner, and the 2CRT bridging party were assembling trestles. At the end of July, a temporary Brigade of construction troops was formed, led by the CO of 2CRT.

On 29 July it was arranged with the XIX Corps water officer for a stand pipe to be erected on the B4 line near St-Jean village by the first day of the battle, to provide drinking water to fill four 1,000 gallon cars, to take forward for troops. On 30 July, A Company of 2CRT moved to an advanced camp on the B1 line between Atherley Junction and White Pole Corner.

The Ypres Salient front during the Third Battle of Ypres
31 July 1917 to 10 November 1917

Many books have been written about the very complex Third Battle of Ypres. This Battle is also often referred to as 'Passchendaele', but officially the Battle of Passchendaele was only the final phase. The battle is legendary for the appalling conditions, especially the mud, and for the casualties.

The plan of the Allied High Command was to take the high ground to the south and east of Ypres, building on the capture of Messines Ridge on 7 June, and then rapidly break out east and north towards the Belgian Coast. Following the breakout there would be naval landings on the coast east of Nieuport, and attacks by the Belgians and by the British Fourth Army on the coast at Nieuport (see below). In practice the offensive became bogged down, but did eventually succeed in taking the high ground, removing to some extent the ability of the Germans to observe and bombard all activities in the Salient area. The battle was fought by the British Fifth Army, with the Second Army involved at the south end, and the French First Army taking part at the north end. Later the Second Army took over more of the southern part, and the Belgians advanced north of the French.

The front lines at the beginning and the end of the battle are shown on Figures 7.1, 7.4, and 7.6. In the afternoon of the first day, 31 July, it started raining, and it rained continuously until 4 August. This caused many streams to overflow, and with the damage to drainage systems, and the constant churning by shelling, began the mud for which the battle was notorious.

Railways
Standard gauge railways

Works during this period are included in Table 7.6.

Proven–Boesinghe (Northern Line) and Boesinghe–Staden

On 27 July, the XIV Corps of the British Fifth Army had already taken up positions to the east of Boesinghe railway bridge across the Yser Canal. On 31 July, Pilckem Ridge was captured which made the path of the standard gauge railway from Ypres to Staden via Boesinghe available as far as the Steenbeek, about 2½ miles (4km) from Boesinghe bridge. XIV Corps captured Langemarck, where there was a station on this line, on 16 August. By the end of the battle, the path of the line had been captured to just south of the Houthulst Forest, 2 miles (3.2km) beyond Langemarck.

Reconstruction in the Boesinghe area by 1CRT was greatly delayed by constant shelling, even though there was no longer direct observation from Pilckem Ridge. A single track from Elverdinghe reached Boesinghe Bridge on 8 August. Work was temporarily abandoned by 1CRT because of the shelling. Work restarted on 1 October, and the new bridge was completed on 6 October. However, the line was only used as a standard gauge railway for a short distance onto Pilckem Ridge. On 23 August it had been agreed that from a point near Pilckem village the formation could be used by a light railway line. By 10 October, 1CRT had laid track up to that point, about 2 miles (3.2km) from Boesinghe Bridge.

Operating the Bergues–Proven and Proven–Boesinghe lines

The main operating Company was the 60th Australian BGROC (60(6)Aus BGROC), based at Bergues Exchange. 112RC operated in the forward areas. The line from Mendinghem Junction to Adinkerke was under Belgian control, and trains were handed over at Waayenberg to Belgian motive power. The line east from Heidebeek junction was under French control, where similarly trains were handed over.

Midland Railway extension – Reigersburg to St-Jean and Wieltje

Construction of this extension was undertaken by 264RC and 268RC. Track was laid across the Yser Canal bridge on the first day of the battle. By 18 August, track had reached St-Jean station and sidings were being laid. Working parties were withdrawn in early September because of heavy shelling. Work resumed on the extension of the line to Wieltje on 1 October. Construction of the St-Jean yards, where there were eventually eleven sidings, and at Wieltje (four sidings) continued into November.

The original Yser Canal bridge for the extension was an embankment. The Canal had not been filled with water at this time, and the embankment was provided with 2ft 6in timber culverts to prevent water collecting upstream. When the Fifth Army light railways had widened the bridge to take the B9 line in July they had only put in two pipes, so with the October rain water was collecting on the Ypres side, and by 28 October was flooding some dugouts. Light railways improved the openings, but, in early November, first the standard gauge bridge and then the light railway bridge were reconstructed with central piled sections, which solved the problem. Pile driving began 3 November (1CRT) and was finished on 22 November. For the light railway bridge, eighteen piles were driven, complete with stringers, by 18 November.

Operating the Midland Railway and extension

The operating company was still 59(5)Aus BGROC based at Peselhoek. In the more forward areas, 264RC operated or piloted the trains. On 9 October 59(5)Aus BGROC were able to start operating over the Canal in daylight, and traffic to St-Jean increased.

Ypres–Roulers line
With the front line moving east during the Third Battle of Ypres it was possible to bring part of this line back into use at the Ypres end. In August, parties from 10RC worked on the line, but on 18 August it was fit for traffic less than half a mile from Ypres station, and by 27 September the line was still 1,000 yards short of Hell Fire Corner. From 2 October, 262RC also worked on the line. By 16 October, the line had reached 500 yards beyond Railway Wood. On 26 October 10RC were ballasting and finishing for 1 mile (1.6km) beyond Railway Wood, 3½ miles (5.6km) from Ypres. Attempts to carry on work beyond the Frezenburg Ridge were discontinued on 25 October. The final end of the line in January 1918 was at Potsdam Keep, half a mile (0.8km) beyond the halt on the Westhoek Road.

Operating the Ypres–Roulers line
A small operating and maintenance party was placed at Gordon House by 10RC on 22 October, and 'considerable traffic' was run during that month. This included, with great difficulty, deliveries of ammunition to the light railway crossing at Wild Wood (C2F line).

Ypres–Comines line
About half a mile (0.8km) of this line had been captured from the Germans during the Battle of Messines, and about the same amount again on the first day of the Third Battle of Ypres, to a point just south of the village of Hollebeke. No more was captured after this. From 17 August, parties were working on the line. In the end, the line was not extended into the Hill 60 cutting, but a branch was constructed in August to the Ravine Wood Howitzer spurs.

Metre gauge railways
During the Third Battle of Ypres, the metre gauge lines in French Flanders were operated and maintained by the *10ème* SCFC, assisted by the pre-war civilian companies responsible for these lines. They also ran some lines in Belgium; Poperinghe to Rousbrugge-Haringhe (*Vicinal* line 115), Poperinghe to Oostvleteren (*Vicinal* line 107), Oostvleteren to Elverdinghe (*Vicinal* line 29), and Elverdinghe to Lizerne (*Vicinal* branch line 107), Lizerne being as far as this line was open towards the front line. The French also ran the military link lines from Herzeele (France) to Watou (Belgium), and in Belgium that from Proven to Crombeke. Lines north from Rousbrugge-Haringhe (*Vicinal* line 115) and Oostvleteren (*Vicinal* line 29), and the line from Oostvleteren to Noordschoote (*Vicinal* line 107) were operated and maintained by the SVCFC. The part of *Vicinal* line 29 from Elverdinghe to Ypres was not open.

From 1 to 20 September 1917, 112RC constructed a metre gauge deviation at Elverdinghe for the French. We presume that this was to allow them to run metre gauge trains through from *Vicinal* line 29 to the *Vicinal* branch of line 107 from Elverdinghe to Steenstraat without crossing the standard gauge double track from Proven to Boesinghe (Northern Line). A new timber bridge was built across the Kemmelbeek but work was delayed by flooding. In September 1917, the new standard gauge line to Hoymille and Les Forts was constructed from Bergues Exchange. This crossed the metre gauge line from Rexpoëde to Bergues near the ramparts at Bergues. A signal cabin was placed at this crossing on 12 October, manned by ROD staff.

Light Railways
Construction and maintenance
As before we have separated out the B9 system of lines from the main A and B systems because this very long line was mainly in support of a different Corps. Each of the Corps areas advanced during the battle in general in a north-easterly direction, following the direction of the Ypres to Staden and Ypres to Roulers standard gauge railways. Works on these light railway lines are shown in Table 7.7.

A system lines (including LX, AB, and Z lines)
During the Third Battle of Ypres these lines mainly supported XIV Corps (Fifth Army) until 29 October when they were replaced by XIX Corps. Their line of advance was along both sides of the Ypres to Staden standard gauge railway from Boesinghe, over the north part of Pilckem Ridge, and eventually reaching and taking Langemarck. To the north was the front of the French First Army.

At the beginning of the battle, 9CRT were in charge of maintenance and construction for the rear areas of the A, AB, LX and Z lines (and for the B system lines), up to the Elverdinghe area. The 277RC Group were responsible for the extension of the lines from Elverdinghe into the captured areas east of the Yser Canal. From 11 August 7CRT took over the rear areas from 9CRT.

With the beginning of the battle on 31 July, the 277RC group were unable to start work until the afternoon, when XIV Corps had advanced over Pilckem Ridge to the Steenbeek stream. Construction began from Euston towards Boesinghe (Bushey). With the heavy rain beginning during the night and continuing for three days, preventing the artillery from moving forward, work also started on extending the A6 line from the beginning of the Lunaville branch to Bletchley, and completion of the loop north on the west side of the Yser Canal to Boesinghe (Bushey).

Railways of the Ypres battlefields 7 June 1917 to 8 April 1918

Table 7.7 Light railway works in the Ypres Salient areas 7 June 1917 - 8 April 1918

System	line	Section	Construction work start	finish	undertaken by	notes
Proven-Boesinghe system						
	A1	St-Sixte siding TS	26.06.1917	05.07.1917	112RC	
	A1 new	Bubbles Wood Yard (Swiss Cottage) - Noppe (A1)		06.1917		
	A1	Euston jct - Boesinghe	31.07.1917		277RC	
A lines		Bushey to Rugby & Stoke	08.1917	by 16.08.1917	277RC	and Wolverton branch and sidings
		Rugby to Stafford (just east of Steenbeke River)	16.08.1917	07.09.1917	277RC	
		Stafford to Langemarck	04.10.1917		277RC	
		Loop Warwick to Hanley	04.10.1917		277RC	
		Ondank Yard	10.09.1917	14.09.1917	7CRT	
	A6	Woking jct to branch to Lunaville	22.06.1917	25.06.1917	2CRT	
		Lunaville branch to Bletchley	31.07.1917		277RC	and loop north to Bushey (Boesinghe) (A1 extension)
		Bletchley to Tamworth & Broad Street	08.1917	by 16.08.1917	277RC	and loop north to Rugby (A1 extension)
		Broad Street to Burton (just east of Steenbeke River)	16.08.1917	07.09.1917	277RC	and loop north to Stafford
		Burton to Wigan (Alouette Farm, Langemarck) & Pheasant Farm	04.10.1917		277RC	Wigan - junction with B9 from late 11.1917
		Cannock to Rudolphe Farm	04.10.1917		277RC	joined Kitchener's Wood line from B9 10.1917
		Swiss Cottage loop and engine sidings	13.08.1917	22.08.1917	7CRT	
	A5	Woking jct to Euston	bef 06.1917	25.06.1917	4CRT 2CRT	
		Waanebeke TS, salvage spur and extension	05.12.1917	29.12.1917	7CRT	
	A7		by 06.1917			
	A8		by 14.07.1917	03.08.1917	9CRT	
	A10			25.07.1917	9CRT	
		Huddleston Locomotive yard	26.11.1917	08.12.1917	7CRT	
		Wolverton ARP	03.12.1917	06.12.1917	7CRT	
AB lines	AB	A1 Swiss Cottage to B12 (Westonhoek)	18.07.1917	09.08.1917	9CRT	
		Vox Vrie locomotive yard	23.07.1917	05.09.1917	9CRT 7CRT	platform completed 10.01.1918
	AB1	branch to Peselhoek RE Park TS & sidings		by 09.08.1917	9CRT	
	AB2	branch to Gric House		by 09.08.1917	9CRT	
	AB3	Branch to Steentje Mill TS and sidings	31.07.1917	by 09.08.1917	9CRT	
	Z1	Appledore A1 new to near De Blauwepoort Farm	07.1917	by 09.08.1917	9CRT	
	Z2	near De Blauwepoort Farm to Noordhoek	07.1917	by 09.08.1917	9CRT	
	Z3	near De Blauwepoort Farm to Zuydschoote	by 20.07.1917	by 09.08.1917	9CRT	branch to Noordhoek SG and MGTS
		Noordhoek branch and stone siding TS	09.01.1918	25.01.1918	109RC 1CRT	

109

Narrow Gauge in the Ypres Sector

System	line	Section	Construction work start	finish	undertaken by	notes
Former French Army system (December 1917 onwards)						
		Link Wolverton to Het Sas	06.12.1917	13.12.1917	109RC	
		Link A10 to Lizerne	14.12.1917	18.12.1917	7CRT	
		Mendoza tractor depot	27.12.1917	28.12.1917	7CRT	
		Link Z1 to Kruisdoorn	18.12.1917		7CRT	
		Charpentier siding	27.12.1917	29.12.1917	7CRT	
		Het Sas revision	01.01.1918	10.01.1918	7CRT	
B9 system	B9	Mission jct (B1) - Yser Canal bridge	by 06.06.1917	07.1917	7CRT 9CRT	pr complete to NE of Brielen by 06.06.1917
	B9	Yser Canal bridge to Bourne jct		08.1917	17NF	
		Battle locomotive yard	22.11.1917	14.12.1917	7CRT	
		Reconstruction of Yser Canal bridge		05.12.1917	1CRT 7CRT	
	Fusilier Loop			08.1917	17NF	
	B16	Admiral's Road loop - B9 to Wieltje	03.09.1917	14.09.1917	7CRT	
		Buff's Road ammunition siding TS and battery spur	13.12.1917	16.12.1917	7CRT	
	B9	Bourne jct to Canopus junction	02.09.1917	28.09.1917	7CRT	
		Kitchener's Wood line - to Rudolphe Farm	17.09.1917	11.10.1917	7CRT	continued to Cannock (A system) by 277RC 10.1917
		Bochcastle (Bochcastel Estaminet) branch and sidings	24.09.1917	28.09.1917	7CRT	Bochcastle ARP 07.11.1917 - 13.11.1917
		Bochcastle loop - B9 Bourne jct - Kitchener's Wood line	10.10.1917	25.10.1917	7CRT	Welch Farm - No Man's Cot
		Bochcastle - Fusiliers loop link	27.10.1917	30.10.1917	7CRT	
	B9	Canopus Junction - Wigan (Alouette Farm, Langemarck)	01.10.1917	30.11.1917	7CRT	junction with A6 line extension (A system) on branch off B9 near Brielen
		Brielen new ammunition dump	01.10.1917	11.10.1917	7CRT	
		Reigersburg spur	01.11.1917	28.11.1917	7CRT	
		St-Julien ARP spur	07.11.1917	28.11.1971	7CRT	
B9 related tramways						
		Near Canadian Farm (B9) to Langemarck	by 15.09.1917			probably connected to B9 again at Langemarck
		Kronprinz - Bellevue		by 31.12.1917		extension by 4ATC begun 31.12.1917
Westonhoek system	B1	DT from Triangle (Hagle) to Atherley jct	08.06.1917	04.07.1917	2CRT	
		Brandhoek branch		by 06.1917		
B lines and extensions		new ammunition spur (Muskoka Spur)	25.06.1917	04.07.1917	2CRT	
		Barrie spur extension		03.07.1917	2CRT	
		Culloden locomotive yard	23.07.1917	30.07.1917	2CRT	
		Pottenhoek rail head and TS	27.08.1917	14.09.1917	7CRT 2CRT	
	B8	Stanley salvage dump	02.09.1917		2CRT	
		Machine Gun Farm new diversion	03.12.1917	13.12.1917	114RC	
	B4	White Pole Corner - St-Jean		06.1917	7CRT 2CRT	with Midland Railway (standard gauge)
		Manner's Yard	08.10.1917	12.10.1917	2CRT	
		Pittsburgh RH and TS siding	28.11.1917	20.02.1918	113RC	SG branch off Midland Railway extension

110

Railways of the Ypres battlefields 7 June 1917 to 8 April 1918

ID	Description	Date	Unit	Notes
B5	White Pole Corner - St-Jean via Potijze	by 06.1917	7CRT 2CRT	
B6	B5 nr White Pole Corner - E Ypres ramparts via Ypres Station Square	by 03.06.1917	7CRT 2CRT 9CRT	work stopped mid-July to 11 August 1917
B11	B5 N Ypres - Ypres Station Square (B6)	by 03.06.1917	7CRT 2CRT	work stopped mid-July to 11 August 1917
B12	Westonhoek TS yards	29.05.1917	7CRT, 2CRT	
	Oakhanger RE Park extension	16.09.1917	2CRT	
	West spur Westonhoek Yard	28.12.1917	114RC	
B14	Canal Bank line B4 to B9, and Noordhofwijk spur	15.06.1917	2CRT	on west side of Yser Canal
B15	B4 La Brique to B9 Plymouth Junction	05.08.1917	2CRT	links to St-Jean standard gauge yard & Zouave Sidings
	St-Jean yard ballast sidings	10.10.1917	2CRT	TS from standard gauge
Y1	B5 Laydon Junction along Menin Road	by 17.08.1917	2CRT 9CRT	connecting with C2 at Hell Fire jct
Y2	Hexham (Y1) - Y1 Menin Road nr White Château (Menin Loop)	24.08.1917	9CRT	
Y3	Menin Gate (B6) - Ecole jct (Y2)	02.10.1917	5CRT	
	Bedlington line to Cameron Loop	02.08.1917	2CRT	extended 26 September 1917
Y4	Bedlington extension	20.10.1917	6CRT	
	Rotherbury line, along Cambridge Road	16.08.1917	2CRT	extended after advance of 20 September 1917
	Wieltje spur	15.08.1917	2CRT	
	Forest Hall line (Forest Hill line from 01.11.1917)	05.08.1917	2CRT	Wieltje diversion from 28 October, open 13 November
	Wilde Wood line, Grey Ruin jct to C2F	29.09.1917	2CRT	link to C system (C2F)
	Forest Hall line extension (Pommern line from 01.11 1917) (Beck House jct line)	06.10.1917	2CRT	most construction from 01.11.1917
Y6	Gravenstafel line - Pommern jct to Seine jct Y8	04.10.1917	2CRT 263RC	called Y6 from 01.11.1917
Y8	Pommern Castle to Seine jct Y6	31.10.1917	2CRT 260RC	called Y8 from 01.11.1917
Y9	Pillbox jct (Y8) - near Zonnebeke Station	27.12.1917	260RC	
Y10	Kansas Cross line	11.1917	2CRT 260RC	Canadian Corps TW to Y8 at Y9 jct

B system related tramways

	Kansas Cross (Canadian Corps) tramline - connection to Y6	05.11.1917	7CRT	maintenance taken over by 2CRT from 09.11.1917

Busseboom system

ID	Description	Date	Unit	Notes
D4	ARP	05.07.1917	7CRT	
D6	Harrow Yard (D6D)	26.07.1917	7CRT	
D, C & most	Harrow Yard D6D dismantling and salvaging	28.08.1917	9CRT	
F lines				
D14	B8 link to D11 (Asylum line)	13.08.1917	2CRT	
D8	nr Yale (D1) - Pocklington	05.1917	7CRT	
C1	Brisbane (D2 D4) - Bedford (C2 V5)	29.06.1917	7CRT	
C2	Bedford (C1 V5) - Hell Fire Corner jct	29.06.1917	7CRT 9CRT	part also known as H12 initially
	Hell Fire Corner jct - Birr Cross Roads	15.08.1917	9CRT	
	Birr Cross Roads - Westhoek	29.09.1917	9CRT 5CRT	
	Birr Cross Roads marshalling yard	02.10.1917	5CRT	
	Transport Stone Spur - C2 to Zillebeke Lake SG RE siding TS	02.10.1917	5CRT	dismantled 26-27.08.1917 9CRT
	Westhoek - Zonnebeke	04.10.1917	5CRT	

Narrow Gauge in the Ypres Sector

System	line	Section	Construction work start	finish	undertaken by	notes
	C2A	Leinster Farm and Halfway House spurs	12.08.1917	27.08.1917	9CRT	
	C2B	Birr Cross Roads - Yeomanry jct (C4)	14.08.1917	27.08.1917	9CRT	
	C2C	C2 jct - Railway Wood	14.08.1917	28.09.1917	9CRT	
	C3	Zillebeke jct (C2) - Dormy jct C4	01.08.1917	08.08.1917	7CRT	
		Dormy jct C4 - Essex (C5)	08.08.1917	27.08.1917	7CRT 9CRT	
		Grange siding C3 line		25.08.1917	9CRT	
	C4	Dormy Junction C3 - Sanctuary Wood	07.08.1917	08.08.1917	7CRT	
	C4A					
	C4 extension					
	C5	Tunnel jct (C2) to Essex (C3)		31.08.1917	9CRT	
	C6	Essex (C5) to Bluff jct (V8)	by 31.08.1917	25.09.1917	9CRT	
	C7	short line near Zillebeke Lake		08.09.1917	9CRT	
	C8	short line near Zillebeke Lake	04.09.1917		9CRT	
	C9	Lambert Spur at end ?	by 06.09.1917	25.09.1917	9CRT	
	C10	branch off C3	25.09.1917		9CRT	
	C11	Dearborth jct (C2) - C2B jct (near Sanctuary jct C4)	28.09.1917	04.10.1917	9CRT 5CRT	work abandoned 4 October
	C2D	C2 'short cut' from Lake Farm spur	bef 07.10.1917	11.10.1917	5CRT 8CRT	
	C2F	C2D to Crecy	07.10.1917	15.10.1917	5CRT	connected to Wilde Wood line, B System

C system related tramways

Birr tramways - Birr Cross Roads (C2B spur) south and East
Surbiton tramway. Near C4 line (Sanctuary) to Surbiton Post & trackway, Menin Road (aka Surbiton Light Railway)

			11.09.1917	30.09.1917	1APB	
Hooge Crater - Château Wood monorail			12.09.1917		2 AFC 1APB	later extended to Glencorse and Polygon Woods
Westhoek towards Zonnebeke (Corps Tramway)			by 04.10.1917	23.10.1917	1 APB 2APB 5APB	
branch to near Molaarelsthoek (Divisional Tramway)			10.10.1917	20.10.1917	5APB	

L System lines

	LX	Swiss Cottage yards (north end) - Heidebeek		07.1917		
		Crombeke ARP	15.02.1918	23.02.1918	109RC	beside *Vicinal* 107 (MG), possible TS
	L4	Quintin (Remy) D1 - Boeschepe Quarry	03.03.1918	04.04.1918	10CRT	further work see Chapter Eight

MG	metre gauge	TW	tramway
SG	standard gauge	aka	also known as
TS	transhipment	pr	probably
DT	double track(ing)	jct	junction

AFC	Australian Field Company (Engineers)	ARP	ammunition refill point
APB	Australian Pioneer Battalion	RC	Railway Company (RE)
ATC	Army Tramway Company (RE)	RE	Royal Engineers
CRT	(Btn) Canadian Railway Troops		

The main (northern) line (A1 extension) was then extended across the Yser Canal, followed by the more southern line (A6 extension) and the loop completed from Tamworth north to Rugby. The northern crossing, 250 yards north of the standard gauge bridge on the Staden line, was made on a mattress of fascines with an open span of 20 feet (6m) supported on crib piers, with the sides of the embankment faced with broken brick from Boesinghe Village. After shell damage, the span was filled in and culverts were placed. The southern crossing was made as an embankment with culverts from the start. This was much higher and took longer to complete.

Some extensions were made by the time the Germans were driven back to Langemarck on 16 August, after which both lines were constructed to the Steenbeek Stream, to get guns and ammunition into the Steenbeek Valley. The main (northern) line was extended along the old Ypres–Staden standard gauge line. This was constructed entirely at night, and covered with camouflage netting and debris of RE material which had been on it previously. Rough earthwork and platelaying of the other line was done at night and finished off by small parties during the day. It was only possible to pack this line lightly, and it was purposely left open to observation. Open span bridges on crib piers were constructed at both crossings of the Steenbeek. The lines were linked up (link about 500 yards) just east of the Steenbeek and were ready for traffic on 7 September. Neither line was much damaged by shell fire until there was evidence of use.

As there was no prospect of another offensive in the near future, the men were put on maintenance in the back areas, and to rest, and only small parties were kept in forward areas. At the same time, the 277RC group took over maintenance of all light railways in the XIV Corps area, about 50 miles in all. During September 277RC were reduced to 80 men, because men were 'combed out' for the infantry, and the York and Lancaster Pioneers similarly lost 400 of their 580 men. This was a severe handicap to construction. By November 277RC had 110 men, the reinforcements being good railwaymen.

Extension of the lines was resumed on 4 October. The southern line was extended to Wigan (Alouette Farm) just south of Langemarck, where it joined the B9 line which reached there by the end of November, and on to Pheasant Farm. This line could not be ballasted because of shell fire under direct observation from the Goudberg Heights and was laid on planks for tramway type operation until the Germans retreated.

The other line was extended along the path of the Staden railway through Langemarck station towards Carre Ney (Ney Cross Roads). Both extensions were very close to the front line in mid-October. Another loop was constructed between the two lines near Pilckem (Warwick to Hanley), and a line south from the southern line at Cannock joined up with the Kitchener's Wood line from the B9 system. Also, part of the line using the standard gauge formation was shifted to the north side to allow the extension of the standard gauge line towards Langemarck. With the end of the Third Battle of Ypres all plans for further extensions were dropped.

From September, shellfire was very heavy and concentrated on the light railways. In forward areas, patrols were required day and night, at night large patrols were required with a smaller party with each convoy of trains. Further back, with telephone communication, a repair party travelled around with a light tractor. Damage was also caused by troops using the formation of tracks under construction, or completed track, as footpaths or mule tracks. Barbed wire fencing and even the use of police was unsuccessful, and the only effective solution was to provide duckboard tracks alongside the line.

B9 system lines
During the Third Battle of Ypres these lines mainly supported XVIII Corps (Fifth Army). They occupied the middle part of the front between Boesinghe and Ypres, and advanced over the southern part of Pilckem Ridge, capturing St-Julien and later Poelcappelle. From 31 October, this part of the front was taken over by II Corps (Second Army).

At the beginning of the battle, 9CRT were in charge of maintenance and construction for the rear areas of the A and B system lines, including the B9 line, replaced by 7CRT from 11 August. The 17NF Group were responsible for the extension of the lines across the Yser Canal. During August 1917 they built the B9 line to Bourne junction (Civilization Farm), and the Fusilier (Hindenburg) loop to the north of this. At the end of August 7CRT took over the B9 construction and maintenance from 17NF. During August, three men of 17NF were killed and nineteen wounded.

In September, 7CRT extended the B9 line on to Juliet Farm, which included a cutting 250 yards long, and a connection was made to the trench tramway from near Canadian Farm. In October, the line crossed the Steenbeek but was not completed to Langemarck until November. At Wigan (Alouette Farm) just south of Langemarck, reached by the end of November, it joined the southern A system line from Burton.

In all, between 26 September and 26 October just over 8 miles (13km) of track were laid. As the other forward construction units found, this was dangerous work. In October, eight men of 7CRT were killed

7.3 Men of the Royal Garrison Artillery loading shells onto light railway wagons at the Arrival Farm ammunition refilling point near Brielen on 3 August 1917. The arrangement of the railway tracks was unusual, with the light railway at right angles rather than parallel to the standard gauge line. The dump was on a branch of the Midland Railway, at the end of the light railway branch south from Reading on the B9 line (location 18, Figure 7.4). (*collection Sandra Gittins*)

and two officers and twenty-five men wounded, plus fourteen killed and sixty-four wounded from attached labour units. The CO reported how much shelling and bombing there was, especially on the B9 extension to Langemarck and the Kitchener's Wood line. Planes dropping bombs had pursued the working parties constantly. A lot of track material was delivered by mule transport, allowing several track groups to be kept busy simultaneously. Overall 'the continuous wet weather, the difficult nature of the ground, the long walk to and from work, the insufferable bombing and intermittent shelling' made this month the most difficult they had yet experienced.

B system lines

These lines extended around Ypres and then east along the north side of the Ypres to Roulers standard gauge railway. During the Third Battle of Ypres these lines mainly supported XIX Corps (Fifth Army). From 20 September they were replaced by V Corps, and from 4 October by II Anzac Corps (Second Army), who captured Gravenstafel and the Abraham Heights. From 15 October this sector was taken over by the Canadian Corps (Second Army), who captured the Passchendaele Ridge in the last stages of the battle. At the beginning of the battle, 9CRT were in charge of maintenance and construction for the rear areas. The B6 and B11 lines in Ypres were handed over to 2CRT on 8 August, and from 11 August 7CRT took over the rear areas from 9CRT.

2CRT were responsible for the maintenance and construction of the lines east of the Yser Canal. On 31 July, reveille was at 2.30am for A Company of 2CRT at the Advanced HQ Camp, and with the King's Own Yorkshire Light Infantry Pioneers they marched to the rendezvous by the Yser Canal. At 3.50am a terrific bombardment opened along the whole front. The morning was spent repairing shell breaks on the

B4 line, and damage from tanks crossing the line. At noon they were relieved by B Company, who continued laying track, and grading towards Wieltje. On the following days work continued on lines forward from St-Jean and Potizje, and the B5 line from North Ypres was completed and linked to the B4 line on 5 August. A bridge 75 feet (23m) long over the Bellewaardsbeek was completed very quickly using pre-constructed bents. However, the greater part of the line ran over very bad ground, and the heavy rainfall caused much of the line to 'sink out of sight', so that it had to be lifted and re-ballasted.

Also on 5 August, 2CRT were forced to put up a fence, and call for police assistance, to stop ammunition columns from destroying the grade by crossing it unnecessarily. On 6 August, five 60 pounder guns were moved about a mile on the B1 and B4 lines to near La Brique, after it proved impossible to move them by road because of the mud. Other guns were later moved by light railway.

The main line east from St-Jean became the Bedlington line. The line further north, just south of Wieltje village, was soon known as the Forest Hall line. By 16 August this had reached Lytham Cottage, where there was a trench tramway. Construction continued during daylight even though under observation and considerable shell fire, and by 19 August this phase of the work was done except for the branch off B5 along the Menin Road, which was connected up with the II Corps line towards Hell Fire Corner (Y1 line) on 23 August. Work resumed on the B6 line through Ypres, suspended in July, at first mostly replacing damaged track. Throughout August, damage from tanks and guns crossing the lines continued. On 29 August, C and D Companies of 2CRT arrived at the Watou Rest Camp from the Belgian Coast, and shortly after took over forward work.

7.4 A train loaded with ammunition headed towards the front along the Y1 line, in the Menin Road between Ypres and Hell Fire Corner, on 3 October 1917. The light railway had taken the path of the metre gauge line to Gheluwe, leading to the C system. This was the day before II Anzac Corps took the Abraham Heights and I Anzac Corps took Zonnebeke and Broodseinde Ridge. (*Australian War Memorial E00913*)

In September, the work was mainly maintaining and improving existing lines. A large amount of brick rubble was brought out of Ypres for use as ballast. On the B5 and Bedlington lines steel sleepers were replaced with wooden ones on curves, and guard rails were put in at curves and crossings. The forward lines were still under German observation and there was much shell fire. Early in the month the forward parties were provided with one petrol electric tractor for ballast trains. Later day parties were provided with this locomotive, and one Simplex (probably a 20hp), which was found inadequate as the locomotive attracted shell fire. On 25 September there were no tractors and steam had to be used, attracting shelling on the B4 and B5 lines.

In the renewal of the offensive on 20 September, British and Australian troops captured Glencorse Wood and Inverness Copse, and on 26 September they took Polygon Wood and Zonnebeke. The Bedlington line was extended, and construction started of the Wild Wood line to link with what was now V Corps under Second Army, south of the Roulers standard gauge line.

For the next stage of the battle, beginning 4 October, the B system area was taken over by II Anzac Corps. The attack on 4 October led to the capture of Broodseinde Ridge and Gravenstafel. On 3 October, the CO of 2CRT conferred with the General Commanding the Artillery of II Anzac Corps and conducted him over the forward lines. He asked for a date when light railways could be built to Spree Farm but considered the given date of 8 October too late, so said a plank road would have to be used instead to bring ammunition up to the guns. In the event, the extended Forest Hall line, which became the Gravenstafel line, reached Spree Farm and was bringing up ammunition by 8 October and was also bringing planks for the uncompleted road. In the early part of October, rolling stock was so over-committed to moving supplies and ammunition that 2CRT men often had to walk to work on the forward lines, which could be 4 miles (6.4km) each way. On 14 October, A Company moved to Manner's Junction (St-Jean village) to be nearer the work, and on 22 October B Company moved to Wieltje. Work was often held up for lack of track and ballast. Also owing to the state of the roads, many convoys of pack mules, carrying field gun ammunition, persisted in using the grade and the Bedlington line and later the Forest Hall line were partly fenced.

Construction around Pommern Castle was very unpleasant. This had been the German Hindenburg defence line. The ground was covered with dead bodies, and the block houses were also filled with dead Germans. Also the heavy rains had made the lines in a 'fearful condition'. In some cases the mud, which was 'of the consistency of porridge', was over the line for a depth of 18 inches (45cm). Wooden sleepers had to be used because the steel sleepers sank. At a Fifth Army conference on 11 October, it was decided that every aid be given to Light Railways to consolidate their lines as their value had been demonstrated during the recent wet spells. The next day, three additional Simplex tractors were allotted for construction, and for 48 hours from 14 October as much motive power and other rolling stock as necessary was provided to put the lines in running order. With the continual rain and lack of ballast, lines were still fast sinking out of sight. The weather became drier, and the state of the lines improved. On the Forest Hall line extension (later called the Pommern line), a bridge 75 feet (23m) long was put in over the Zonnebeke stream.

The Canadian Corps took over from II Anzac Corps on 15 October, and 2CRT expressed their great satisfaction at this, the first time that they had worked with Canadian infantry since coming to France in January 1917. Canadian units camped around the 2CRT advanced camp near Ypres, making it seem 'like the old days in Canada'. On 20 October, ADLR II took over the B system for the Second Army. The new Light Railway Chief Engineer was a former 2CRT Company Commander. Work on the Bedlington line extension (Y4) was given to 6CRT on 21 October.

In the middle of October, a spur was constructed into St-Jean village, to collect brick rubble for ballast, and on 24 October 80 tons were brought out. Sand and mine earth was being brought from the St-Jean standard gauge yards, and sand from Culloden siding near Mission Junction on the B1 line. Later in October there was a big effort to complete the Gravenstafel line. From 1 November ADLR II changed the name of some of the lines; the Forest Hall line to Forest Hill, the extension of the same to the Pommern line, the Bridge House gun spur to Y7, and the Gravenstafel line to Y6. From 5 to 8 November, the Y6 line was connected to the Canadian Corps (Kansas Cross) tramway. From 9 November, 2CRT worked on improving the tramway, and it is probable that it was taken over for through running from the light railways.

On 26 October, 2CRT handed over the forward lines to the 260RC, and the section from the Westonhoek Yards to Atherley Junction to 114RC. During the Third Battle of Ypres, 2CRT had laid 23 miles (37km) of light railway track and maintained up to 36 miles (58km). 114RC maintained the system and put in new lines at Machine Gun Farm on the B8 line. A new holding siding at Oakhanger and the West Spur at Westenhoek, including a floor for washing out locomotives, were also put in before they handed over to 9CRT on 3 January 1918.

C, D and Y system lines (Busseboom system)

During the Third Battle of Ypres these lines mainly supported II Corps (Fifth Army), who became part of Second Army from 16 August. I Anzac Corps, with

X Corps and IX Corps to the south, took over from 20 September. The area these lines served was around Zillebeke Lake, and then east to the south of the Ypres to Roulers standard gauge railway as far as Polygon Wood and the Gheluvelt Plateau. The lines of the system adjoined the Ouderdom (Voormezeele) lines in the area of X Corps (Second Army), at the north end of the Messines front. They also took part in the battle, and there was some overlap of the use of the light railways. On 20 September, they advanced about a mile across the Gheluvelt Plateau, taking the Menin Road Ridge, and in a major assault on 4 October Broodseinde Ridge was captured.

Construction and maintenance was undertaken initially by 7CRT, but from 10 August 9CRT took over, with 7CRT going to relieve them on the A and B systems. The C2 and C3 lines were considered part of the forward Zillebeke system. Construction of C3 commenced on 1 August and that night some ammunition was transported. Construction on all lines was difficult because of shell fire, killing a lieutenant and five labour company men, and the heavy rain. On 4 August the men were put on maintenance elsewhere. However, on 7 August they were brought back and instructed to have the C3 and C4 lines ready for the night of 8 August to take up ammunition towards Sanctuary Wood. The lines were open at 8pm on 8 August. Also, the C3 line was probably only completed to the C4 junction; it was extended later in August. Construction of the C system continued through August and September, with the usual problems. Later in September work was also hampered by the premature explosions of shells from British guns behind them, and by three tramway crossings (the Surbiton tramway system). From 30 September, 5CRT changed places for two weeks with 9CRT in order to give the latter some rest and sleep.

With the impending attacks of 4 October, the C2 line was completed to Westhoek, and immediately work started extending the line eastward. There was also pressure to make progress with the C11 line, to become the lateral line nearest the front line. On

7.5 The C2F light railway line, near Westhoek, in October 1917 (locations t7 - t8, Figure 7.6), looking towards Westhoek. On the left is a battery of 6in howitzers, covered with camouflage netting; the ammunition was delivered via this line. (*Australian War Memorial C01354*)

2 October, Sanctuary Junction ammunition dump was set on fire, the explosion destroying 100 yards of track. On 4 October, I Anzac Corps captured Zonnebeke village and the Broodseinde Ridge. 5CRT were advised of the immediate need to push the C2 line forward to Zonnebeke. Construction of the C11 line was abandoned, and 1,100 additional labourers were provided. However, on 7 October, 5CRT were instructed by ADLR II to push the grading of the C2D line west to join the work of 8CRT, and to construct the spur line from this north-west behind the Westhoek Ridge to serve batteries. This became the C2F line and was eventually linked to the B System. In any case, track laying on the C2 extension was 'practically at a standstill' owing to the unauthorized use for tramways of the required light railway material by '1st and 2nd Anzac Pioneers' (1 and 2APB). Breaks prevented new material being brought far enough to be of use. From 9 October, every man walking to Westhoek to work was required to carry one light railway sleeper from Birr Cross Roads or further back. Between 13 and 16 October, 5CRT and 9CRT changed places again, with 9CRT returning to the C System lines. Men of 5CRT were awarded two MCs and one MM for their work on the C System. From later October to the end of the battle on 10 November, the work was mainly maintenance and improvement of the lines.

Operating the Ypres Salient lines

As for the period leading up to the battle, there continues to be a lack of information about the light railway operating companies in the Ypres area during the battle. More is known about the C, D and Y system lines (Busseboom system). Activity on Fifth Army light railways for the week ending 17 August 1917 is shown in Table 7.8. Here the B9 System is not separately identified but is included with the B System.

Appreciation for the work of the railways was sent on 11 August by the QMG of the Fifth Army. In late September a letter from the Surgeon-General at GHQ was passed on to the DLR, about the evacuation of wounded on light railways. 'Thanks to arrangements made, we were able to get the wounded back regularly and methodically with minimum discomfort. Featured the running of small trains with only 50 to 100 at a time, avoiding large numbers being dumped at one time, consequently helping them to deal with the wounded without delay'. He was sure that the extension of the system would be an advantage.

A system lines (including LX, AB, and Z lines) and B9 system lines

The main operating company for the A system is likely to have been 10LROC who arrived at International Corner on 10 July. A detachment, probably a platoon, of 1LROC was at International Corner up to 2 March 1918, and was probably there from late October 1917. Other units may have been involved. It is known that the main operating unit for the B9 system in August 1917 was 29LROC. They were based at Poperinghe at the beginning of the battle, but on 2 August moved to near Vlamertinghe. Among the few records for this unit is that they supplied the heavy batteries of XVIII Corps throughout August 1917, 'owing to it being impossible to supply [these] by road'. Upwards of 2,000 tons of ammunition and other materiel were carried by them over the Yser Canal per day during that period, over the single B9 line. It is probable that they remained the main B9 System operators for many months after that. Other units which may have been operating on the A or B9 Systems during the battle were 33LROC, known to have been in the Ypres area in the early stages of the Battle, and 54LROC, known to have been in the area from 20 June 1917 at least to December 1917.

Table 7.8 Goods and personnel carried by Fifth Army Light Railways week ending 17 August 1917 (tons unless otherwise stated)

	A System	B System (Inc. B9 System)	D System (Busseboom)
Personnel (no.)	927	1,968	988
Coal	268	80	129
Ammunition	8,408	10,767	12,328
RE Stores	151	1,176	400
Timber			33
Railway material	1,025	2,087	1,422
Ballast	2,051	4,183	3,280
Salvage	2,686	743	1,123
Roadstone			231
Guns (no.)		35	
Petrol	131	39	1
Other traffic	252		71
Total	15,994	21,077	20,016
Grand total Fifth Army Area	57,087 tons		

The system consisted of loops ('balloons') linking two lines toward the front. This gave an alternative route in case of a break or other obstruction on one forward line, and it was found that the enemy did not generally shell both lines leading to a loop at the same time. This worked satisfactorily for the operational conditions, which were a series of short advances with fairly long intervals between. Later links made the A and the B9 systems effectively a three line system serving two Corps, the loops on each side of the centre (A6 extension) line being 'staggered'. Heavy artillery was based near the lines or branches to avoid long 'push' tram lines, but it was found best not to put main lines less than 200 yards from heavy batteries because they attracted shell fire. The construction companies also found it inadvisable to put in many sidings in forward areas, because controllers were not usually stationed there, and at night time train crews tended to leave trucks in any sidings and return to base with their tractors if there was any difficulty experienced in getting ammunition to its destination. A system of Divisional Tramways was inaugurated to overcome this problem but had not been fully organised by the end of the battle.

B System

The most likely operating companies during the Battle were 7SA LROC and 8SA LROC. As for the other systems, 33LROC and 54LROC may also have been involved. From August to October, 2CRT (the construction and maintenance battalion) were piloting ammunition trains at night in forward areas. As for the period before the battle, 22TCC were probably assisting in operating the B System.

More is known about the use of the lines to bring in wounded for some of the later stages of the battle. On 4 October, the attack began at 6.30am. All returning empty trucks carried wounded, and during the day large numbers of wounded were carried to west of the Yser Canal. At the beginning of the attack on 26 October, the lines in the forward areas were again busy bringing down wounded from Bridge House on the Forest Hall line. About 500 wounded in all were carried over the system. During the attack on 30 October, on the outskirts of Passchendaele, returning empty trains again brought wounded, 300 during the day.

7.6 Wounded men of the 66th Division (II Anzac Corps) being conveyed by light railway on 10 October 1917, after the failure of the Division to advance onto Passchendaele Ridge the previous day. The railway is probably in the B system, and the train is headed by a Simplex 40 tractor. (*National Library of Scotland 74301221 License CC BY 4.0*)

7.7 The light railway yard at Ellarsyde (Location 24, Figure 7.6) in October 1917. There are four steam locomotives, the one on the far right is a Baldwin 4-6-0T. (*Australian War Memorial C01388*)

C and D system lines (Busseboom system)

For the Busseboom system, the main operating companies were 5NZ LROC, based at Quintin (Poperinghe), and 15(1)Aus LROC based at Ellarsyde. Both these units were partly operating also on the Messines front (Ouderdom System). They were joined by 17(3)Aus LROC who arrived at Ellarsyde on 20 September and moved to Mimico (D11 line), near the Ypres Asylum, on 30 September, to operate mainly the northern C lines. From Mimico their main access to the C System would have been through the ruins of Ypres using the D11 and B6 connection, then the Y1, Y2 and Y3 lines to Hell Fire Corner. The 22TCC were also supporting operations in this area.

Operating from the Mimico depot, 17(3)Aus LROC reported great difficulty forward from Hell Fire Corner for the whole of October, because of shell breaks, as many as twenty in 12 hours. Because of this they operated trains in convoys of five between bursts of fire. The eight available steam locomotives (Alco-Cooke 2-6-2T type) operated as far as the Birr Cross Roads forward marshalling yard on the C2 line, and the eight Simplex 40 tractors forward from there. There were problems with water supplies for the locomotives in early October, water from shell holes being unsuitable, and camp water causing scaling. The main traffic was ammunition but some RE stores were carried. Casualties were heavy.

At the end of the battle, 5NZ LROC reported that they were operating to Hanebeke, on the C2 line beyond Westhoek towards Zonnebeke. During the battle (which they refer to as Passchendaele) they carried 45,000 tons of ammunition. Since their arrival at Poperinghe on 27 February 1917, they had carried 245,000 tons of goods, of which 138,000 was ammunition. They had brought in many wounded from Dressing Stations. The principal C System construction companies, mostly 9CRT during this period, continued to maintain and operate the D6 yard at Harrow.

7.8 A working party of soldiers loading stone for road repairs beside the standard gauge railway at Dickebusch. The light railway runs alongside, east of Dunedin Junction (location 39, Figure 7.6). 9 August 1917. (*Imperial War Museum* ©IWM *Q5852*)

7.9 A load of heavy artillery shells being taken to the front line on 14 September 1917, with men of the 5th Australian Pioneers (APB), probably an unloading party. Near Dickebusch (Figure 7.6), exact location unknown, but another line is coming in from the right beyond the locomotive. (*collection Sandra Gittins*)

Rolling stock

At the end of 1917, 15(1)Aus LROC reported the available rolling stock each week since June for the whole operation of the Ouderdom and Busseboom systems. A small sample shows the change over the period (see page 94). In October 1917, 17(3)Aus LROC operating from Mimico had the use of eight Alco-Cooke 2-6-2T steam locomotives, and eight Simplex 40 petrol tractors. These are included in the numbers reported above.

Accidents and incidents

The citation for the bar to his MC for Captain H.P. Burrell of 5 CRT illustrates the type of incident which occurred frequently under shell fire. On 6 October, near Westhoek (C2 line), a light railway train of two Simplex 40 tractors and five wagons was derailed and the track broken by a direct hit from a shell. With a corporal and six men he immediately worked to repair the track and re-rail the train. The corporal was killed, three men were wounded, and the rest concussed. He helped the wounded, gathered another party, and despite further shelling completed the job.

Tramways

Tramlines from light railway group stations to heavy batteries were laid as they moved forward, and salvaged as positions were vacated. During the Third Battle of Ypres, the 4th Army Tramway and Forward Company (RE) (4AT&FC) laid such tramways in the XVIII Corps area, from supply points on the B9 system. On 31 July they laid about 1,200 yards of tramline. They continued this work at least until the end of November. From October, 6AT&FC undertook similar work on the A System. During October, twenty-two tramlines were laid.

7.10 Light railways after a direct hit by shellfire, near a junction, with a derailed Simplex 40 tractor near one line and a loaded bogie wagon on the other. 1 November 1917, on the C2 system, at the junction with the C2B line at Birr Cross Roads (location r8, Figure 7.6). (*National Library of Scotland 74411181 License CC BY 4.0*)

6AT&FC activity October - December 1917

	Tramlines laid		tramlines salvaged		Ammunition moved	
	km	miles - yards	km	miles - yards	rounds	tons
October	3.22	2 - 18	7.20	4 - 880	70,130	7,350
November	1.89	1 - 317	3.41	2 - 229	29,684	2,937
December	0.58	0 - 1,021	0	0 - 0	19,242	908

Between October and December 1917, sixteen men of 6AT&FC attended tractor driving training in various places. This all argues for preparations for mechanical transport on tramways, and inter-running with light railways. However, there are no diaries for this unit from January 1918 to 8 February, when they had moved to the Somme Sector.

The 1st Australian Pioneer Battalion (1APB) arrived in the Ypres area in early September 1917, and began work on roads in the II Corps area on 9 September. From 11 September they began construction of a tramway system east from near Birr Cross Roads. This was later called the Surbiton Tramway, because eventually it extended to near Surbiton Villas. On the 1APB work map for September 1917, it is labelled the 'Surbiton Light Railway', and at its origin there is a clear connection with the C2B line, indicating the intention of inter-running onto the tramway. However, on 29 September 5CRT were ordered to remove this connection immediately.

The 2nd Field Company (Australian Engineers) arrived in the Hooge Crater area on 9 September 1917, and among other tasks surveyed and pegged out the route for a monorail track from Hooge Crater, on the Menin Road, to and through Château Wood. Salvage parties collected steel rails for the monorail along the Menin Road, perhaps from the disused metre gauge railway (*Vicinal* line 121). Strutted posts about 7ft (2.13m) tall, on which to mount the rails, were constructed at the HQ at Kruisstraat and brought forward. We do not know who designed the monorail system, or the rolling stock, but a switch (points) designed by a unit officer was used. Because of shelling and shortage of materials the monorail was not completed for the attack of 20 September, and on 18 September 1APB took it over for completion and extension. By this time it was about 400 yards long, and reached almost to the west edge of Château Wood. Further north, 2APB constructed a tramway 800 yards long from Bellewaard Bend to 100 yards west of Westhoek between 23 and 25 September. We do not have a map showing this tramway.

Before the attack of 4 October, a tramway had been constructed from Westhoek to the Anzac strongpoint, with light (9lb/yard, 4.5kg/m) rails. On 4 October, 1APB were given the task of extending this towards Zonnebeke. Before 6 October, 2APB were improving this tramway, which became the Corps Tramway from Westhoek to Zonnebeke. The tramway was taken over by 5 APB on 6 October, who found it in poor condition and extending only to the Hanebeke stream. Work was difficult because of rain and shelling but by 9 October the formation was complete to Zonnebeke Village, and the track laid to near the Anzac strongpoint. The line was probably finished to Zonnebeke by 23 October. It was intended for mule haulage. By November, 5NZ LROC were operating this as a light railway to the Hanebeke stream.

From 10 to 20 October, 5APB built a Divisional tramway, branching off the Corps Tramway from Westhoek to Zonnebeke near the Hanabeke Loop on the C2 light railway. The line finished on the road south from Zonnebeke near Molaarhelsthoek, with a terminal siding and a rail head and loop at Albania Wood. 5APB were proud to say that it had been constructed entirely with 20lb/yard (10kg/m) rails, that is those used for best quality light railways with heavy traffic. No doubt these were those taken without authorisation from 5CRT at Westhoek, delaying the C2 extension to Zonnebeke. On 24 October, the line was improved by laying planks down the centre to help pushing wagons.

The Ypres Salient front after the Third Battle of Ypres
11 November 1917 to 8 April 1918

Shortly after the official end of the Third Battle of Ypres on 10 November 1917, the British Second Army under General Plumer moved to North Italy. The Fourth Army, who had been on the Belgian Coast since 5 July (see below) took over. Confusingly they were renamed Second Army, but the new name was not consistently applied. From December 1917, the Fifth Army moved south to take over the extended British front on the Somme. Also in November, the French First Army started to leave, so that the British and Belgian forces were once more side by side at the north end of the Ypres Salient. The Second Army returned from Italy on 17 March and took over from the Fourth Army (now named Fourth Army again), who were sent south. The British Second Army was now responsible, as they had been up to 10 June 1917, for the front line from the Hazebrouck to Armentières railway in the south, to north of Boesinghe on the Yser Canal, so that they held the Messines and the Ypres

7.11 A group of soldiers huddle around a brazier just outside the Menin Gate as a horse-drawn transport column passes, 20 December 1917 (location 23, Figures 7.4 and 7.6). The gap in the damaged ramparts behind is the Menin Gate, with the ruins of Ypres beyond. The light railway running out of the gate and off to the right is the B6 line, with the Y3 line branching off nearer the men. (*Pen & Sword Archive*)

Salient front lines. To the south was the British First Army, and to the north the Belgians. There were no major offensives on either side, and very little change to the position of the front line, during this period. The winter of 1917–1918 was marked by periods of intense cold, with frozen ground, interspersed with short periods of thaw, when the ground quickly turned to mud. On the night of 15-16 January there was extremely heavy rain, with flooding.

Railways

During the Third Battle of Ypres the French First Army had been responsible for standard gauge, metre gauge and light railway lines in their section of the front. A *Procès Verbal* (formal meeting for agreement and handover) was held on 14 December 1917 with Belgian, French and British representatives. Details of the agreement are given for each gauge of line in the section for that gauge.

In spite of the dislodgment of the Germans from the ridge to the east of Ypres, and the resulting loss of direct observation from ground level over the Salient, shelling continued to be very heavy. Shell breaks were frequent on all lines back to Poperinghe and Proven, and the construction and maintenance companies spent a great deal of their time keeping the lines repaired and open for traffic. Shelling intensified from the middle of March, around the time of the offensive in the south which began on 21 March 1918. This continued until the offensive in this area which began on 9 April. During April, preparations were made for the destruction of major bridges if the need arose.

Railways of the Ypres battlefields 7 June 1917 to 8 April 1918

Standard gauge railways

During this period the British took over some of the French standard gauge railways, especially the Heidebeek to Steenstraat line. Details of works on these lines are shown in Table 7.6. The operating company for these lines, and for the Northern line, remained 60(6)Aus BGROC based at Bergues Exchange. 59(5) Aus BGROC based at Peselhoek continued to operate the Midland railway and branches, and the extension to St-Jean and Wieltje.

Metre gauge railways

At the *Procès Verbal* held on 14 December the agreement for metre gauge lines was as follows (reference to Figure 3.1 may be of assistance):

The *10ème Section* would hand over to the SNCV the lines in Belgium they were operating which had been laid prior to the war, which were Rousbrugge to Proven (*Vicinal* line 115), Crombeke to Oostvleteren (*Vicinal* line 107), Oostvleteren to Elverdinghe (*Vicinal* line 29), and Elverdinghe to Lizerne (branch *Vicinal* line 107).

The French would also hand over the military link line in Belgium from Proven to Crombeke.

Stocks of permanent way material would be taken away by the French, with the exception of those rails and fittings necessary for maintenance for 6 months.

All of these lines would be operated by the Belgians from 22 December 1917.

Of the two standard to metre gauge transhipments at Heidebeek, that with road access would be handed to the Belgians by 22 December, that without would be used for the final evacuation of French material.

The *10ème SCFC* would hand over to the British for operation two SNCV lines laid before the war, Poperinghe to Proven (*Vicinal* line 115), and Poperinghe to Crombeke (*Vicinal* line 107), the military link line from Herzeele (France) to Watou (Belgium), and in France the line from Esquelbecq to Herzeele. The hand over date was to be fixed later, but before 1 January 1918.

French rolling stock on lines handed over would be handed back to the French within periods to be fixed by a special conference. Apart from Esquelbecq to Herzeele, and that part of the military link line from Herzeele to Watou which was in France, that is the part to Houtkerque, the pre-war civilian French Companies would take over all lines on French territory.

Bollezeele, St-Momelin to Esquelbecq, Esquelbecq to Proven and Poperinghe, Poperinghe to Crombeke

To operate these lines the British were to take over the metre gauge depot and workshops at Bollezeele, from where the *10ème* SCFC had been running these lines. The 85th Canadian Engine Crews Company (85Can ECC) arrived in France at Boulogne on 12 December 1917 and were sent to Audruicq. On 1 January 1918, about half the Company, that is one captain, one lieutenant, and 123 other ranks, travelled from Audruicq to Bollezeele by train via Hazebrouck. They probably changed to a metre gauge train at Esquelbecq. There was accommodation for forty men in part of the workshops, the rest stayed for the moment in the box cars they arrived in.

From 3 January, men were placed at stations along the line to work with the French, and drivers and firemen travelled on the trains with the French to gain knowledge of the lines. It is clear from the records and from an accompanying map that the ROD (for the present 85Can ECC) were in charge of the lines from Esquelbecq to Poperinghe and Proven, and from Poperinghe to Crombeke. They would also operate on the line from St-Momelin via Bollezeele to Esquelbecq. The civilian company, the SE, were in charge of the lines from St-Momelin to Bergues and to Esquelbecq (both via Bollezeele) but would also operate between Esquelbecq and Herzeele. There were metre gauge facilities on the Aa Canal wharf at St-Momelin, which had been extended by the British in early 1917 (see Chapter Six).

The Canadian Company made inventories of all locomotives and other rolling stock, workshop fittings, buildings and stores. They took over the operation of their lines on 9 January. All men were, as nearly as possible, allotted the same employment as their civil occupation. On 9 January there was heavy snow, and it proved difficult to maintain traffic, but after that everything went smoothly.

There were forty-two British engines (War Department) and fourteen Belgian (SNCV), all 0-6-0T locomotives of *bicabine* type, the latter taken over from the *10ème* SCFC. The British ones were from the fifty ordered from Robert Stephenson and Hawthorn Leslie, delivered in 1916 and 1917 (see Table 6.3, Chapter Six). Any remaining locomotives or rolling stock were removed by the French. There were 1,000 British War Department wagons distributed along the line, a considerable number located on the line operated by the Belgians beyond Crombeke, but the *10ème* SCFC arranged to have these returned to the ROD lines. In March 1918 there is a record of the locomotive numbers awaiting repair or under repair. Fifteen are identified, all British War Department stock. One of these, Hawthorn Leslie ROD No. 244, is now preserved at Tramsite Schepdaal, near Brussels (see Chapter Twelve).

The running shed and repair shop at Bollezeele remained the property of SE but was handed over to ROD, with a few SE employees. There were also two SE men at Wormhoudt and at Poperinghe, who were paid by the ROD. The repair shop at Bollezeele continued to be used by SE for repairs and housing of engines with suitable payment to the ROD. There were also running sheds at Herzeele and at Poperinghe with only very limited accommodation for the housing of engines. Registration of civil and military traffic in the yard at Bollezeele was performed by the ROD under the supervision of a Civil Agent. Two station staff were placed at each of the eleven stations and halts from Esquelbecq to Proven and Crombeke. In addition to those at Bollezeele, there were two engine crews (driver and fireman) at St-Momelin, one train crew (driver, fireman and brakeman) at Esquelbecq, and three train crews at Herzeele. Early during operations drivers had to be reminded of the speed instructions, that they must not exceed 10mph (16kph), with 5mph (8kph) round curves, traversing points, and through villages.

Full records are available for activity on these ROD lines from the weeks ending 17 January to 28 March 1918, and this is worth recording (Table 7.9), as we know of no other similar metre gauge records. At the outset, the CO at Bollezeele estimated that there would be work for about twelve locomotives at various points on the line. Table 7.9 shows that the highest weekly average was nine, but it may have been higher on individual days. Note the quite high proportion for engine distance running light (15 per cent in January), likely to represent engines from Bollezeele picking up trains elsewhere. Note also the relatively short average distance per wagon movement (column 23) compared with the line length operated (58km, 36 miles).

The civil traffic (column 20) was the train run daily from Monday to Saturday, from Bollezeele to Poperinghe and back, which among other things carried women employed at the camouflage works at Wormhoudt. This must have consisted of two passenger carriages. In addition, SE ran a daily civilian service each way between Bollezeele and Herzeele. This was likely to have been the same as that on the timetable of 1 July 1917 (Table 7.3).

After a frosty winter there was a thaw from 21 January, and by 2 February there were snowdrops, crocuses and primroses in the farms. A British Army circular of 8 February predicted a world shortage of potatoes and green vegetables after June, so with the reduced work the Company acquired plots adjacent to the lines at no rent, and men tended the vegetables. On 4 March, forty-eight men were declared surplus to requirements for these lines and were transferred to assist 58Can BGROC at Merris. Later in March the Company was advised that men over 55 could apply to go home if they wished.

Instructions were received on 7 April to turn over six locomotives and 100 wagons to the ROD at Vendroux. These were transhipped from Poperinghe. Further instructions were that a lieutenant and sixty men were to remain at Bollezeele, the rest to proceed to Merris to reinforce 58Can BGROC. All men not required at out stations were brought in to Bollezeele. The remaining men were attached to 6(60)Aus BGROC, who took over on 9 April, the first day of the German offensive. There was now more traffic again, with considerable troop movements to and from Poperinghe, and shelling at Poperinghe and Proven.

Heidebeek to Steenstraat
At the Noordhoek sidings, 1CRT were instructed in January 1918 to convert the standard gauge siding to 3-rail dual gauge (metre gauge added). This would enable the transhipment to and from the light railway to be used by the metre gauge *Vicinal* line 29 as well as by the British standard gauge line.

Bergues to Hoymille and Les Forts branch
From 13 March 1918 the blockmen at the crossing of this line with the metre gauge line from Rexpoëde to Bergues were withdrawn. Fixed signals would indicate 'line clear' to all metre gauge trains and 'on' or 'danger' to all standard gauge traffic, unless changed using a key obtained from one of the Bergues Exchange block cabins. The scheduled metre gauge train times were noted, indicating that there were now two trains each way daily between Rexpoëde and Bergues. The times at this crossing were, to Bergues 7am and 3.30pm, and to Rexpoëde 8.30am and 6pm. On 1 July 1917 there had only been one each way (Table 7.2).

Light Railways
At the *Procès Verbal* on 14 December the agreement for light railways was as follows; the British would take over the lines:

Kruisdoorn to Woesten branch line
Athenee to Noordhoek branch line
Lizerne to Het Sas line and the branch towards Boesinghe
All lines in the British zone east of the Yser Canal

Railways of the Ypres battlefields 7 June 1917 to 8 April 1918

Table 7.9 85th Canadian Engine Crews Company metre gauge activity based at Bollezeele, January to March 1918

W/E	Line length km	Engines (1)	(2)	(3)	(4)	engine km (5)	(6)	(7)	(8)	engine hours (9)	(10)	Personnel (11)	(12)	(13)	(14)	(15)	Traffic (wagons) British (16)	(17)	(18)	Other (19)	(20)	Total (21)	(22)	(23)
17 Jan	58	56	12.1	9.8	0	844	602	623	308	433	223	144	16	43	26	9	63	63	369	2	24	521	7593	14.6
24 Jan	58	56	16.1	6.5	0	1363	693	947	619	573	187	154	15	46	25	16	126	39	543	4*	24	736	13144	17.8
31 Jan	58	56	15.8	8.7	0	1081	670	744	450	400	216	145	14	46	25	17	168	63	387	0	24	642	10320	16.1
total Jan from 10 Jan						3288	1965	2314	1406															
7 Feb	58	56	11.7	6.1	0	912	640	799	94	313	207	140	14	46	25	17	129	62	266	0	24	489	9278	19.0
14 Feb	58	56	6.7	7.4	0	1111	417	691	95	251	197	141	13	43	18	17	331	55	94	0	24	504	9902	19.6
21 Feb	58	56	5.8	6.1	0	734	244	314	148	175	198	142	15	43	28	17	41	28	149	0	26	244	4794	19.6
28 Feb	58	56	6.1	6.4	0	722	336	220	116	183	188	157	15	36	28	17	130	69	100	0	24	323	5179	16.0
total Feb						3479	1637	2024	453															
7 Mar	58	56	5.4	6.7	0	611	131	18	36	112	121	143	17	25	26	18	62	33	66	0	24	185	3502	18.9
14 Mar	58	56	4.4	4.7	0	623	81	0	50	104	82	122	13	26	26	14	59	41	52	0	24	176	2594	14.7
21 Mar	58	56	4.8	4.1	0	655	269	133	104	155	65	122	12	26	25	13	61	36	162	0	37	296	4658	15.7
28 Mar	58	56	5.0	3.4	0	662	388	251	92	185	58	122	12	28	25	14	98	27	45	0	24	194	4547	23.4
total Mar to 28 Mar						2551	869	402	282															

(1) Daily average engines
(2) Daily average engines in steam (%)
(3) Daily average engines under or awaiting light repairs (%)
(4) Daily average engines under or awaiting heavy repairs (%)
(5) Engine km loaded
(6) Engine km empty
(7) Engine km assisting
(8) Engine km light
(9) Engine hours trains
(10) Engine hours shunting
(11) Personnel total strength
(12) shed duties
(13) running
(14) ground staff
(15) repairs

Traffic, loaded wagons per week
British military traffic
(16) miscellaneous
(17) supplies, ordnance
(18) construction

Other
(19) French military
(20) Civil

Totals
(21) total wagons
(22) wagon km
(23) average km/wagon movement

* French troops

The remaining lines further north were taken over by the Belgians.

The British and the Belgians agreed to notify the French as soon as possible of lines they did not wish to retain, so that the French could take them up. The remaining lines would be handed over with the obligation of repayment in kind to the French. Stocks of permanent way material would be taken away by the French, with the exception of the proportion of rails and fittings necessary for maintenance for six months. All rolling stock, installations, workshops, tools and gear would be taken back by the French, who would provide the transport necessary for the British and Belgian armies until 15 January, leaving ten tractors and fifty flat wagons on the system with the necessary personnel (one construction group and one operating group). Connections of the lines taken over by the British were made in later December 1917 (see A System below, and Table 7.7).

At a transportation meeting at Second (Fourth) Army HQ at Cassel on 28 December 1917, to discuss the programme of construction during the winter, it was decided to make the 60cm light railways the bulk distribution system. In discussion of new work on the standard gauge system from Proven to Steenstraat, taken over from the French, it was thought that a light railway ballast siding would have to be constructed in this area. This became the Noordhoek branch, transhipment and stone siding, constructed by 109RC in January 1918 (see metre gauge section).

Construction and maintenance

From 21 November 1917 the CRCE lent additional Railway Companies (RE) to Second Army light railways for construction and maintenance under ADLR II(S) or II(N). These were:

263RC	from	23.11.1917	to	26.01.1918	to work for ADLR II(S)
114RC		25.11.1917		04.01.1918	ADLR II(S)
260RC		27.11.1917		26.01.1918	ADLR II(S)
269RC		30.11.1917		04.01.1918	ADLR II(S)
109RC		05.12.1917		24.02.1918	ADLR II(N)

A system lines (including LX, AB, and Z lines) and B9 system lines

Shortly after the end of the Third Battle of Ypres, 7CRT took over responsibility for all the A system lines, as well as continuing with the B9 system. The 17NF, and the 277th Railway Company RE (277RC), both already working on the A system, were attached. 277RC were allocated responsibility for an area of lines, as was each of the four companies of 7CRT, and 17NF assisted with the advanced construction work. 7CRT were now responsible for a vast area, with 3-4,000 men as attached labour.

With the end of the battle, construction work slackened off. In all only 6.3 miles (10km) of track was laid between 26 October and 26 November. The final ¾ mile (1.2km) of the B9 line was completed to Langemarck during November, where it connected with the A lines extensions. The final sections of the line, close to the Langemarck to Zonnebeke road, were protected by screens to prevent observation. There was great difficulty constructing the grade in the mud, with more embankments than usual, and about three times as much ballast as under normal conditions.

On 5 December, 277RC left the district, and were replaced by 109RC, again working under the CO of 7CRT. Reconstruction of the Yser Canal bridge for the B9 line was completed. During December, 7CRT took over the agreed part of the system formerly used by the French, and links were constructed. On 13 December, 109RC connected with the French system, from Wolverton to Het Sas. Between 14 and 18 December, the A10 line was connected to the French system near Lizernes, and on 18 December work started on a connection from the Z1 line. 7CRT were also assisted by 269RC, who arrived at Reigersburg on 30 November 1917 and left again on 4 January 1918.

The end of December 1917 and early January 1918 brought a hard frost, making construction and maintenance difficult. Traffic was light, and 7CRT undertook a lot of salvaging. In the very heavy rains of the night of 15-16 January the HQ camp of 7CRT, by the Poperinghe Canal, was flooded out, but by next day the flood waters were subsiding. Some track was sunk and bridges had to be repaired. On 18 January the A system lines were handed over to 109RC and the B9 system to 17NF, and 7CRT moved south for a rest period, travelling by train from Peselhoek. They never returned to the Ypres Sector. On 25 January the bridge for the A1 new line over the Poperinghe Canal collapsed following the floods but was fully repaired by 109RC. Meanwhile, the old A1 line remained in place for use. On 19 February, 109 RC handed the A System lines over to the 17NF, and by 25 February they had returned to standard gauge work. On 1 January 1918 17NF had moved their HQ from St-Jean to near the B12 and AB lines just north of the Westonhoek Yards, where they remained until 8 April 1918, on the eve of the German advance. They were employed on the maintenance of light railways and tramways of the 'Northern System', that is on the A and B9 Systems. There were no major construction projects and casualties were light.

B system lines

The Westonhoek, Oakhanger, Edwaarthoek and Peselhoek yards are shown in Figure 7.5, as in April 1918. 2CRT continued as the main construction and maintenance unit until 27 November, when they went south to the Arras area. Between the end of the

Third Battle of Ypres and their departure, D Company continued construction on the Pommern line, Y8 line, and the Y10 (Kansas Cross) line, and maintained the Forest Hill line and lines around Wieltje. With the extension of the standard gauge facilities at Wieltje, the Forest Hill line was diverted around the western end of the yards. The diversion was opened on 13 November and the old track taken up. When they left, the Pommern line was connected to the Y4 line at Beck junction, and the Y8 line was laid to Delva loop,

The Pittsburgh railhead and transhipment sidings are a slight puzzle. Commenced on 28 November 1917, the standard gauge branch from the Midland Railway extension, with two sidings forming a loop, is shown on all later maps. The standard gauge work, and the light railway earthwork, was undertaken by 113RC. However, no map shows the light railway sidings. These were probably a branch of the B4 line just east of the Yser Canal, which the standard gauge sidings almost reach. The only other light railway line in the area was the B9 line at the north end of the sidings. RCE II refers to the Pittsburgh 'railhead and transhipment' as being complete, confirming the presence of light railway track, but not until 20 February 1918.

On 23 November 1917, 263RC moved into the Canadian camp near the St-Jean yards and on 27 November they were joined by 260RC who established their HQ in dugouts at Manner's Junction. Between them they took over from 2CRT construction and maintenance of the forward part of this system. On 25 November, 114RC moved from the Arras Sector to near Vlamertinghe and took over from 2CRT construction and maintenance of the back areas, from Westonhoek Yard to Atherley Junction.

260RC continued the construction of the Y8 line eastwards, to the north of Zonnebeke. 263RC continued the Y6 (Gravenstafel) line which joined the Canadian Corps (Kansas Cross) tramway and used the tramway for about 800 yards. It was then also extended eastwards just to the south of Abraham Heights. The Y6 and Y8 lines were linked up on 29 December about 800 yards south-east of Abraham Heights, at

7.12 A working party of Royal Engineers on a light railway train on a passing loop in the Ypres area, probably returning from work nearer the front line, on 7 January 1918. Note the control point building on the left, the telephone wires overhead, and the excellent state of the track despite the surrounding mud and shell holes. (*Collection Sandra Gittins*)

Seine Junction. During January 1918 both companies continued maintenance and some construction, often under heavy shelling. In the rains and flooding of 16 January, three sections of the Pommern Line were washed away between the Y8 (Iberian) and Y4 (Beck House) Junctions. A few days later, the same section of track was broken in many places by shell fire. In both cases traffic was restored very quickly.

Although only recently moved from standard gauge work, 260RC and 263RC were congratulated by ADLR IVS on their light railway work. On 31 December, 263RC were warned to be prepared to move, and dismantled their huts and went under canvas. However, the move was delayed nearly four weeks, and they had to continue work in the frost and rain from this poor accommodation, before moving to Maricourt in late January. On 26 January, 260RC entrained at St-Jean (standard gauge) and moved to Péronne. 114RC, working on the back areas of the B System, had already moved to Saigneville (Somme Sector) on 4 January 1918. All three companies went back to standard gauge work, and none ever returned to Ypres.

9CRT, who had left the C System on 21 November, returned to the Ypres Sector on 21 December 1917, and established their HQ near the Culloden locomotive yard on the B1 line. Advanced camps were in the Ypres and St-Jean areas near the B4 and B5 lines. During January their CO complained that they were under-employed, no doubt because of the continuing presence of 260RC and 263RC on the forward lines. Nevertheless, they undertook maintenance work, constructed the Oxford ammunition refill point, probably on the Forest Hill line at Wieltje, and did some pile driving at the B4 bridge over the Yser Canal, possibly as part of a reconstruction. They regraded and improved the Wild Wood line and the linked C2F line (C System). After the rains of 16 January, they repaired washed away track on the Wild Wood line and at Grey Ruin Junction (Bedlington Line). From early January they took over all the back areas from 114RC, and from 26 January all the forward lines from 263RC and 260RC. During January they had laid 2.9 miles (4.6km) of track but maintained 70 miles (112km). On 1 February 1918, 9CRT left the Ypres area for the last time, entraining at Vlamertinghe for Achiet-le-Grand, south of Arras. The entire B System was handed over for construction and maintenance to 10CRT, already responsible for the C System and the northern part of the Messines front.

On 3 March, 10CRT began construction of the L4 line from the D1 line at Quintin (Poperinghe), following close to the Poperinghe to Hazebrouck standard gauge line. By 4 April 4 miles had been constructed, as far as Boeschepe Quarry, just north of Godewaersvelde. After 9 April this was continued as part of the extensive L System (see Chapter Eight). Apart from this, 10CRT did not undertake any major new construction on the B System and Y lines. It is confirmed by later maps that Seine Junction (Y6 and Y8 lines) was the most eastern extent of this system. Trench maps in July 1918, following the German re-occupation of this area, show a loop from the Y6 line up onto the ridge past Crest Farm to Passchendaele village, then to Passchendaele Station (Roulers line) and back to Seine Junction. This is shown on a German photograph with a sign in English pointing to Crest Farm, but the sign probably relates to a duckboard track rather than the railway, and it is most likely that this was a German line linked to the captured British systems.

C, D and Y system lines (Busseboom system)

After the end of the battle, 9CRT remained the construction and maintenance unit, until they moved to the Arras Sector back areas for rest and training on 21 November. They returned to the Ypres area from 21 December to 1 February 1918 (see B System). The work was taken over by 10CRT, already responsible for the more southerly C System lines (C5, C6 and C10) and the north end of the Messines Front. Because of continued shell fire, 9CRT found it impossible to work on the C2 line beyond the crest of the Westhoek Ridge. Little construction was undertaken, with less than 1 mile of new track being laid between 27 October and 19 November. Construction and maintenance were continued by 10CRT until the beginning of the German advance on 9 April 1918. They did not undertake any major new construction in this period. Bombing and shelling continued. On later maps, the C11 line, on which work was abandoned on 4 October, remained incomplete, and there is no evidence for the extension of the C2 line east from Zonnebeke. Indeed, some later maps do not show the C2 line east of Westhoek.

Operating the Ypres Salient lines

In general operations were less than during the battle, especially towards the end of March 1918.

A system and B9 system

As before, we know little about the operating companies in this area during this period. In January 1918, 7CRT had an average of fifteen men daily acting as drivers and guards under the direction of the LROCs, and an average of ninety-five attached troops daily on special work, such as building dugouts and unloading coal, for the ADLR and the operating companies. Also, by January 1918, 7CRT had emergency parties standing by during the night from 6pm to 6am. A party for XIX Corps was based at Huddleston locomotive yard for the A system, and one for II Corps at Battle locomotive yard for the B9 system. Each was in the charge of an NCO and had a tractor and truck of emergency tools available to ensure completion of any track repairs, and were available to personally conduct ammunition deliveries to very forward areas. COs of the LROCs were asked to get touch with Officers in charge at these locations to expedite forward deliveries of ammunition.

B system

When 263RC took over the B system in late November 1917, two locomotive drivers were transferred to them, one each from 22TCC and 2LROC, indicating that these companies were the operating units for the B System at that time. On 13 February 1918, 12LROC arrived at Westonhoek and took over operating the B System from the 92nd South African BGROC. 12LROC moved from Romarin (Messines front) to Westonhoek and based operating detachments at Manner's Yard (B5 line, St-Jean). Manner's Yard was heavily shelled on 16 March, with many casualties among 10CRT and attached Labour units. Also one man of 21TCC, attached to 12LROC, was killed, and five men of 14LROC were wounded. On 18 March, Manner's was again shelled, and three men of 12LROC were wounded. Finally, on 21 March, the control cabin at Manner's was completely destroyed by a direct hit, with one man wounded. The camp was evacuated back to Nordhofwijk, just west of the Yser Canal, but operating continued. Clearly, 14LROC and 21TCC were operating with 12LROC on the B System at that time. It is known that 14LROC had been in the Ypres area from June 1917 and remained until August 1918. In March 1918, the award of a medal to one of their men was mentioned by 3(17)Aus LROC.

C and D system lines (Busseboom system)

At the end of the battle, the operating companies were 5NZ LROC, based at Quintin (Poperinghe), 15(1)Aus LROC based at Ellarsyde, and 17(3)Aus LROC based at Mimico. On 15 February 1918 15(1)Aus LROC moved to Savy-Berlette (Arras Sector), and their work was taken over, in part at least, by 17(3)Aus LROC.

For the week ending 4 January 1918, 15(1)Aus LROC reported that 41,540 tons of goods had been carried on the whole system, reducing to 32,290 for the week ending 18 January. From January to March, 17(3)Aus LROC reported traffic busy but easily managed with the available resources. All traffic ceased for 24 hours on 17 January after the previous day's rains washed out sections of track. Towards the end of the month, there was an increase in the carriage of personnel, mainly reliefs for front line troops. In February 17(3) Aus LROC carried 39,000 and in March 36,200 front line and working party troops. They also carried brick rubble from the ruins of Ypres, 2,784 tons in February and 1,134 tons in March, for use as ballast. Towards the end of March, traffic was very low, and in early April was mainly salvage from forward areas. Shelling remained heavy, and became heavier in February, with more gas shells. Fire on marshalling yards increased in March, but there were few casualties and little damage. In spite of seventy line breaks in February and sixty-one in March, traffic was not affected.

From 13 January 1918, 17(3)Aus LROC ran a new section of trench trains operated with Simplex 20 tractors with 3 ton box (four wheel) wagons for the transport of RE materials and 'front line' rations. The lines operated were probably those forward from Westhoek. On 16 February they took over the operation of another section of trench train line of 2¼ miles, this section running through Glencorse and Polygon Woods. This was the Surbiton Tramway, which had been extended from Clapham Junction to Polygon Wood. The records are not specific as to whether trains ran through from the light railways, or whether goods were transhipped at the tramway beginning. The former is much more likely, especially as by this time inter-running had been discussed as a policy.

From 5 March 1918, the new HQ of Australian Railway Troops renumbered the 15th, 16th and 17th (Australian) LROCs as the 1st, 2nd and 3rd Australian LROCs. For clarity in this book, we have abbreviated the first of these 15(1) Aus LROC up to 4 March 1918, and 1(15) Aus LROC after that, following the same convention for the others. The Australian Broad (Standard) Gauge Railway Operating Companies were also renumbered, and we have applied the same convention.

Rolling stock

In January 1918, 17(3)Aus LROC reported that 137 steam locomotives, 70 Simplex tractors, 42 PE tractors, and 835 bogie wagons were in use on the C and D System lines. We do not have any figures for the other systems. Following previous neglect of wagons, a special staff was authorised by 17(3)Aus LROC from 19 January 1918, and within ten days many of the difficulties had disappeared.

The British Fourth Army on the Belgian Coast
20 June 1917 to 28 November 1917

From 20 June 1917, British troops moved to the North Sea Coast of Belgium and France. They were under General Rawlinson's Fourth Army, which set up their HQ at Malo-les-Bains (Dunkerque) on 5 July 1917. The army consisted of one Corps, XV Corps, with two Divisions. The front line was at Nieuport, and inland bounded by the flooded area around the Yser River as far as Dixmude. At first there was a lightly defended bridgehead on the east side of the Yser estuary between Nieuport and the sea but following a German attack on 10 July only the Palinbrug *redan* and limited other areas remained in Allied hands on the north-east bank of the Yser (see Figure 7.7).

The Fourth Army took the area over from the French and Belgians, and were separated from the British Fifth and Second Armies at Ypres by the Belgians, and south of them the French First Army. The plan was that if the Third Battle of Ypres was successful, with a break out from the Ypres Salient, the Fifth Army would swing round east and north to the coast and capture the ports of Ostende and Zeebrugge. At the same time there would be naval attacks and coastal landings east of Nieuport, and the Fourth Army would advance north-east along the coast to link up. The Fourth Army went into training for these operations and were to a large extent isolated to maintain secrecy, as part of 'Operation Hush'.

Railways

While on the coast, the British Fourth Army undertook considerable construction of standard gauge and light railways in preparation for the planned offensive, as well as taking over and maintaining existing lines. Railways of all gauges in the coastal area from Adinkerke to Nieuport in September 1917 are shown in Figure 7.7.

Standard gauge railways

For the construction and maintenance of standard gauge railways, a new Railway Chief Engineer, RCE VI, was appointed, with staff and supporting railway units (RCC VI). The HQ was at Malo-les-Bains, with or close to Fourth Army HQ, by 23 June. Although we have put the start date for the Belgian Coast presence as 20 June 1917, the day the first British troops moved into the front line at Nieuport, 4CRT had moved from Poperinghe to Ghyvelde-Bray-Dunes standard gauge station on 15 June and started work the next day. On 20 June they were joined at Ghyvelde by 114RC. The line from Dunkerque was converted to double track as far as Furnes. Other work was at the St-Idesbald sidings and light railway transhipment, which were close to the main line between Adinkerke and Furnes, and a long way from the coastal village of that name.

Metre gauge railways

The existing metre gauge railways are shown on Figure 7.7. The British Fourth Army played no part in any metre gauge operations, and clearly relied upon the standard gauge and light railways for their needs. From early 1917, Baron Empain's SVCFC were responsible for running *Vicinal* line 2 between Furnes and Nieuport for military purposes. On 29 July 1917, 2CRT were maintaining the diamond crossing of a light railway (60cm gauge) with *Vicinal* line 2 beside the road between Oost-Dunkerke and Nieuport town. This suggests that there was metre gauge traffic.

Light Railways

There is considerable difficulty identifying some places on the Fourth Army Light Railway systems on the Belgian Coast in 1917. Of the units on construction and maintenance, 2CRT in July identify most lines with numbers and map references, but with almost no place names. In August they mention some place names with map references before they departed on 7 August. 10CRT, who arrived on 2 July, provide almost no details of the location of work done in any form. 4CRT, who worked from late September, identify many place names, most linked with line numbers, but no map references. Of the operating companies, the main unit, the 13Can LROC, record many place names but usually with no line numbers or map references. The available maps between them show all the lines but give no line numbers and relatively few place names. The place names and line numbers shown on Figure 7.7 are those which can be confidently identified from these sources.

Construction and maintenance

Light railway works are summarised in Table 7.10. Two Companies, C and D, of 2CRT arrived at Coxyde on 20 June. Their move from Bapaume was chaotic, causing the Battalion CO, based with A and B Companies near Poperinghe, to complain in writing to the HQ of Canadian Railway troops (CRT). The move was a surprise to C and D Companies when notified on 16 June. The CO sent lorries on 18 June to transport stores. They were told that they were to entrain at noon on 19 June. The Battalion CO went to Dunkerque himself to sort out the arrangements. He asked CRT HQ, the DGT and GHQ for the location of XV Corps (Fourth Army HQ was not yet established). At 2am on 20 June he found XV Corps HQ at Dunkerque, and that morning he was given a 'chit' for a Camp Commandant at Coxyde, who was a French Officer with a British interpreter. The camp allocated was on the coast between Coxyde-Bains and Oost-Dunkerke-Bains. Up to 2pm, the Railway Transport Officer was unable to say at which rail head they would arrive – Coxyde, Furnes, Adinkerke, or Ghyvelde? At 11pm he was told that they would be likely to arrive early morning 21 June, at Ghyvelde-Bray-Dunes. At 6.30 am he found their train on an ammunition siding at Ghyvelde. They were requested to off-load at once as an ammunition train was due in 20 minutes. They marched 10½ miles (17km) to the camp. In the afternoon of the same day they were told that the camp ground was needed for an Infantry Division moving in the next day, and they would have to move again. Up to 22 June they had no information as to what work was required, and where, and the Companies had been occupied from 16 to 21 June on a move which

Railways of the Ypres battlefields 7 June 1917 to 8 April 1918

Table 7.10 Light railway works on the Belgian Coast 20 June 1917 - 28 November 1917

System	line	Section	Construction work start	finish	undertaken by	notes
Coxyde system						
A lines	A	Coxyde - Yser Estuary (Nieuport Harbour)		bef 20.6.1917	French	
		repaired/rebuilt	24.06.1917	20.07.1917	2CRT	
		E line	19.07.1917		2CRT	
		Lloyd yard		01.08.1917	2CRT	
	A1	A (nr Oost-Dunkerke) - Nieuport-Bains		bef 20.6.1917	French	
		repaired/rebuilt	24.06.1917	19.07.1917	2CRT	
		B spur	16.07.1917	19.07.1917	2CRT	
		C spur	16.07.1917	20.07.1917	2CRT	
		Diana gun spur	24.07.1917	26.07.1917	2CRT	
		Langley gun spur	24.07.1917	26.07.1917	2CRT	
		A jct (Suicide Corner) - Nieuport-Ville		bef 20.6.1917	French	
		repaired/rebuilt	24.06.1917		2CRT	
	A2	A (nr Witte Burg) - A1 (nr Zebbern)	27.06.1917	14.07.1917	2CRT	
		D line (spur)	19.07.1917		2CRT	
		Coxyde to St-Idesbald TS	02.07.1917		2CRT	
		Coxyde (additional) Yards	02.07.1917		2CRT	
		Angus locomotive yard		by 21.07.1917	2CRT	
		Winnepeg Marshalling Yard		by 21.07.1917	2CRT	
		St-Idesbald TS to Duinhoek Sablière		07.1917	pr 10CRT	
	A3					
	A4			by 09.07.1917	2CRT	
	A5	Oost-Dunkerke to Nieuport-Ville line		by 27.07.1917	2CRT	
		Lancaster Dump (A line) to A5 jct	08.1917	by 09.09.1917	?10CRT	'Oost-Dunkerke southern bypass'
	A13	St-Idesbald TS to near Oost-Dunkerke	10.1917	03.11.1917	4CRT	'Coxyde bypass'
Associated tramways						
		Adinkerke to La Panne (horse drawn, before 1914)				taken over by British Army for mechanical traction 1917
Avecappelle system						
		Avecappelle (SG line) to Boitshoucke area, and branch	10.07.1917	10.08.1917	10CRT	later called the B line, with branches B1 and B2
		Avecappelle (SG line) to Oostkerke, and branch	30.09.1917		4CRT	

pr	probably		SG	standard gauge
jct	junction		TS	transhipment
CRT	(Bttn) Canadian Railway Troops			

should only have taken one day. Finally, on 23 June, D Company were established at Coxyde, and C Company moved to near Oost-Dunkerke-Bains, on the coast. The CO of D Company met the CO of 13Can LROC, who had arrived on 21 June to operate Fourth Army light railways, and together they reconnoitred the line from Coxyde to Nieuport.

2CRT took over the three lines already constructed by the French. These were labelled, probably by the British, the A and A1 lines, the third being the branch from the A line into Nieuport town. From 24 June, 2CRT were repairing and where necessary rebuilding all three lines. D Company moved six parties each of six men to a maintenance camp halfway from Coxyde to Nieuport town. Repairs were not without problems, including heavy shelling, and in one case replacement light rails (9.5 lb/yard, 4kg/m) were more distorted than those they were sent to replace.

2CRT began additional spurs and sidings at once, and on 25 June surveyed additional yards between Coxyde and the standard gauge main line between Adinkerke and Furnes, and a new line through the dunes east of the A1 line linking that line and the A line. This became known as the A2 line, and construction started from both ends on 27 June. This line supported several gun positions, but construction was difficult because of the soft sand and some tall dunes. In places embankments had to be supported with canvas and wire.

In early July, 2CRT began construction of a line south from Coxyde, known as the Coxyde-Furnes line, although it only went as far as the St-Idesbald transhipment. On 2 July, instructions were received for an associated major new yard near Coxyde, to be completed by 26 July. This caused difficulty because there were standing crops on the site, and neither sickles nor a mower could be obtained. Finally, on

6 July, an officer went to Dunkerque to purchase sickles or scythes, successfully, and in the evening a party cut down the crops.

On 10 July, the A2 working party of C Company reached their work site but were then forced to withdraw by an intense bombardment. This continued all day and that night, 'sweeping the area behind the lines and along the coast' for 7 miles (11km). An ammunition dump exploded at Furnes rail head, where some 2CRT men were helping with unloading, but there were no casualties. That night, D Company moved camp, spreading men out among the dunes. This was the day of the German attack, taking the whole British bridgehead east of the Yser by the sea, except for the Nieuport fortifications (the *Redan*). The next day extensive repairs were required to all lines. The A2 line was ready for operation, but no traffic was possible because of shelling.

During July, yards were constructed on the Coxyde to St-Idesbald line. The Winnipeg Marshalling Yard had five tracks, with about 400 yards of sidings, The Angus locomotive yard (not to be confused with that of the same name at Ellarsyde on the Ouderdom system) had one loop 166 yards long, three more tracks with about 100 yards clear space in each, and a Y track for turning. These yards were ready for operation on 21 July. A further set of sidings nearby, on a loop (Mansell Loop), was also constructed in July 1917, adjacent to the Coxyde to Furnes road and the Metre gauge *Vicinal* line 2. This is likely to have been a transhipment to one or both of these.

Also on 21 July, 2CRT received permission, after frequent requests, to place five wooden sleepers to each rail length on curves, and the same day 135 sleepers (27 curved rail lengths) were laid. The benefits were seen by the next day, with no derailments despite a marked increase of traffic. At one point in the dunes, where there was a considerable gradient, this allowed the immediate addition of 60 per cent to train load in that section. On 30 July, there was difficulty on the A line between Suicide Corner and the RE dump where the line crossed the metre gauge loop to Nieuport-Bains. The track was no sooner repaired than it was blown out again, and the party took cover in a nearby trench. At 10.30am they moved to other work, returning in the afternoon, when things were no better. The light railway operating department were notified that traffic was impossible. Perhaps it was not coincidence that this was the eve of the Third Battle of Ypres further south.

10CRT arrived on 2 July, from Chaulnes (Somme Sector). Initially at Coxyde, they soon moved their HQ to La Panne. Companies were based at Oost-Dunkerque (A) and Coxyde (D). During July, they constructed the line west from the St-Idesbald transhipment to the Duinhoek *sablière* (sand quarry), also building a spur to the 1st Canadian Casualty Clearing Station at Oosthoek, where there was probably another transhipment from standard gauge lines. It was probably at this time that the pre-war 60cm gauge horse tramway from Adinkerke to La Panne was taken over by the British. 10CRT may also have built the line around the west and south of Oost-Dunkerke village, linking the A line at Lancaster Dump to the A5 line. On 7 August, C and D Companies of 2CRT handed their work over to 10CRT and entrained at Coxyde for Bergues, for standard gauge work. A plan of work done between 27 June and 8 August, records that they had built and then maintained 14.8 miles (23.7km) of light railway and maintained a further 16.4 miles (26.2km). In August, 10CRT laid 13 miles (21km) of track, in the Coxyde and Avecappelle systems, and built an aerial cableway ¼ mile (400m) long. 10CRT continued to work in the Coxyde area until on 25 September they were moved back to the Ypres area (La Clytte).

Some line numbers are known to have existed which we cannot attach to any line in Figure 7.7, even though the lines shown must include those to which they were attached. Thus, the A7, A8, and A11 lines are not identified, and presumably there were A6, A9, A10 and A12 lines as well, which are never mentioned in the unit diaries.

On 23 September, the CRCE (Fourth Army) agreed to lend one company of 4CRT to light railways for two weeks, mainly to construct a line 3½ miles (5.6km) long required by Fourth Army. A Company was lent from 29 September and remained on light railway work in the Coxyde area until 13 November. During October, they built the A13 line, linking the St-Idesbald transhipment to the Oost-Dumkerque 'southern bypass'.

By early August 1917, 2CRT had constructed a spur from the A5 line to the Swiftsure naval artillery position. CORCC had a small detachment in the Nieuport area from July 1916, mainly undertaking work with Royal Navy gun positions. They remained until July 1918. In November and December they were building light railway track at the Swiftsure gun positions. Guns were moved around by light railway, and on 28 and 29 December 7.5in guns were loaded onto light railway trucks and moved to Coxyde, even though by this time operating these lines had been handed back to the French or Belgians.

The Avecappelle system

C and D Companies of 10CRT moved to near Pervyse on 9 July to work on the first line of the Avecappelle system. This separate light railway system, in the Belgian sector, began about 2 miles (3.2km) east of Furnes on the main line towards Dixmude. The first line was 5 miles (8km) long, and by 27 July was sufficiently complete to be taking traffic, but the final canal bridge on one of the two eastern ends was not finished until 10 August.

A second company of 4CRT was lent to light railways on 29 September, when B Company moved to near Avecappelle to build the second line of that system, with Belgian labour. Starting from the same place as the first line, it went south and east to near Oostkerke station, with one branch. Three companies of 4CRT moved to Sorel (Somme Sector) on 13 November, leaving C Company on maintenance of the Coxyde and Avecappelle light railways, as well as standard gauge work. Later C Company also moved south, leaving a detachment until 12 December.

Operating the Belgian Coast lines

The main operating company was the 13th Canadian LROC (13Can LROC) who arrived at Coxyde on 21 June 1917. On 18 July an officer and sixty men (one platoon) of 17(3)Aus LROC also arrived.

13Can LROC had left Aldershot on 9 June and this was their first deployment. On 25 June, they took over the existing light railways from the French, and the next day had their first experience of hostile shell fire. They spent the first few days unloading rolling stock and preparing for operating the railways. On 3 July they handled ammunition for the first time, and during the month carried ammunition to the artillery in increasing quantities. At the end of July, 2CRT increased the level of forward area maintenance because of the effect of heavy traffic, and on 29 July a petrol tractor spur with 275 feet (84m) of siding was opened at Lloyd Yard.

From the beginning of August 13Can LROC give much more detail of their work. There were many derailments, and the track suffered in the heavy rains of early August. On the evening of 2 August, all trains were cancelled because of the bad condition of the track, and on 3 August the main line was washed away just east of Coxyde, and ammunition was held up. Other hold ups were caused by shellfire. Over a period of four weeks in August, 13Can LROC carried 45,137 tons of supplies, the main part ammunition; average 1,612 tonnes per day, maximum 2,454, minimum 798. At the end of August, a 2-6-2T Alco-Cooke locomotive derailed seven times between Winnipeg yard and Coxyde, after heavy shelling. Early September was much quieter with little ammunition carried. On 11 September, the company entrained at St-Idesbald rail head for Savy-Berlette (Arras Sector).

The 17(3)Aus ALROC detachment worked wholly with Simplex 20 and 40 petrol tractors, delivering ammunition on forward lines, which fed batteries further forward than had been usual in their Somme Sector operations. On 13 August, tractor 2138 (a Simplex 40) fell in a shell hole. The Australian driver and guard were badly wounded. The Australians also suffered gas casualties. This detachment was still at Coxyde on 30 September, but we do not know for how long they stayed. We also do not know who took over the operating from 13Can LROC after 11 September. 32LROC had one man killed on 13 October and one 19 October, and both are buried at Coxyde, so it is possible that they may have taken over. The Avecappelle System was in the Belgian sector, and we presume was taken over for operating by them.

Rolling stock

More is known about motive power than in many areas, because in August 1917 13Can LROC noted the British Light Railways serial numbers of locomotives and tractors involved in derailments and other incidents. From this we know that they had at least sixteen Alco-Cooke 2-6-2T steam locomotives. One of these was No. 1275, now *Mountaineer* on the Ffestiniog Railway, and another was No. 1257, formerly preserved at Froissy (CFCD) but now at Tacots des Lacs (see Chapter Twelve). No other types of steam locomotives are recorded.

They also had at least two PE tractors (one Dick, Kerr and one British Westinghouse), at least four Simplex 40 tractors, and at least one Simplex 20 tractor, the last probably used in the yard at Coxyde. The 17(3)Aus LROC detachment were operating with Simplex 20 and Simplex 40 tractors. The PE tractors were not popular. On 27 July, 2CRT blamed these, and heavily loaded wagons, for increased track maintenance. During the heavy rains of early August, they were forbidden to go out on the track because of frequent derailments. Finally, on 14 August they were loaded onto standard gauge wagons and taken away to another section of the front.

Tramways

The 2nd Army Tramway Company RE (2ATC) arrived at Coxyde on 25 June 1917. XV Corps had two Divisions, and half of the Company were allocated to each. The 1st Division held the front line on the coast around Nieuport-Bains, and were allocated 2ATC Left half, hereafter 2ATC(L). The 32nd Division were inland in the area of Nieuport town, which was in ruins, and were allocated 2ATC Right half, hereafter 2ATC(R).

Liaison between the tramways and light railways seems to have been closer than in some areas, with the CO of 2ATC meeting the ADLR as well as working with Divisional Transport Officers. He obtained from light railways some track and rolling stock material and started negotiating for tractors. There were already tramways from the light railway rail head at Nieuport-Ville towards Nieuport-Bains along the Yser estuary and harbour side, and through the town to the east of the Yser. These were improved, with the intention of allowing through running.

In early July, they were sent light railway material. Designs for trucks, and regulations for tramway working were agreed. On 5 July, a light railway train was able to reach the centre of the ruins of Nieuport, and on 6 July the tramway to the *briquetterie* (brick works) was re-opened after repairs. Wooden rails were replaced with steel on the lines beyond the bridges over the Furnes Canal and the canalised Yser River, to the moated fortifications and to *Maison Blanche* (White House). The Chief Railway Engineer for 32nd Division agreed that both these lines would be made fit for tractor traffic.

On 10 July the dune area on the coast was taken by the Germans, but XV Corps retained the *Redan* and the *briquetterie* area to the north of the Yser. 2ATC took over maintenance of the line from Oost-Dunkerke to Nieuport on 16 July. Towards the end of July, there was pressure to urgently complete the lines as far as the *Cinq Ponts* (Five Bridges), which implies that the lines north of these were not yet open, but it may just have been the upgrading to tractor traffic which was not complete.

At the beginning of August, 2ATC undertook work in the Groenendijk area, including battery lines. Track was salvaged from the Rue d'Ypres at Nieuport, and used to clear and renew the line in the Rue Longue. On 13 August, after gas attacks, work in Nieuport town was discontinued, and 2ATC(R) was transferred to other work with 148th Army Transport Company RE. Work continued in the Groenendijk and Nieuport-Bains areas. On 9 September, 2ATC(R) moved with the Company HQ to near Adinkerke. 2ATC(L) worked at the Fourth Army workshops at Y Corp Dump (Adinkerke Chaussée), practicing laying light track, and later heavier track on wooden sleepers. On 30 September they were replaced at Adinkerke by 2ATC(R), who undertook the same work and training.

2ATC(L) moved to Ramscappelle on 2 October and took over work in the Nieuport area from the NF Pioneers. 2ATC(L) worked on Divisional artillery lines south-east of the A5 light railway line. From 14 October they were laying light track at *les Cinq Ponts*, Nieuport. On 19 October the two halves swapped, with 2ATC(L) at Adinkerke again, maintaining the lines at Fourth Army workshops, and training on trestle bridging. 2ATC(R) were laying heavy 60cm track at Triangle Wood, on or near the main light railway line into Nieuport. On 7 November, the whole Company moved to Arras to work for the Third Army.

The end of the Belgian Coast deployment

On 13 October 1917, Field-Marshal Haig conceded that with the aims of Third Ypres now limited to the capture of Passchendaele Ridge, the strategic plan was no longer possible. The troops in training were stood down on 20 October. Fourth Army HQ remained near Dunkerque until 5 November, then moved to Cassel to take over the Ypres area from the Second Army. The office of RCE VI at Malo-les-Bains closed on 28 November. Only the exploitation of the Ghyvelde Sand Quarry continued, under a CORCC party, a programme independent of and pre-dating the Fourth Army presence. The remainder of the standard gauge and light railway lines were handed over to the French or Belgians, probably the latter.

The eve of the Battle of the Lys

There was a general increase in shelling on the Ypres and Messines fronts around the time of the German advance on the Somme which began on 21 March. Shelling increased again in early April 1918, leading up to the Battle of the Lys River (also known as the Fourth Battle of Ypres) which began on 9 April 1918. This is described, for the Ypres and Messines areas, in the next Chapter.

Colour Plates

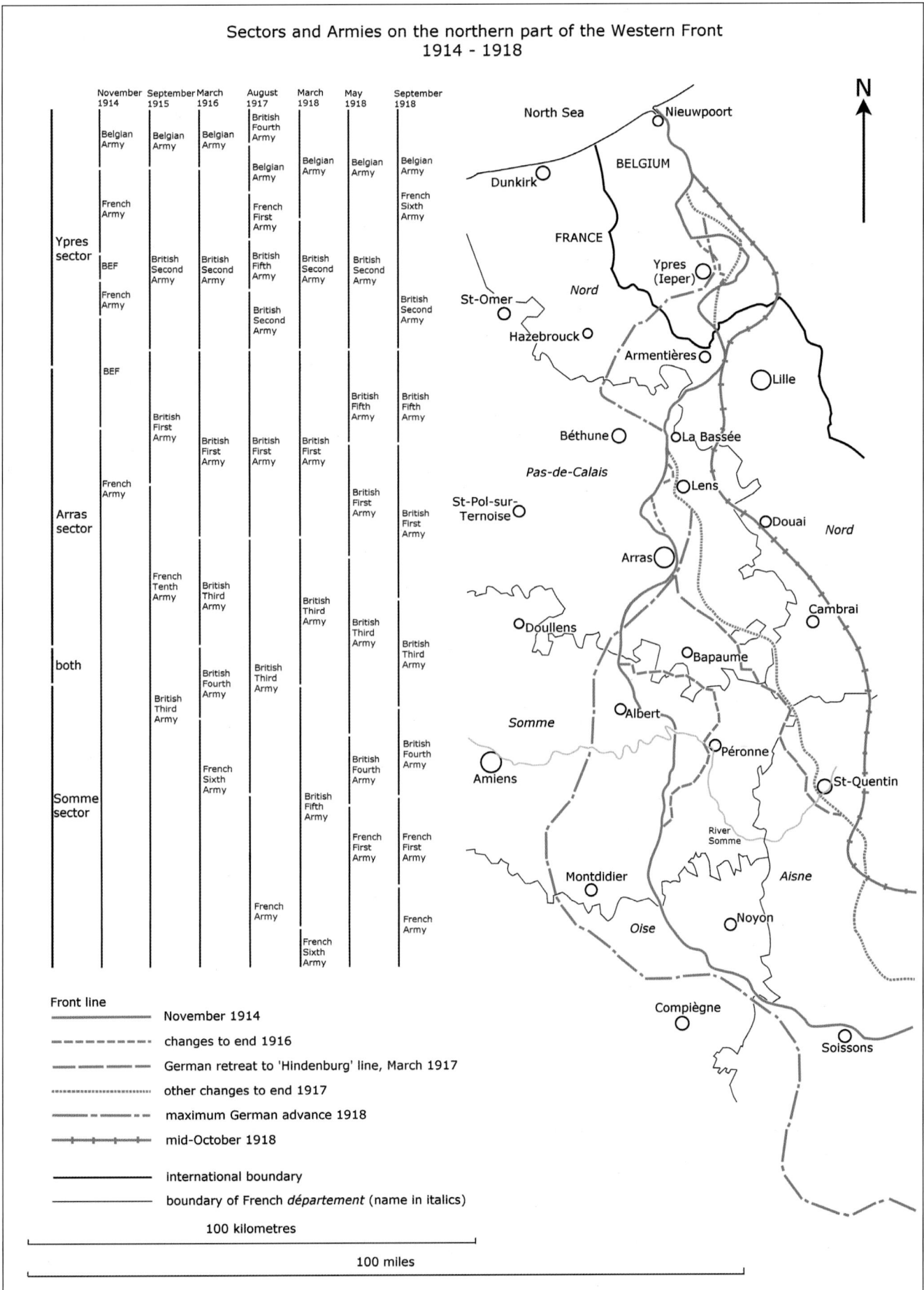

Figure 0.1 Sectors and Armies on the northern part of the Western Front 1914–1918.

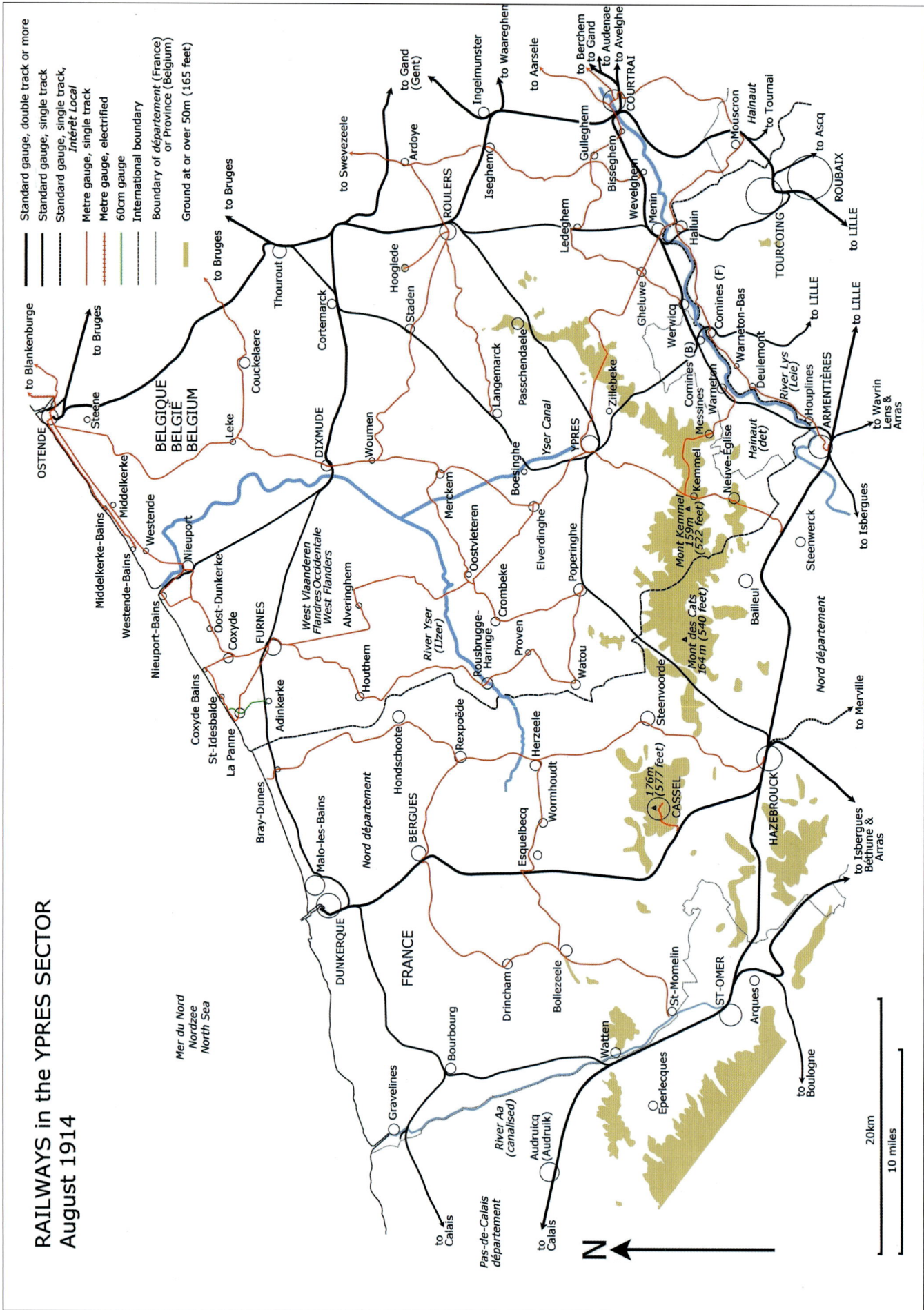

Figure 1.1 Railways in the Ypres Sector August 1914.

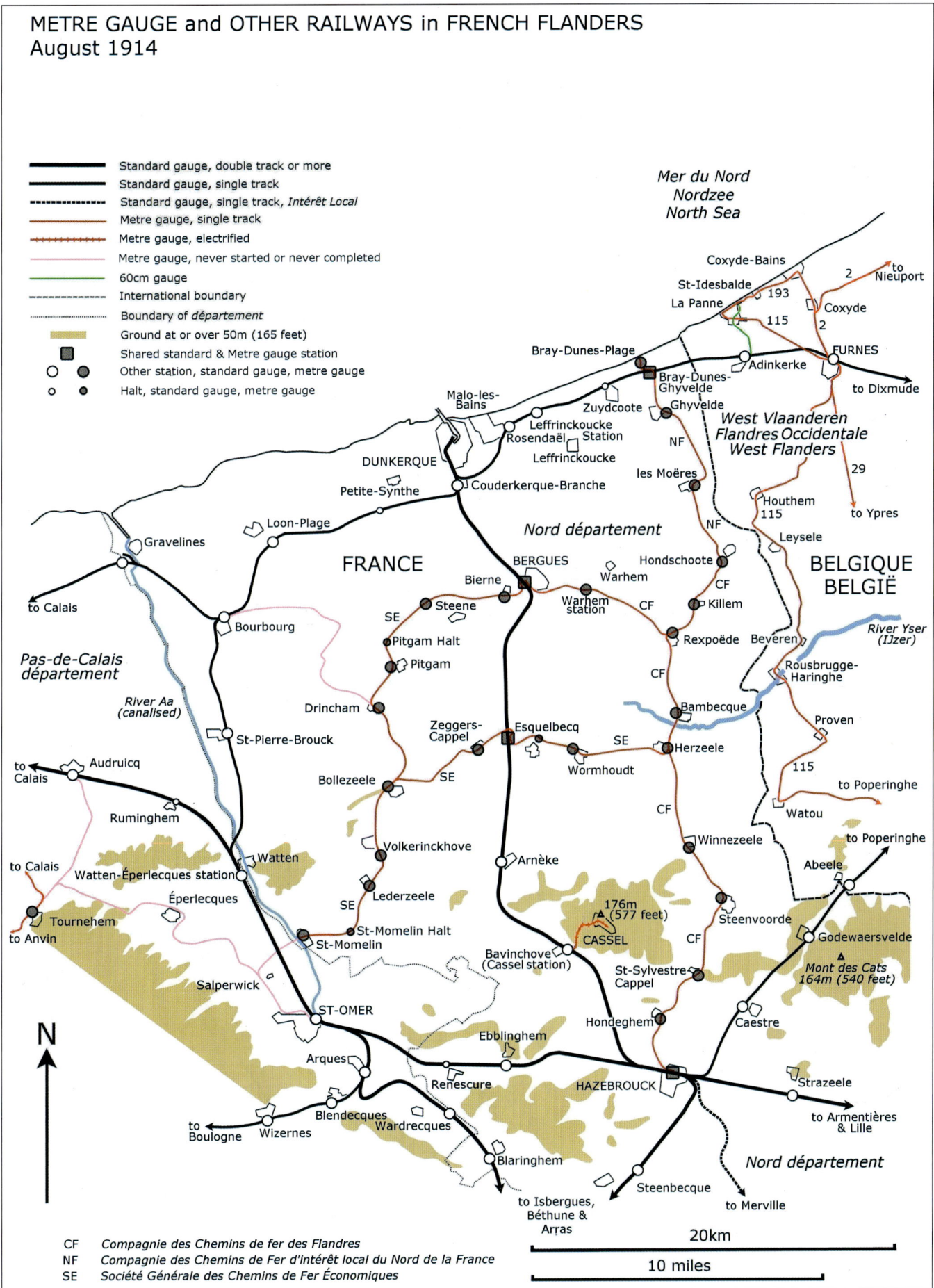

Figure 2.1 Metre Gauge and Other Railways in French Flanders, August 1914.

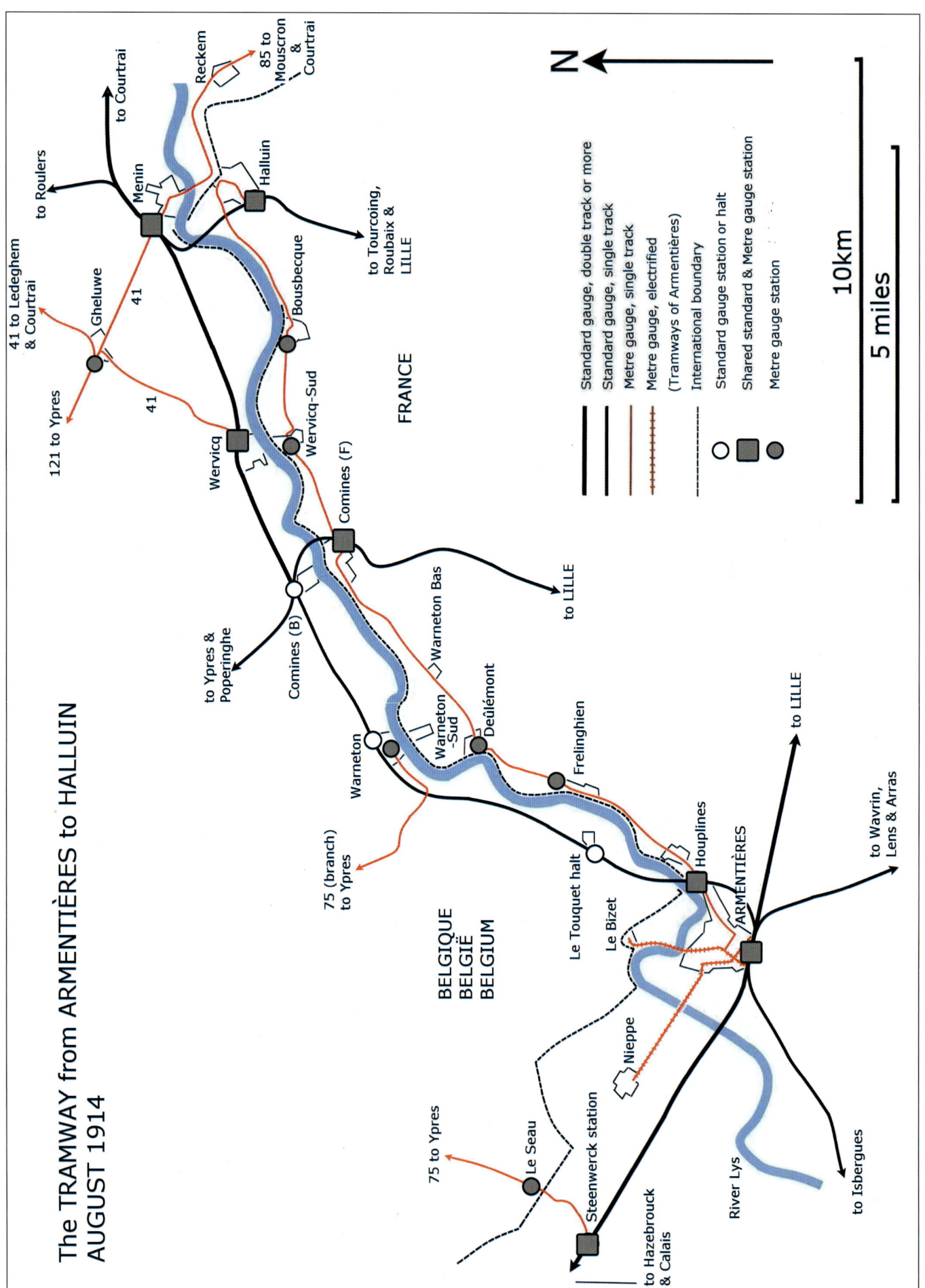

Figure 2.2 The tramway from Armentières to Halluin 1914.

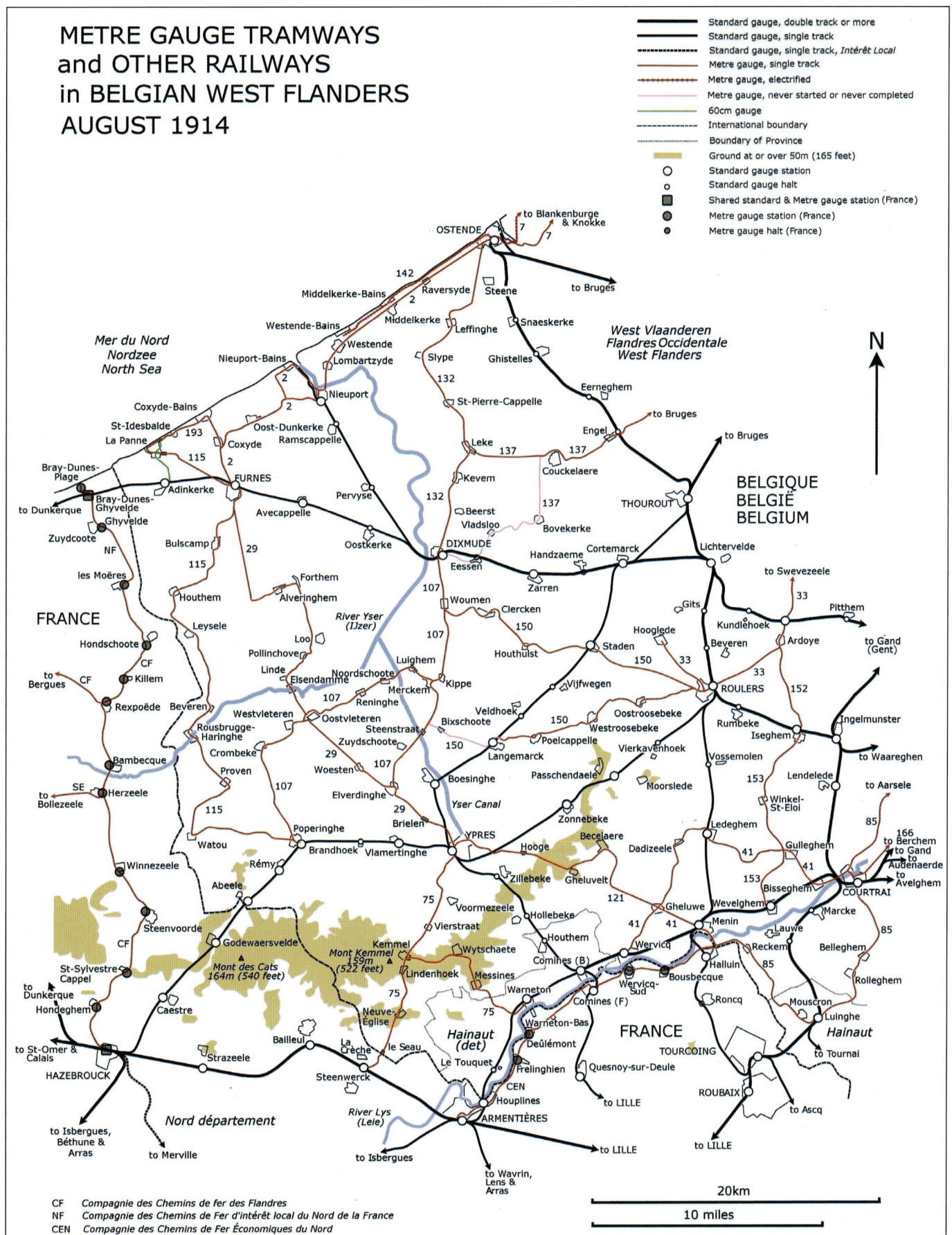

Figure 3.1 Metre Gauge Tramways and Other Railways in Belgian West Flanders, August 1914.

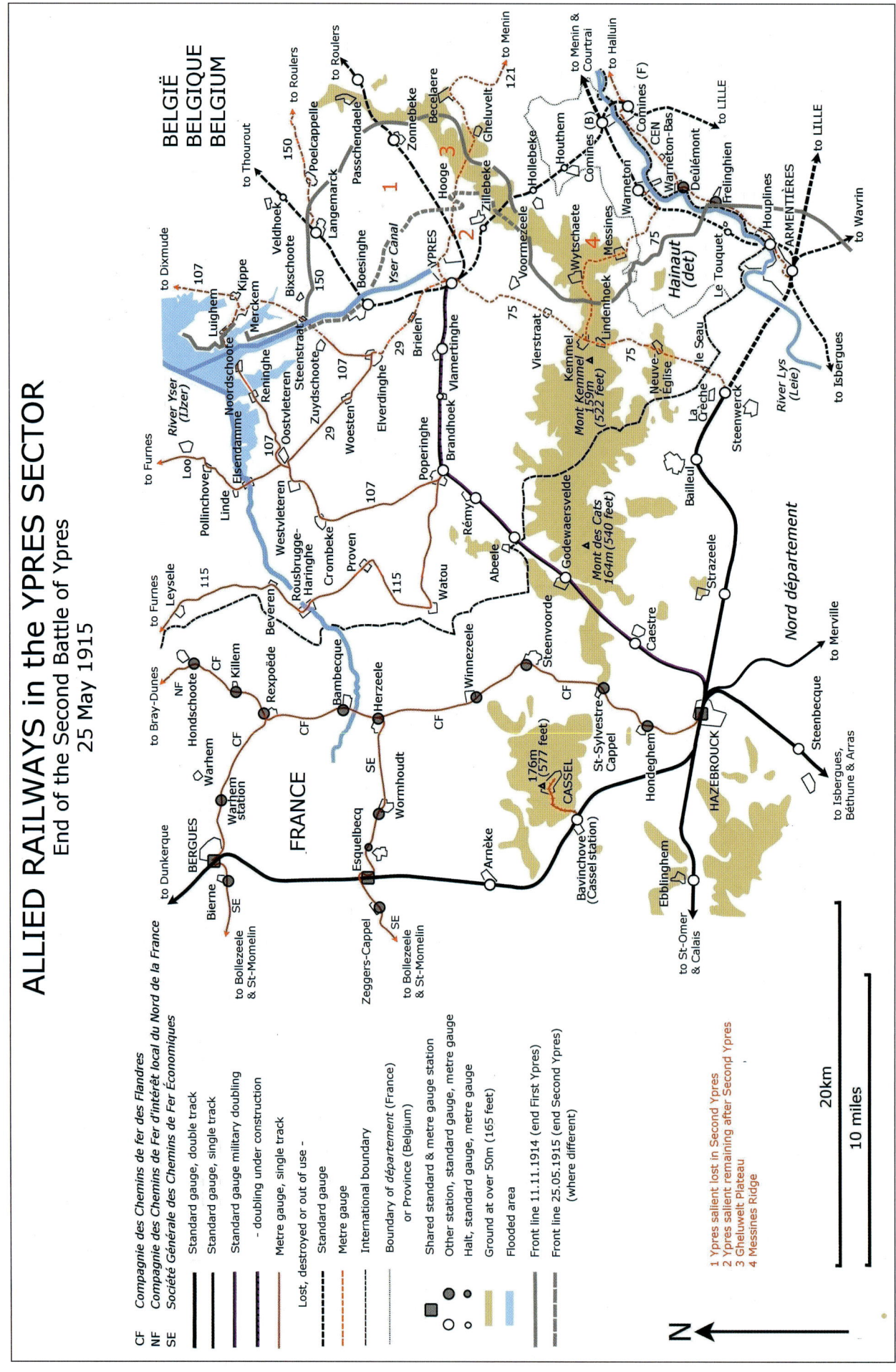

Figure 5.1 Allied Railways in the Ypres Sector. End of the Second Battle of Ypres, 25 May 1915.

Colour Plates

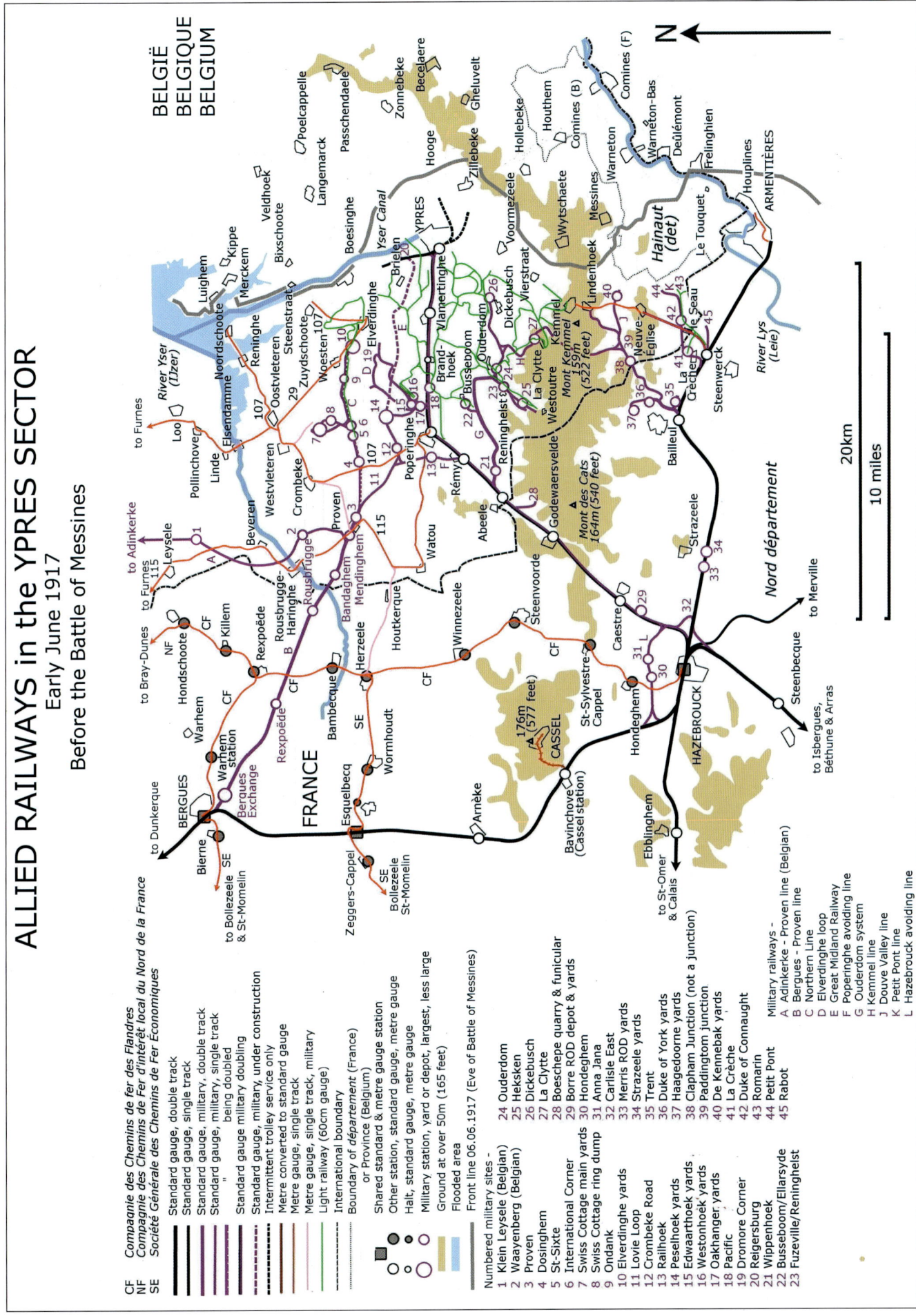

Figure 6.1 Allied Railways in the Ypres Sector, early June 1917. Before the Battle of Messines.

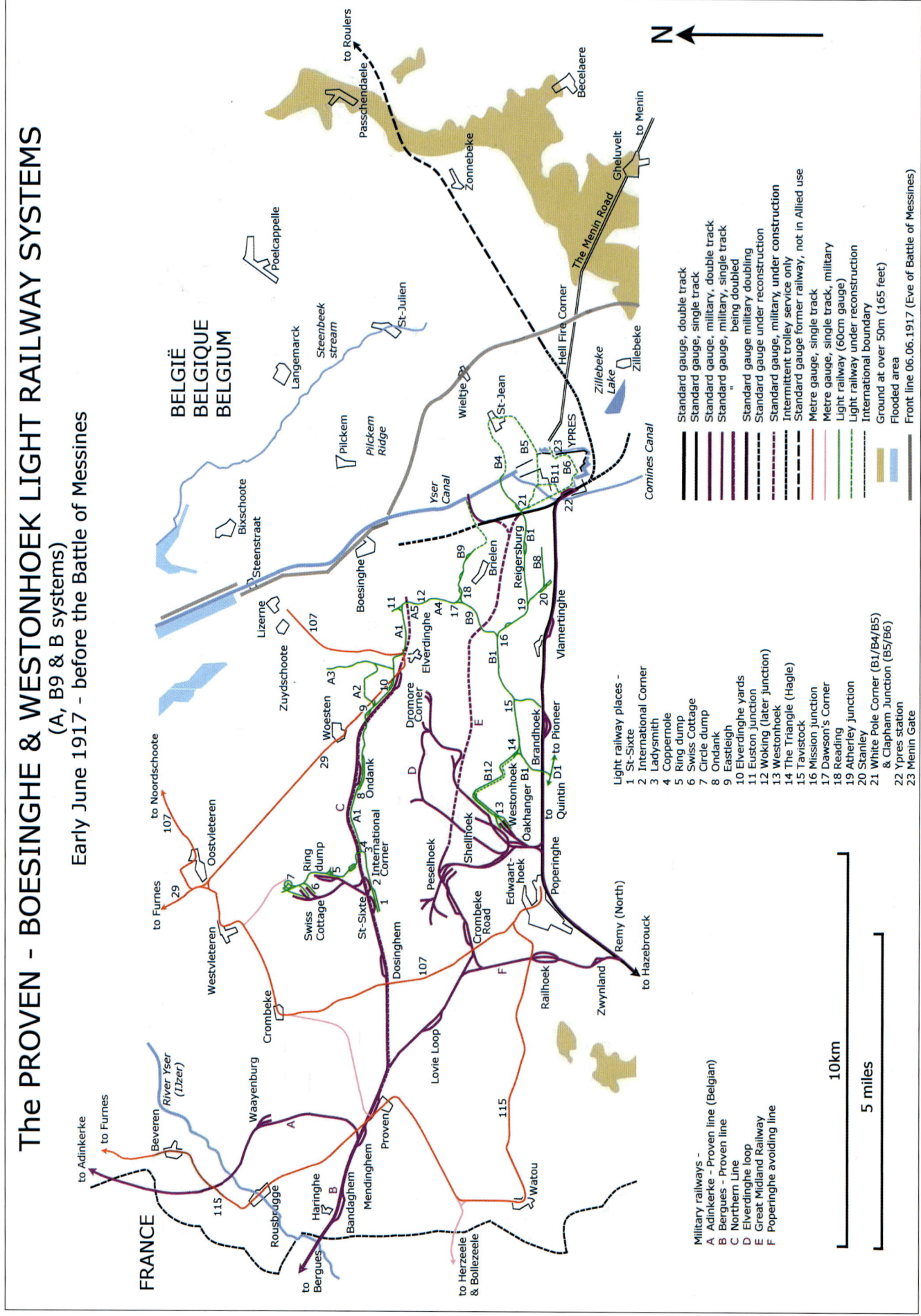

Figure 6.2 The Proven–Boesinghe & Westonhoek light railway systems (A, B9 & B systems). Early June 1917–before the Battle of Messines.

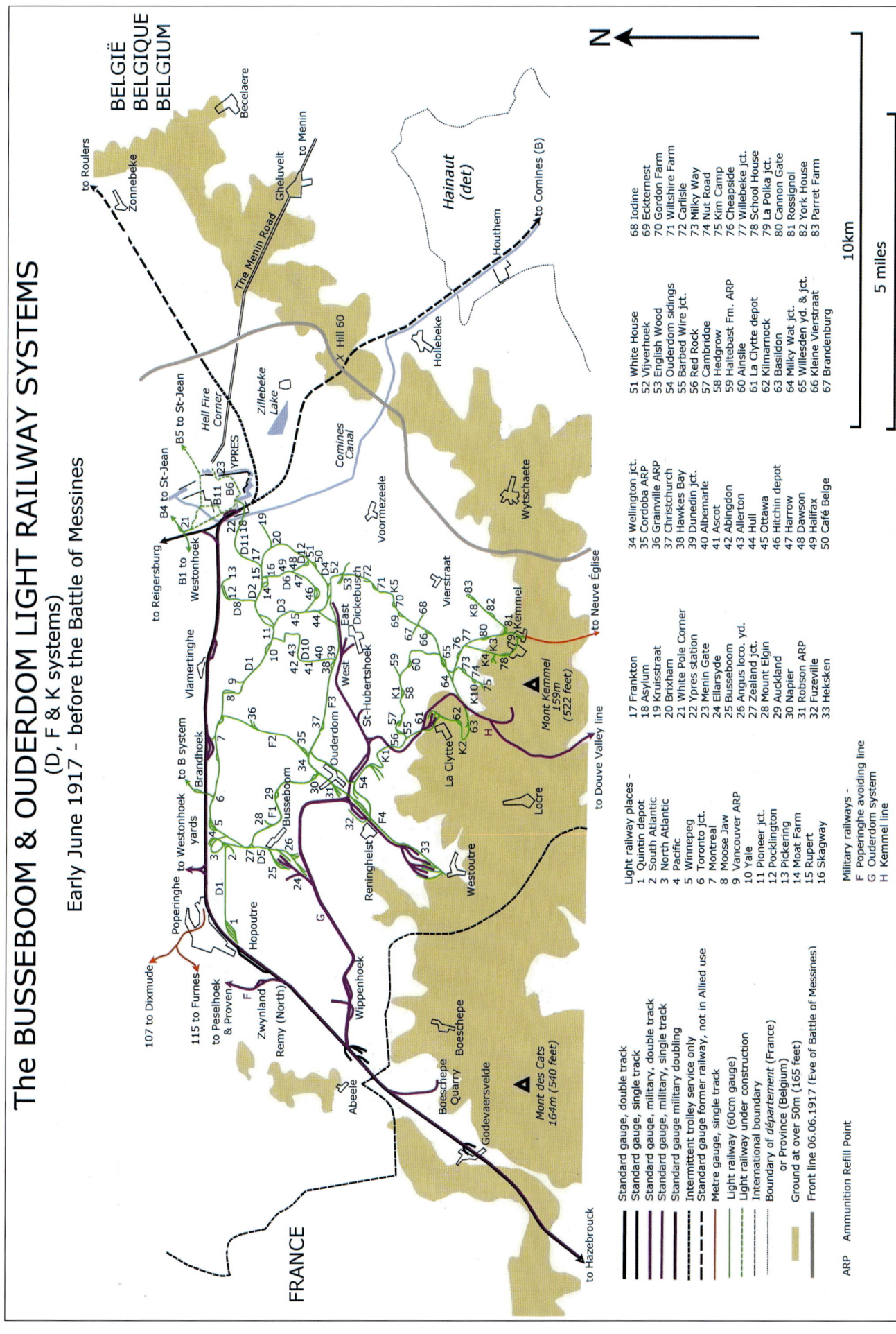

Figure 6.3 The Busseboom & Ouderdom light railway systems (D, F & K systems), early June 1917–before the Battle of Messines.

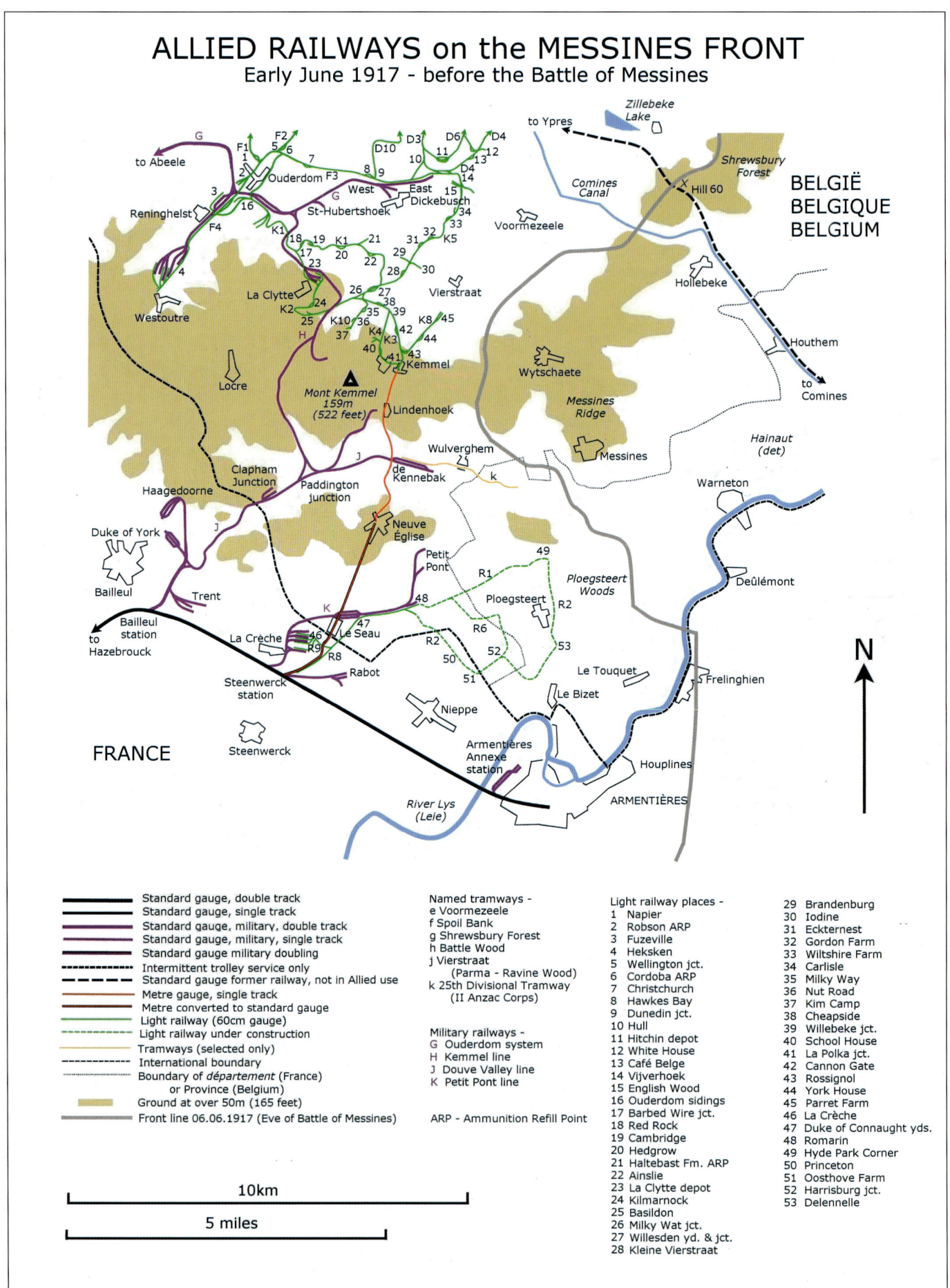

Figure 6.4 Allied Railways on the Messines Front, early June 1917–before the Battle of Messines.

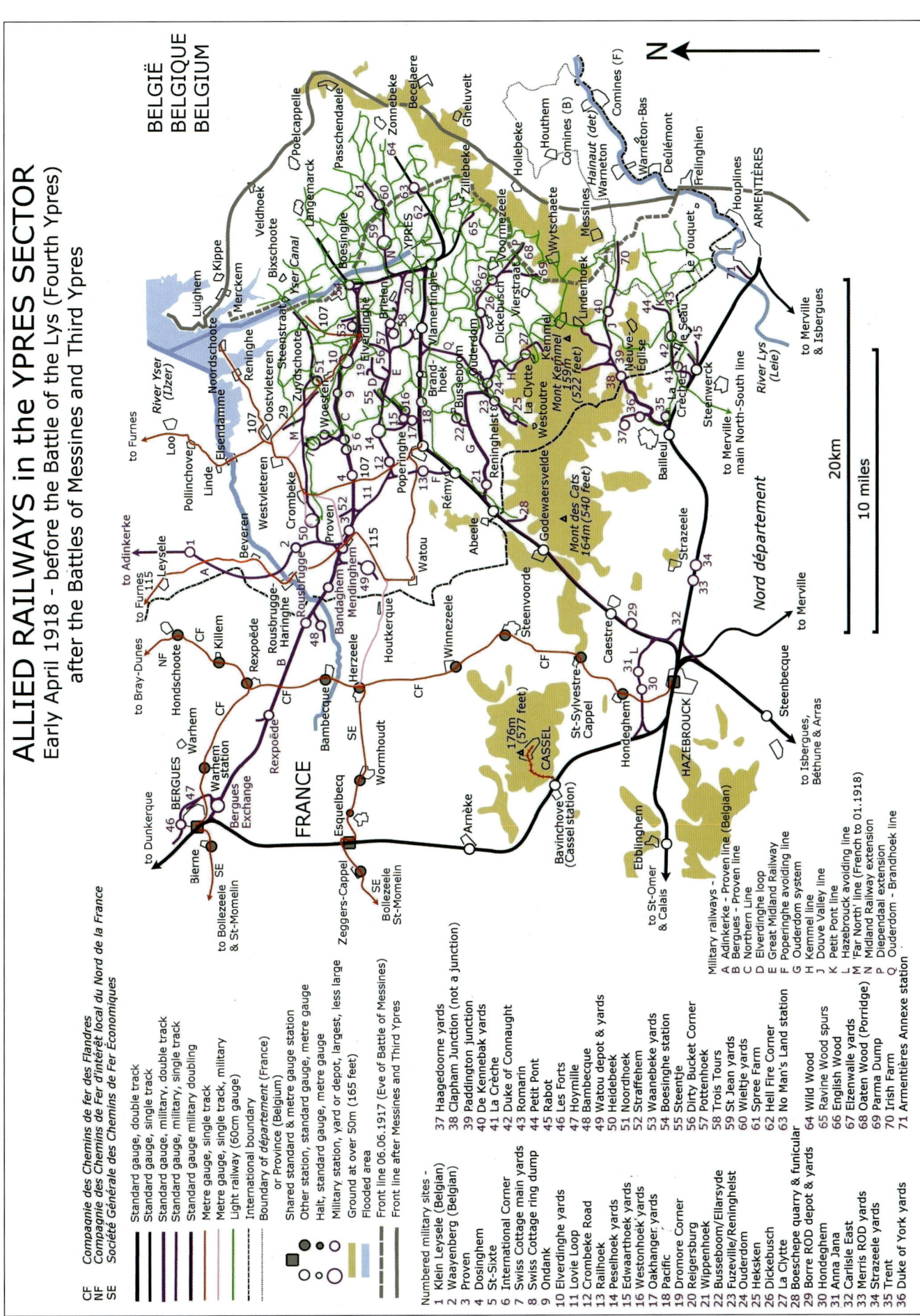

Figure 7.1 Allied Railways in the Ypres Sector. Early April 1918–before the Battle of the Lys (Fourth Ypres), after the Battles of Messines and Third Ypres.

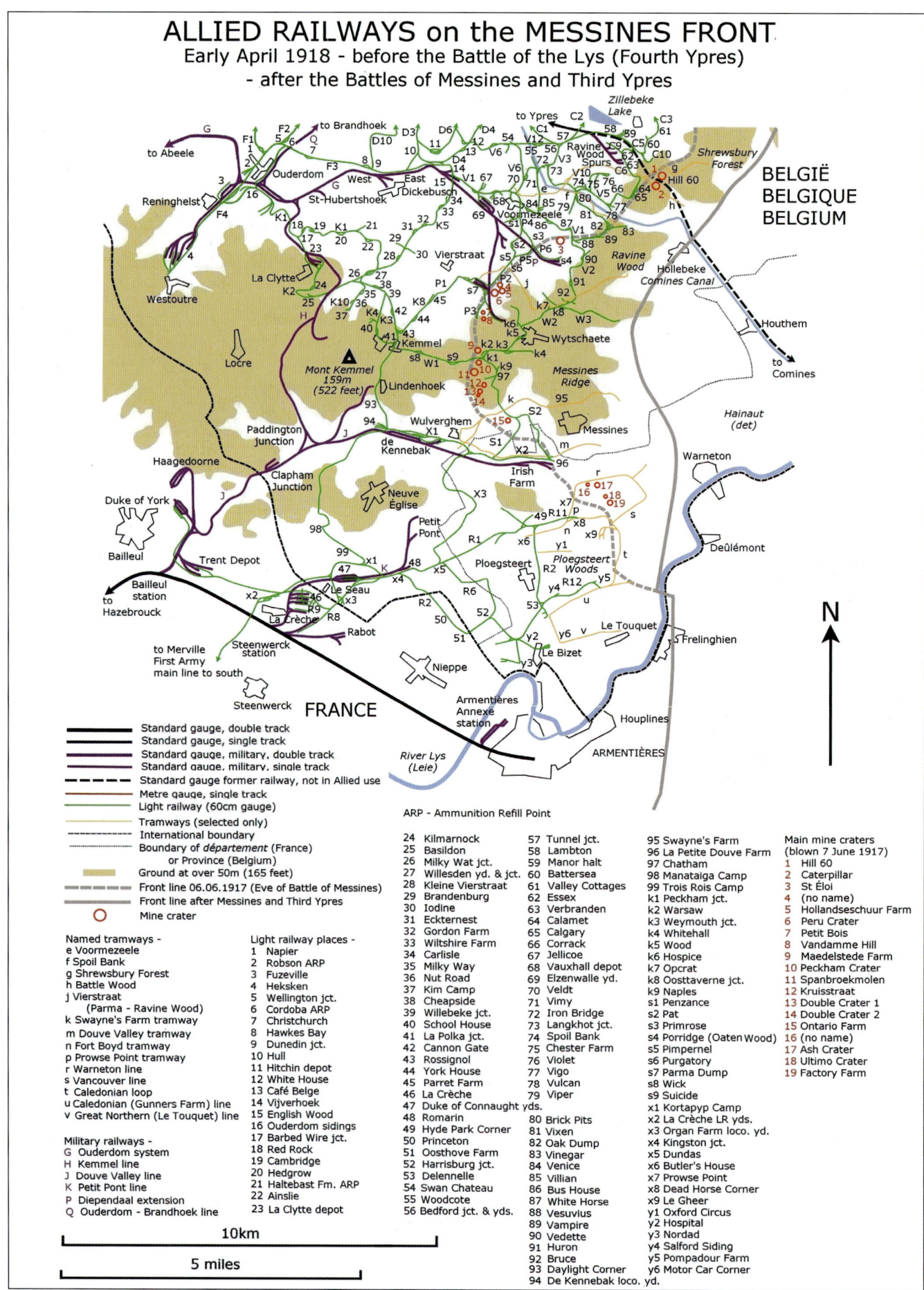

Figure 7.2 Allied railways on the Messines Front. Early April 1918–before the Battle of the Lys (Fourth Ypres), after the Battles of Messines and Third Ypres

Colour Plates

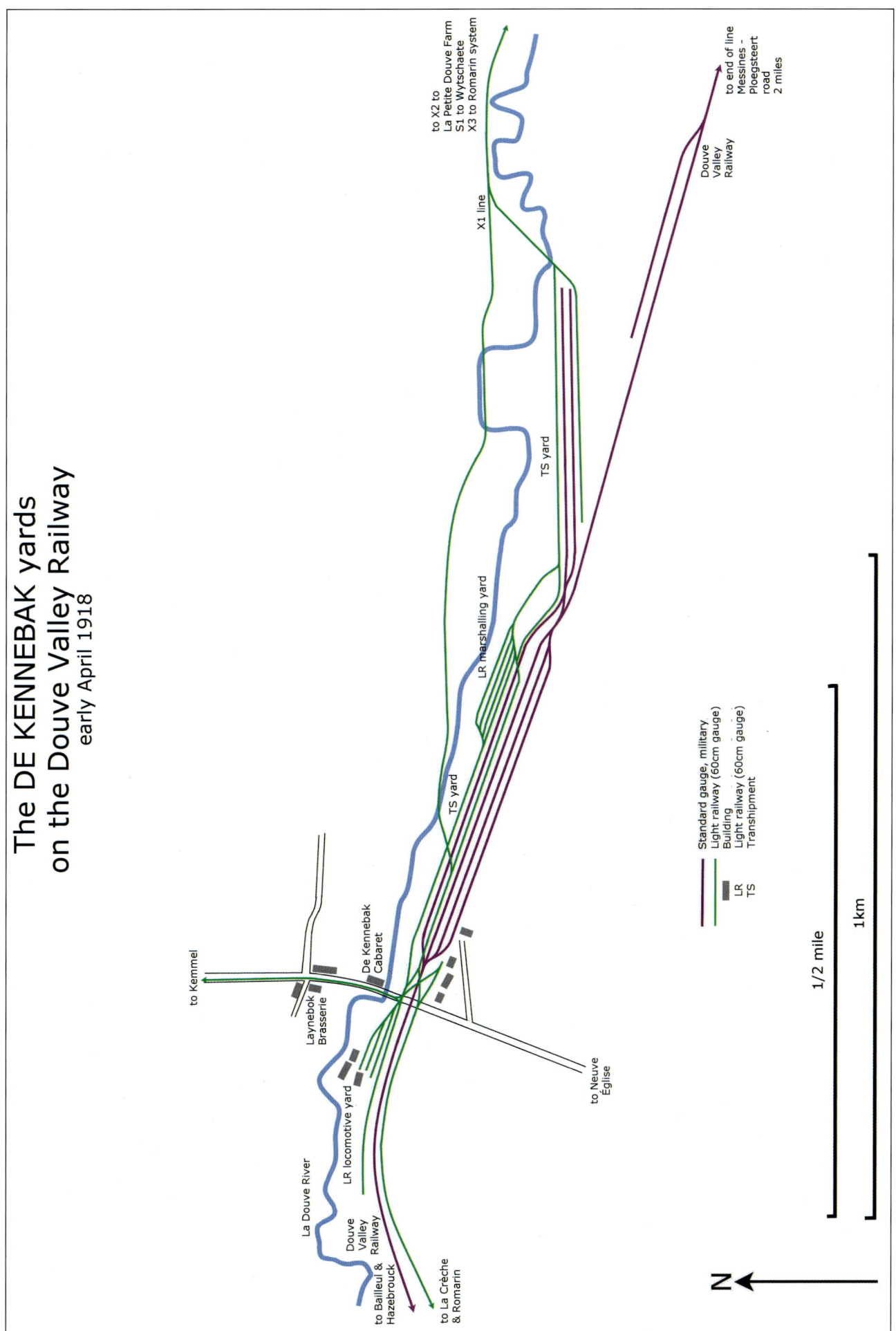

Figure 7.3 The De Kennebak yards on the Douve Valley Railway. Early April 1918.

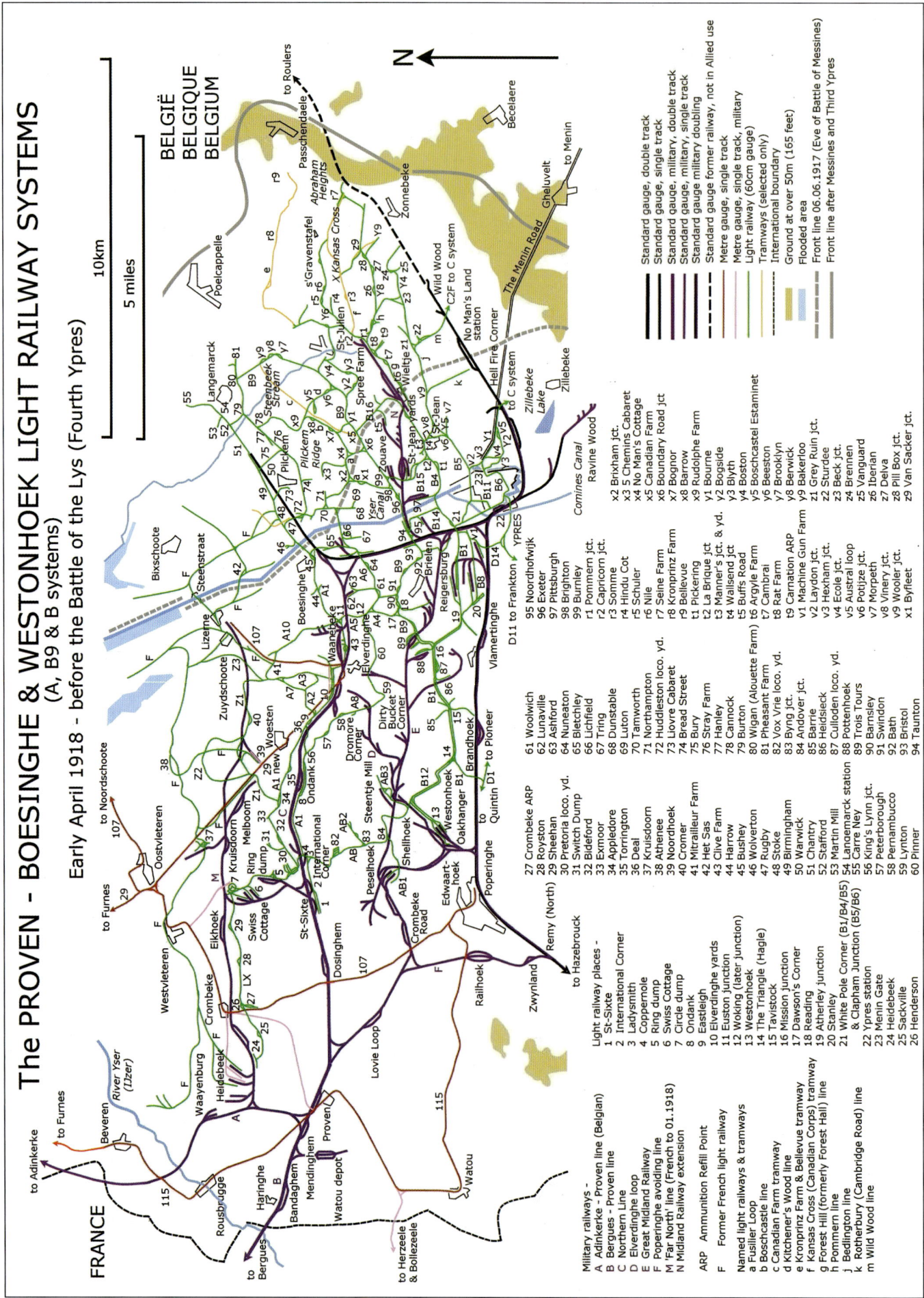

Figure 7.4 The Proven–Boesinghe & Westonhoek light railway systems (A, B9 & B systems). Early April 1918–before the Battle of the Lys (Fourth Ypres).

Colour Plates

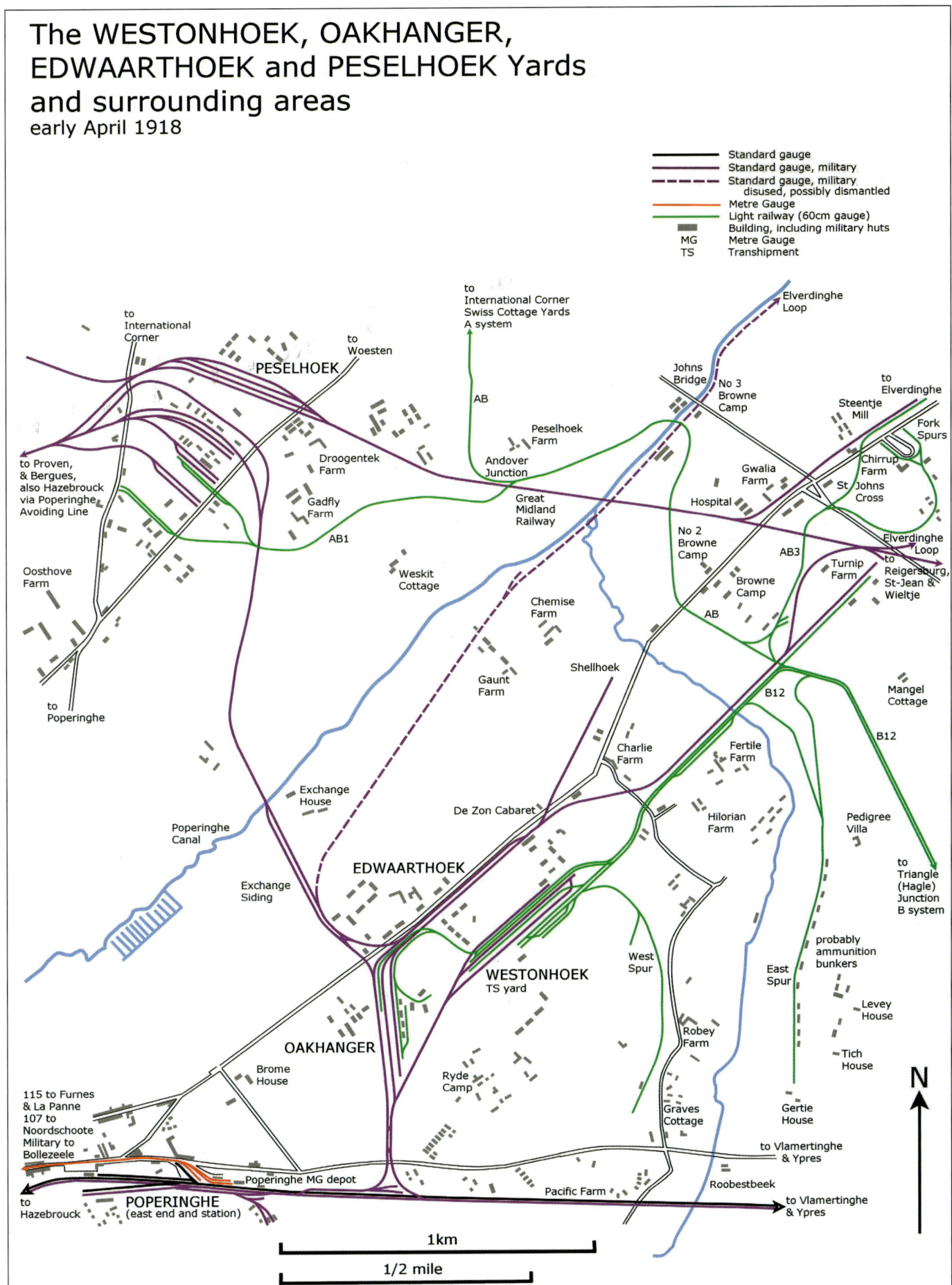

Figure 7.5 The Westonhoek, Oakhanger, Edwaarthoek and Peselhoek Yards and surrounding areas. Early April 1918.

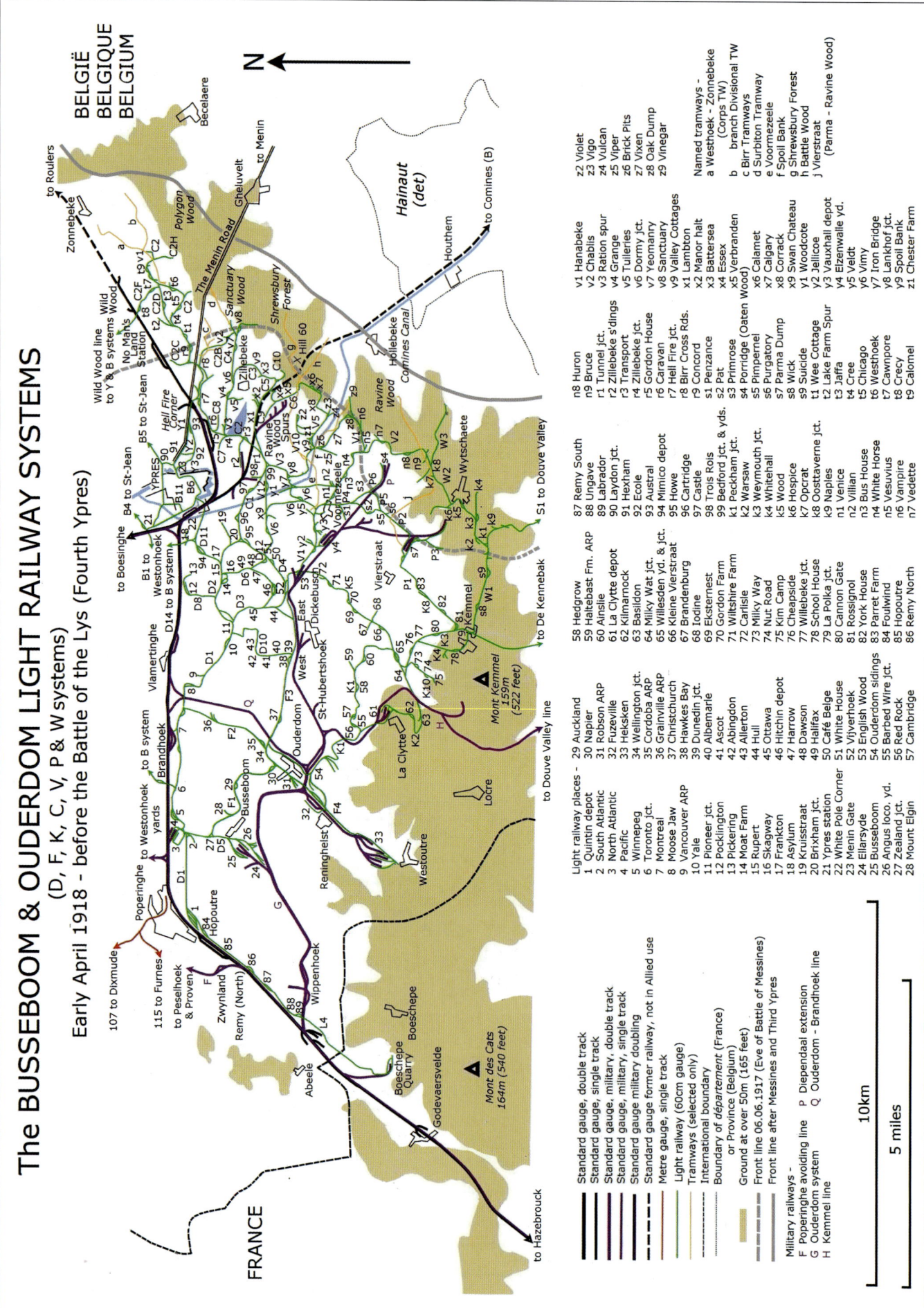

Figure 7.6 The Busseboom & Ouderdom light railway systems (D, F, K, C, V, P & W systems). Early April 1918–before the Battle of the Lys (Fourth Ypres).

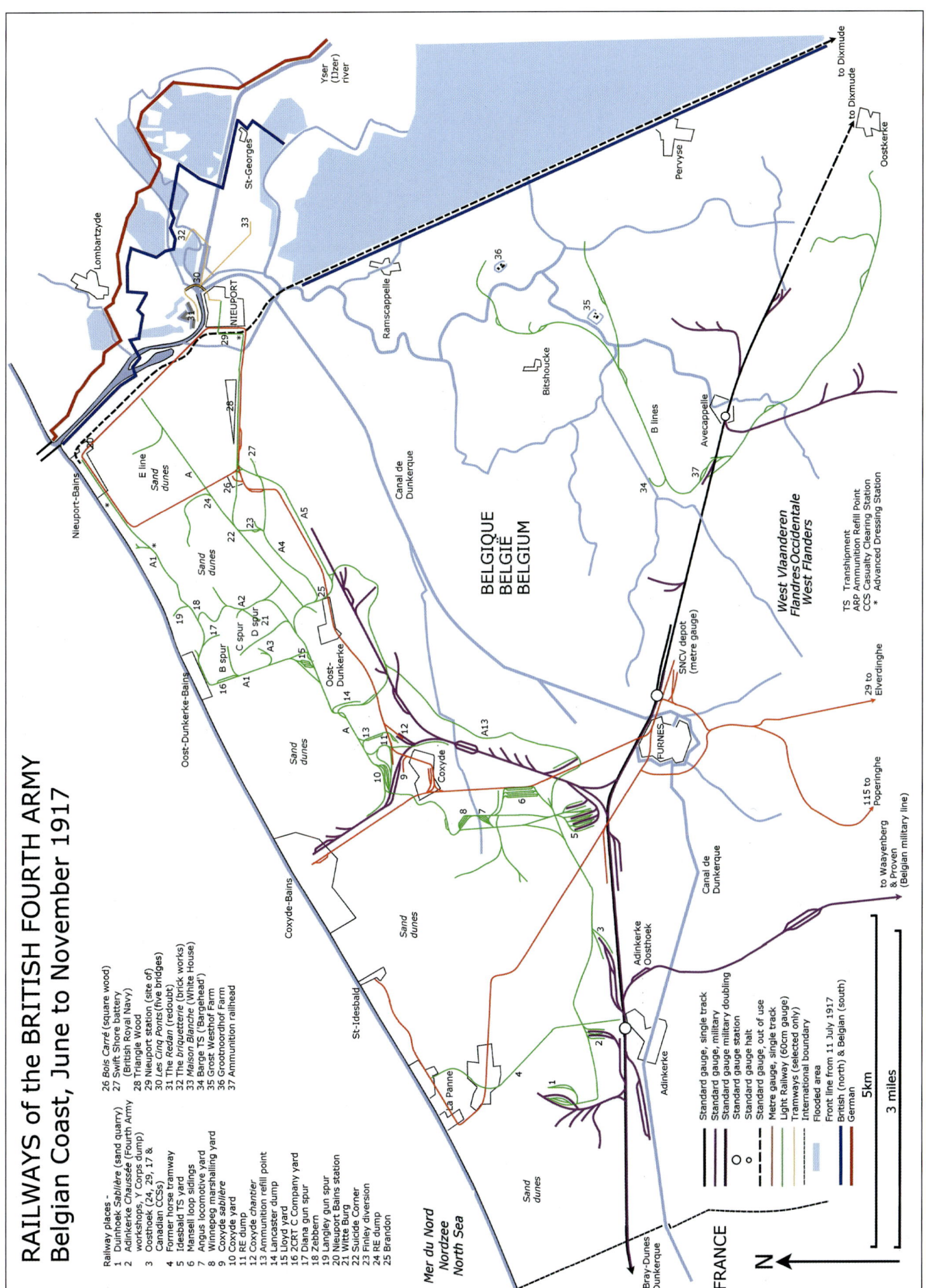

Figure 7.7 Railways of the British Fourth Army. Belgian Coast, June to November 1917.

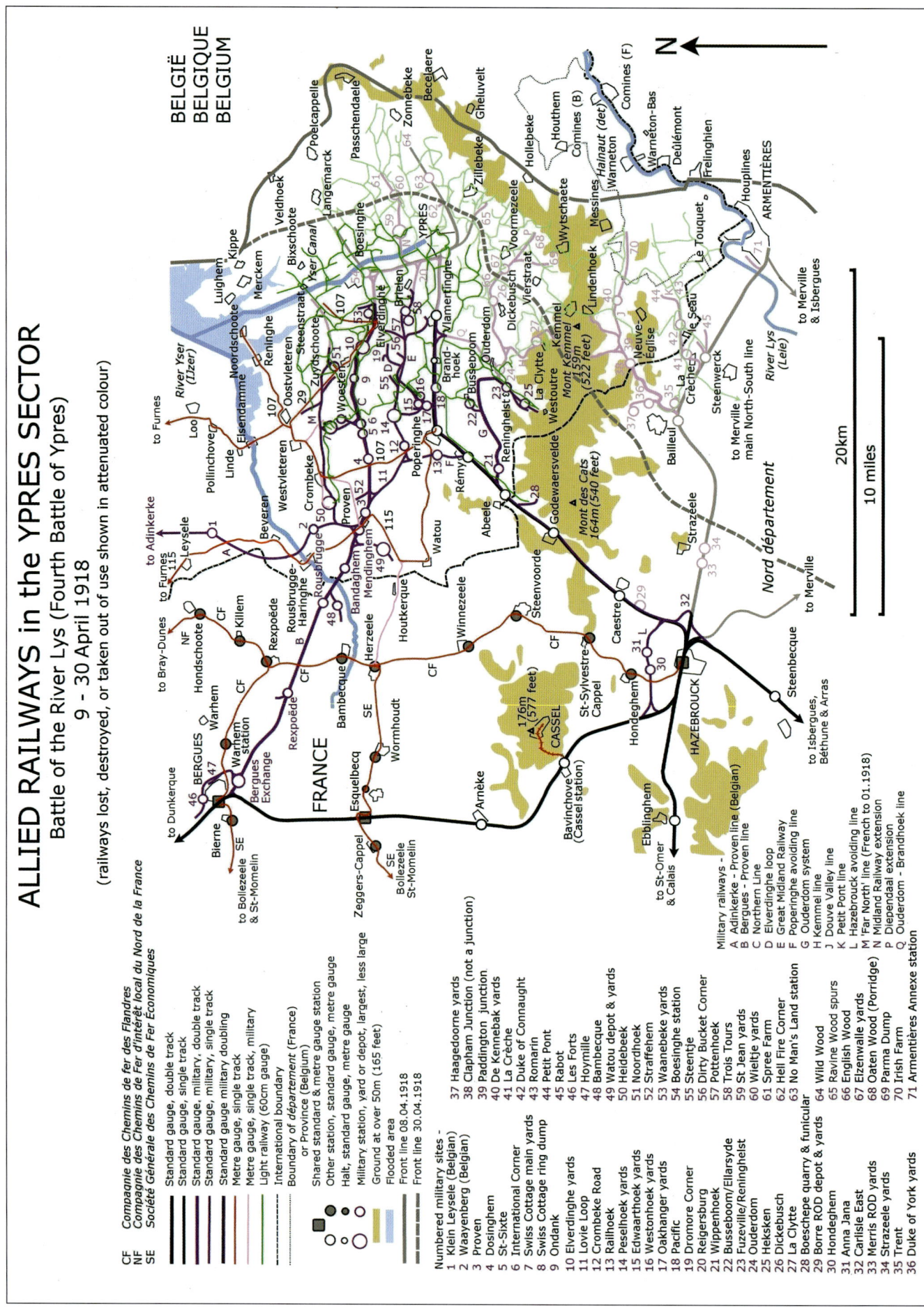

Figure 8.1 Allied Railways in the Ypres Sector. Battle of the River Lys (Fourth Battle of Ypres) 9–30 April 1918.

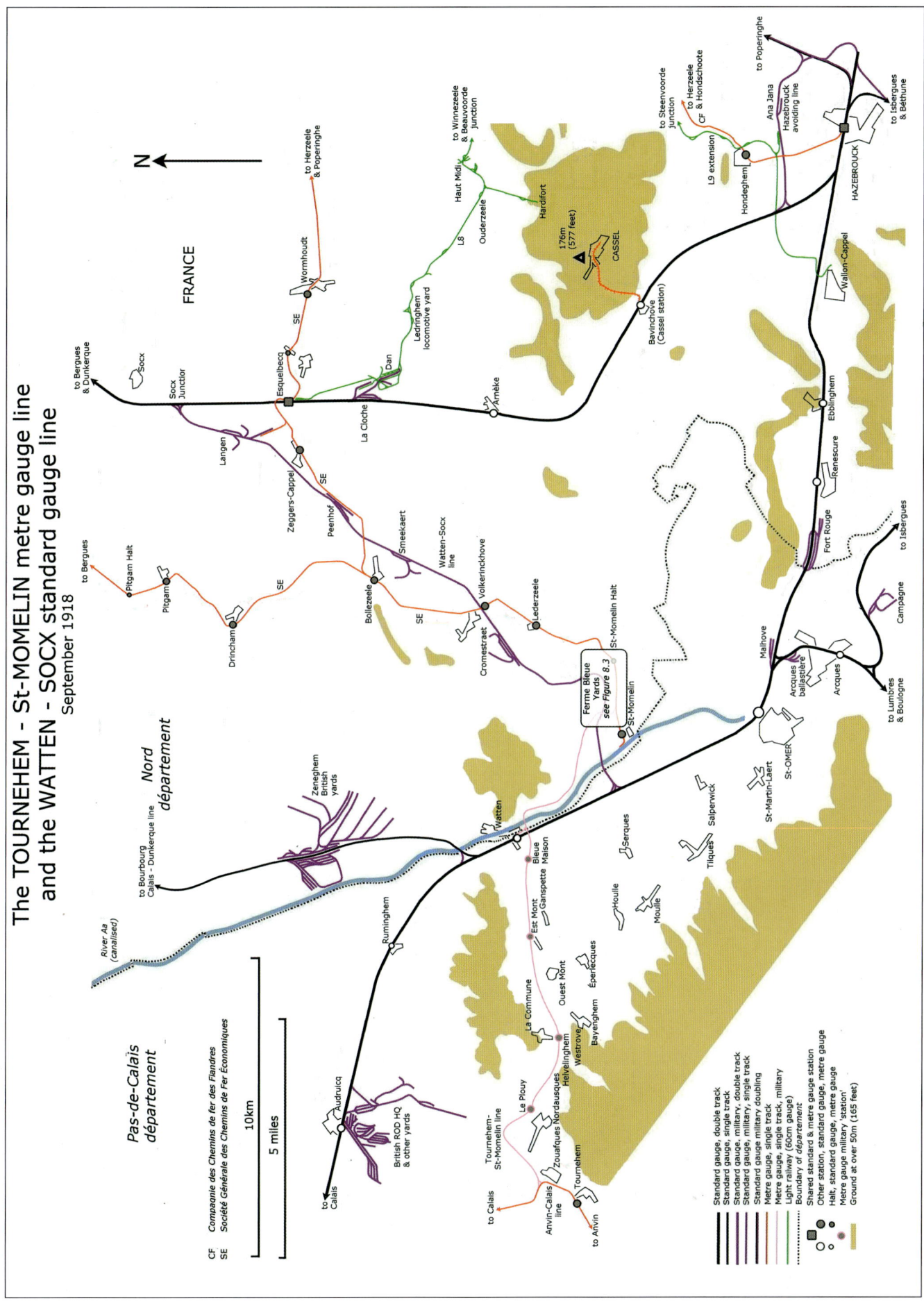

Figure 8.2 Tournehem–St-Momelin metre gauge line, and the Watten–Socx standard gauge line, September 1918.

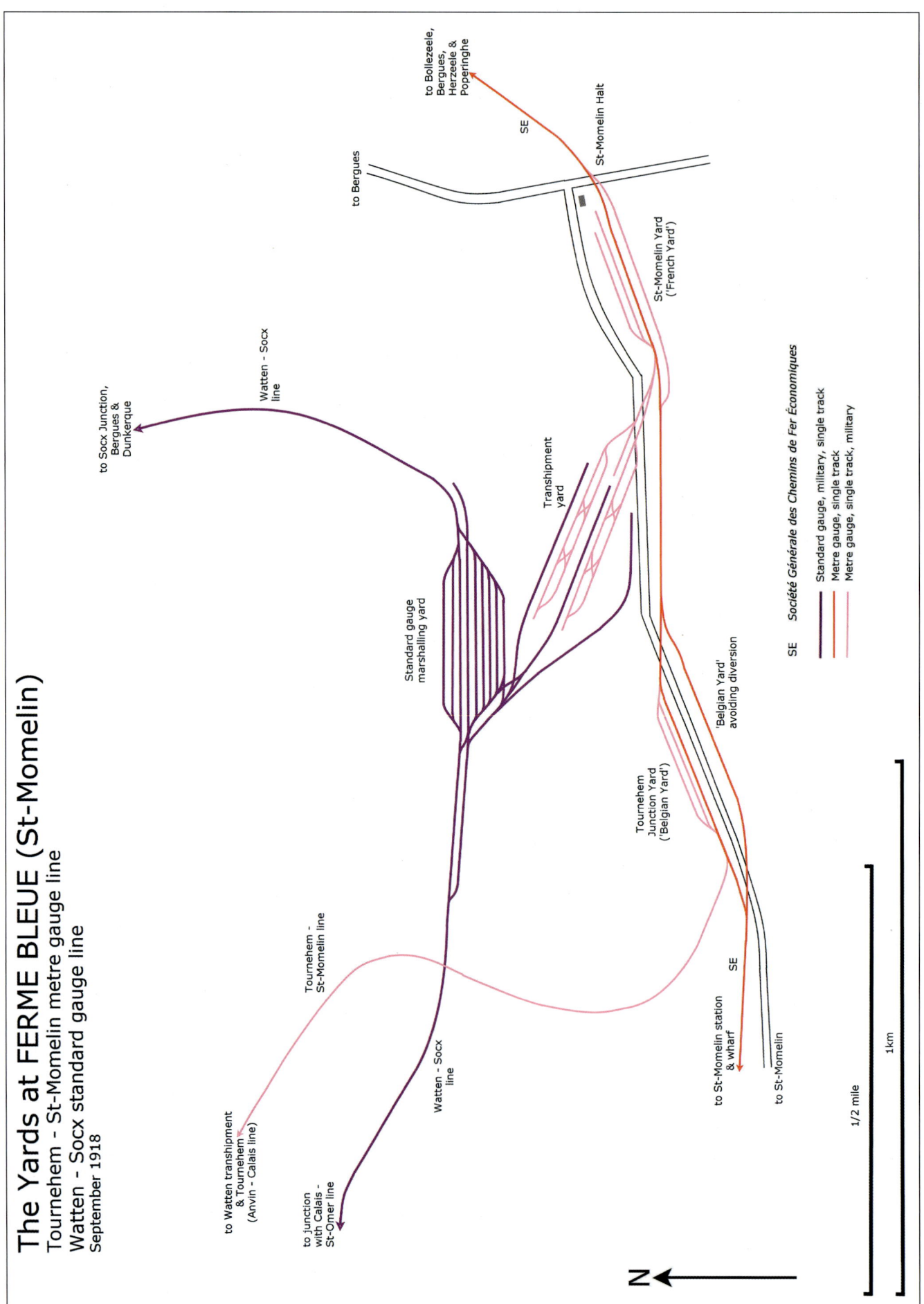

Figure 8.3 The Yards at Ferme Bleue (St-Momelin). Tournehem–St-Momelin metre gauge line, Watten–Socx standard gauge line September 1918.

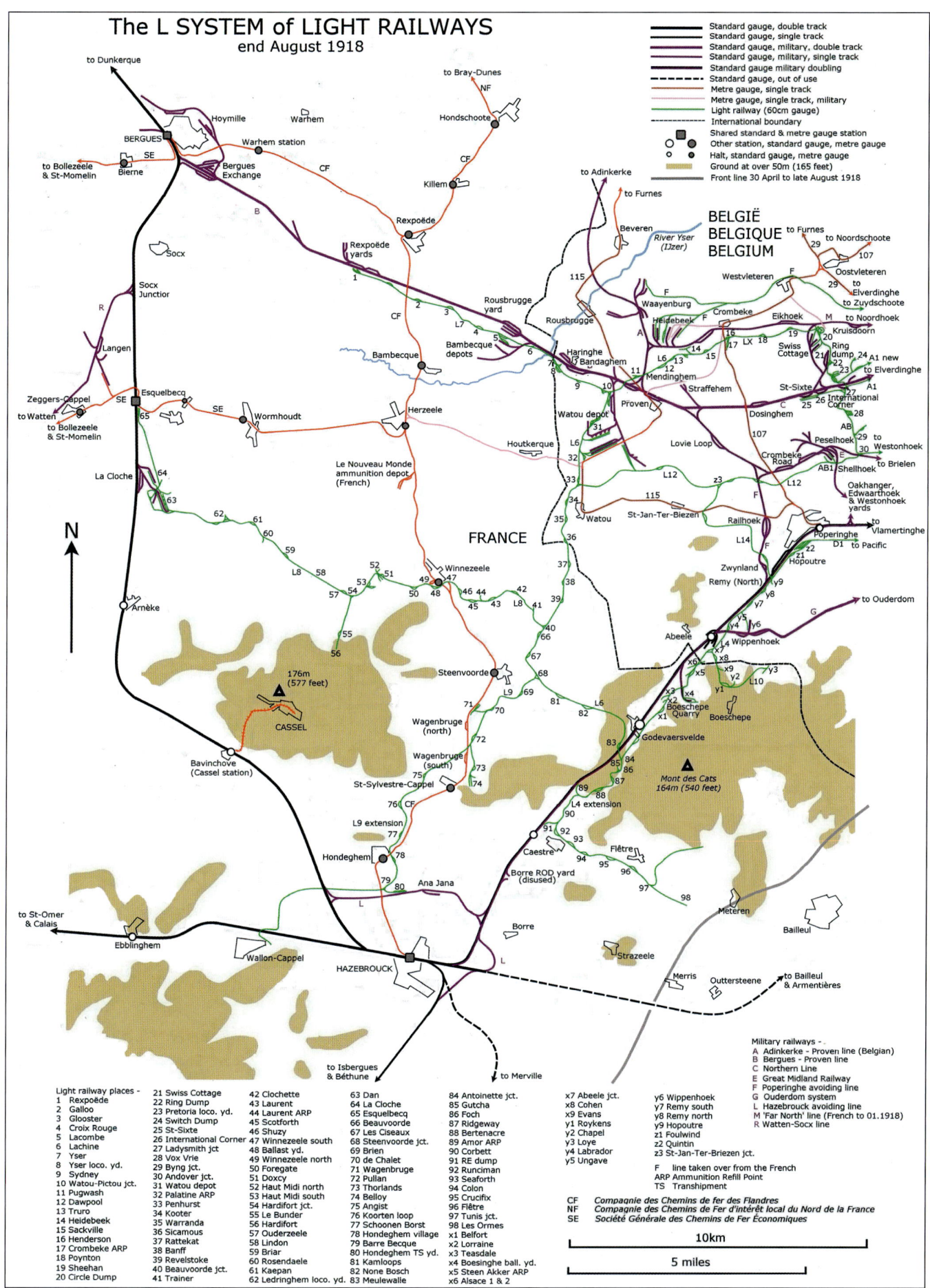

Figure 8.4 The L System of Light Railways, end August 1918..

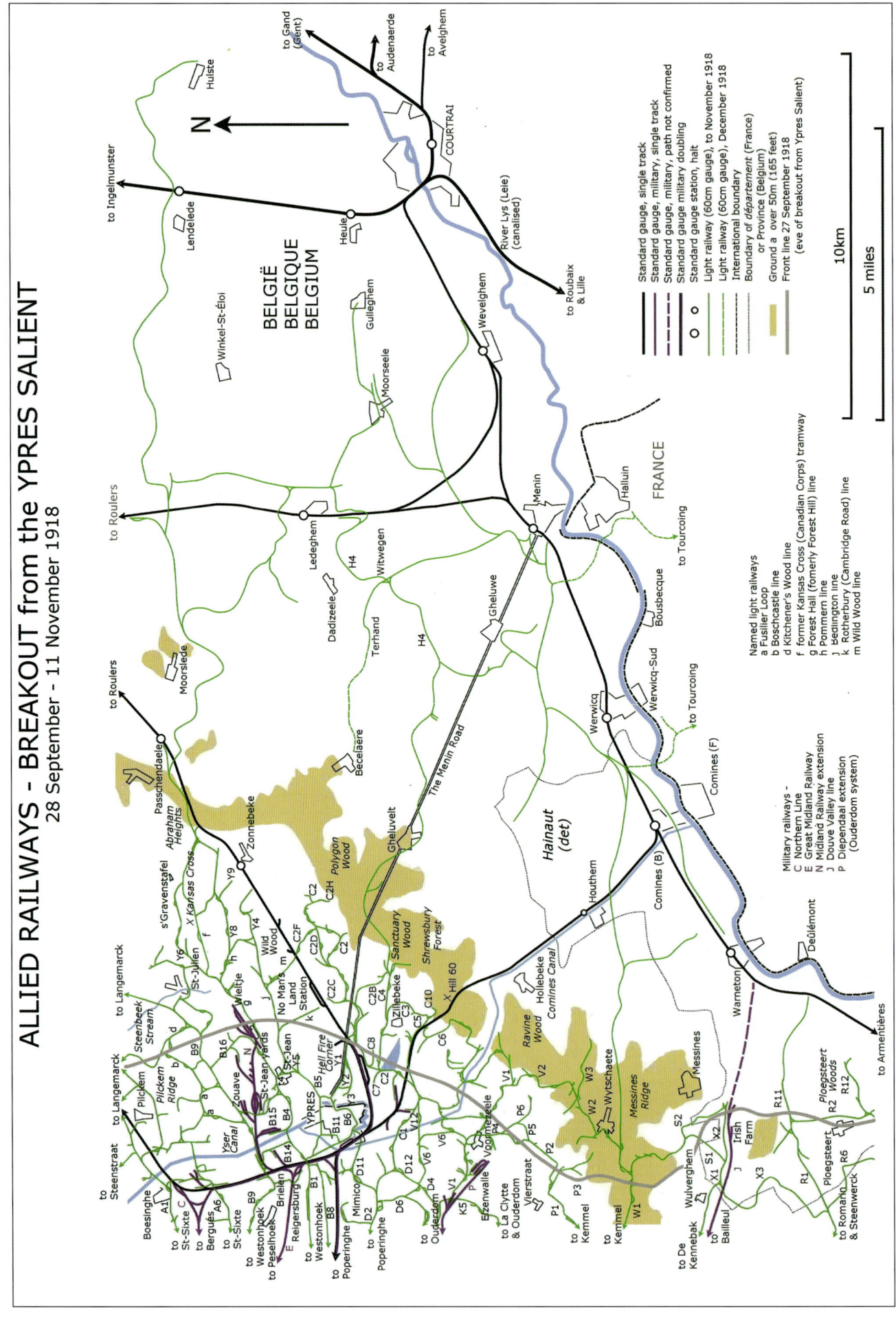

Figure 8.5 Allied Railways – Breakout from the Ypres Salient. 28 September–11 November 1918.

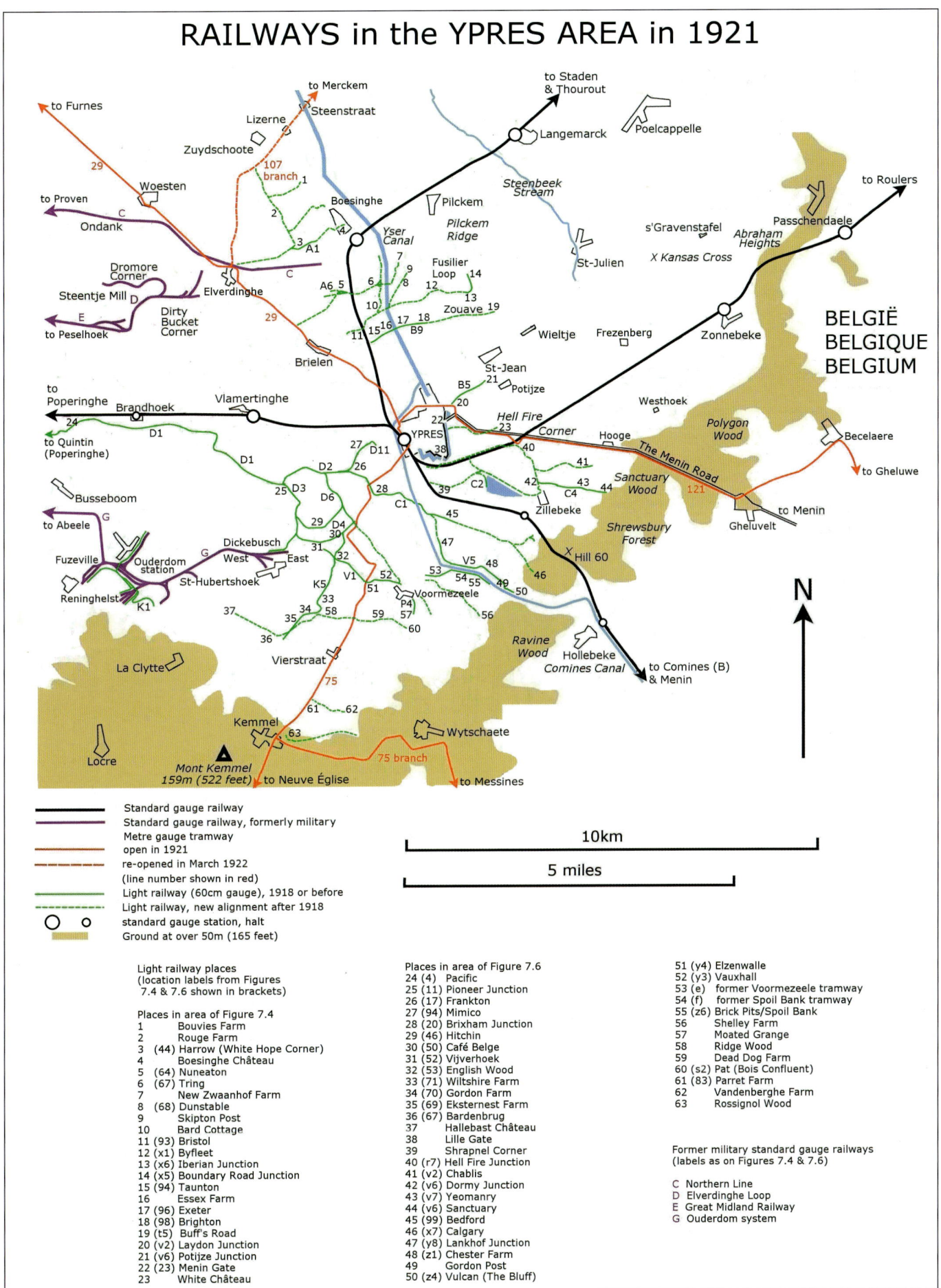

Figure 9.1 Railways in the Ypres Area in 1921.

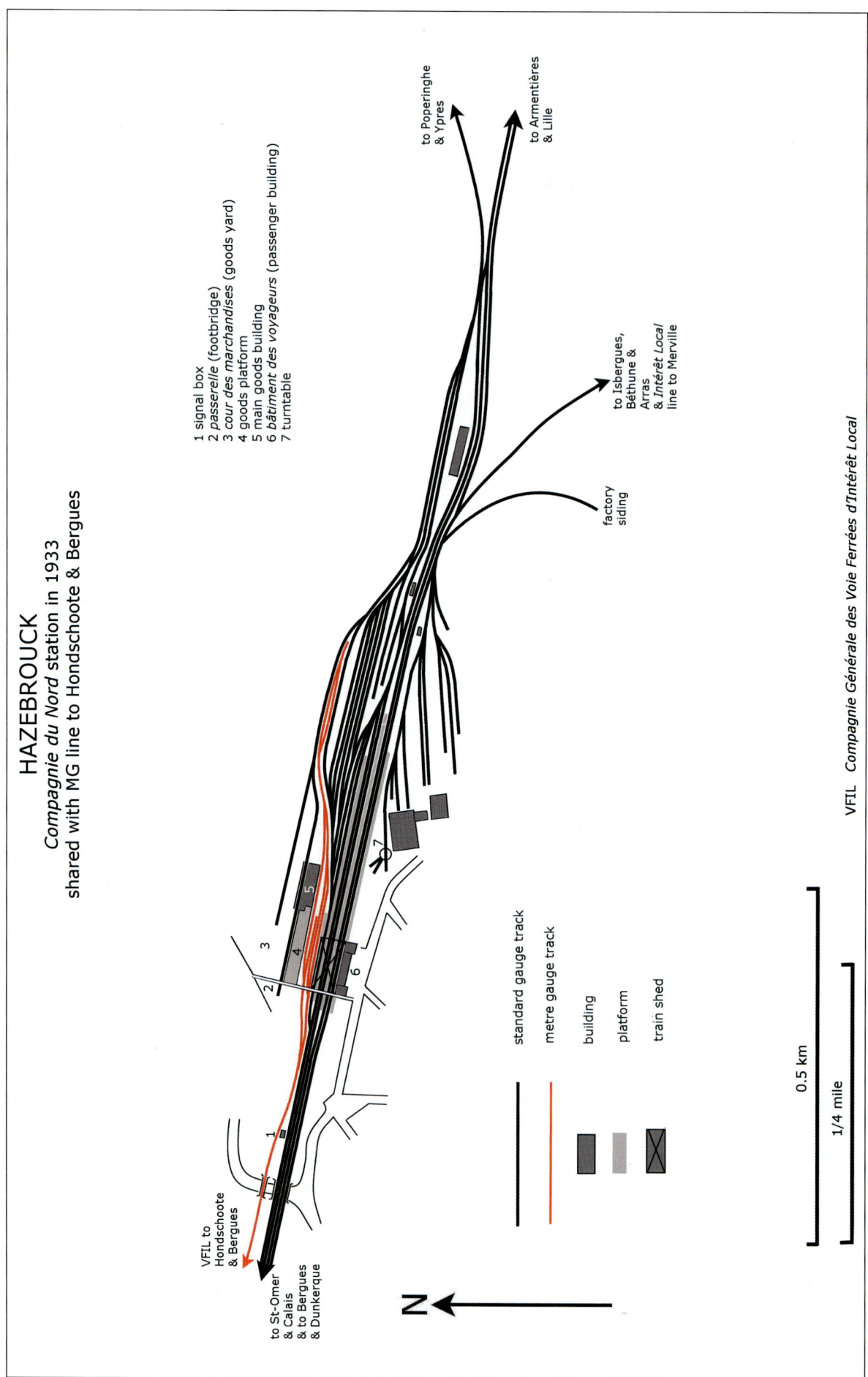

Figure 10.1 Hazebrouck *Compagnie du Nord* station in 1933, shared with MG line to Hondschoote & Bergues.

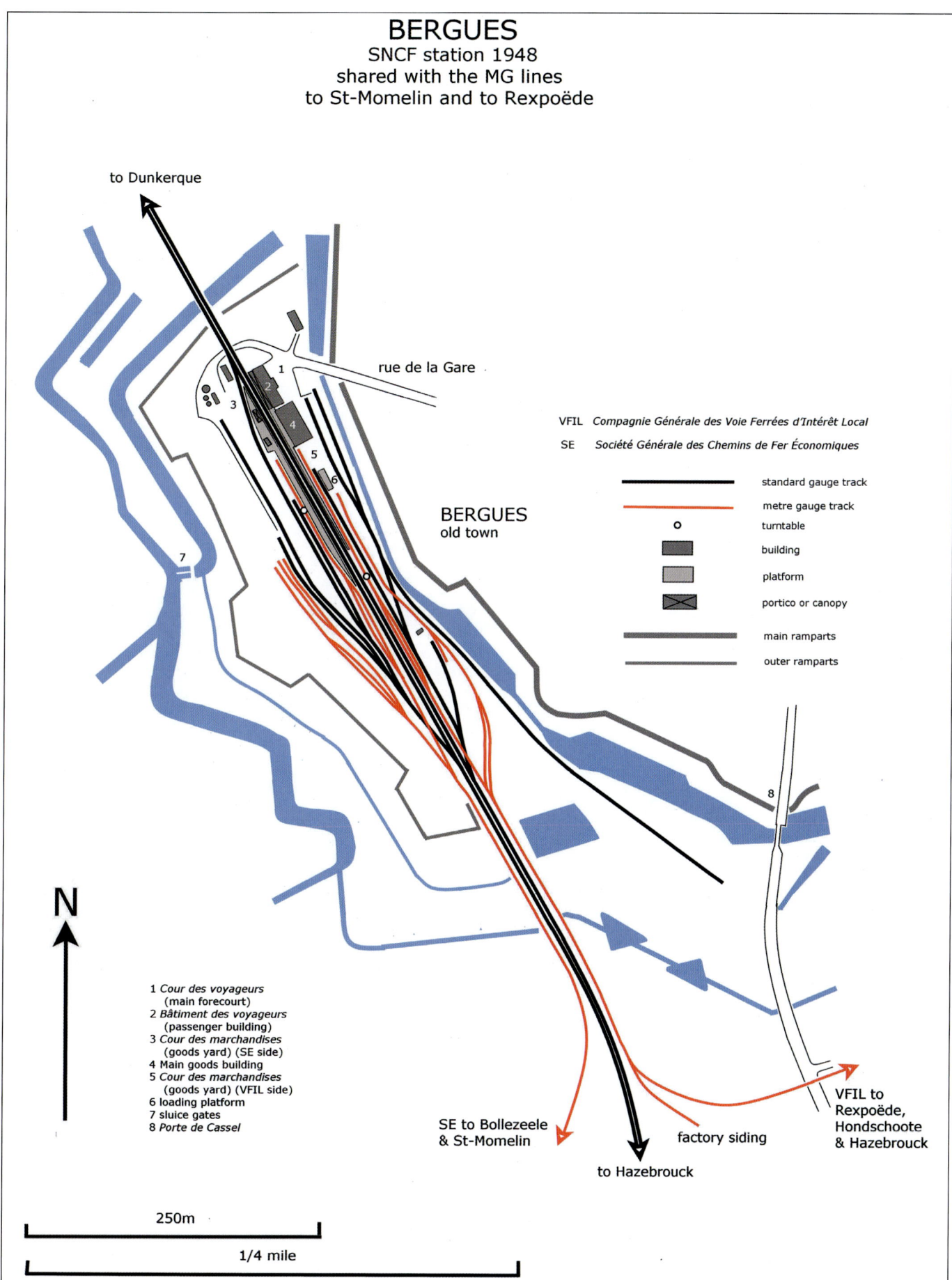

Figure 10.2 Bergues SNCF station 1948, shared with the MG lines to St-Momelin and to Rexpoëde.

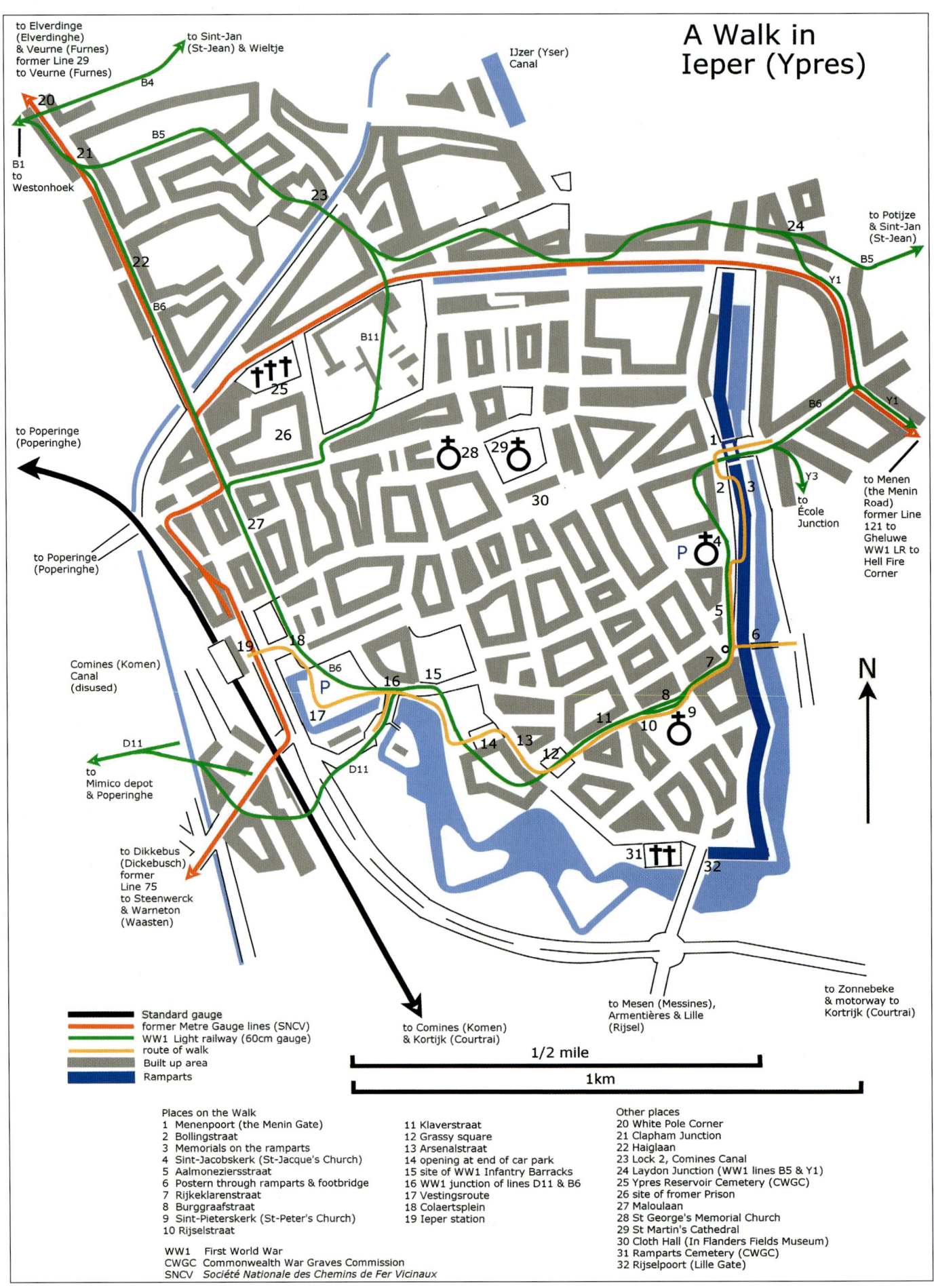

Figure 12.1 A walk in Ieper (Ypres).

Chapter Eight

Railways of the Ypres battlefields 9 April 1918 to 11 November 1918

This chapter covers the period from the beginning of the final German offensive and advance in the north to the Armistice.

Battle of the Lys River (Fourth Battle of Ypres) 9 to 30 April 1918

This was the second major German offensive of spring 1918. The first in the south began on 21 March when the German advance on the Somme nearly reached Amiens. Their aim was to reach the sea and cut off the armies in the north. These were mainly British and Belgian troops. This German attack officially ended on 5 April. On 26 March, at an emergency conference at Doullens, it was agreed that Marshal Foch should be in overall command of the Allies.

An offensive in the north had been expected for some time. The Battle of the Lys River opened on 9 April 1918. The German aim was to reach the channel ports. On the British First Army front south of Armentières, they broke through and advanced five miles. On 10 April they also attacked north of Armentières, the British Second Army front, up to the Ypres–Comines canal. By evening they had almost taken Messines and later they also attacked north of the Comines canal.

On 11 April, Haig sent his famous message saying that in the three weeks following the attack in the south, the enemy had made little progress in his objects – to 'separate us from the French, to take the Channel Ports, and to destroy the British Army'. There must now be no retirement. 'With our backs to the wall … each one of us must fight on to the end'.

The main advance was halted by 17 April but the Germans had advanced 15 miles (24km) on the plain of the Lys, threatening Béthune, and, less closely, Hazebrouck. They had taken the whole of Messines ridge and the Allies had lost all the gains of the Battle of Messines. In the Ypres salient they took Passchendaele Ridge, all the territory gained in the Third Battle of Ypres, and a bit more. Their maximum advance was 5 miles (8km). Things were going better in the north, the Belgian sector. There the Germans failed to advance except a little just north of Boesinghe. On 26 April they also took Mont Kemmel. This gave them a commanding position overlooking the Ypres Salient back areas west of the Yser Canal. The Germans reached 2 miles (3km) west of Mont Kemmel on 27 April. On 30 April they had reached their final position and called off the offensive. They were in Hell Fire Corner and the front line was only 1½ miles (2.5km) from the Cloth Hall in the centre of Ypres. The front lines at the beginning and end of the battle are shown in Figure 8.1.

Railways

After the attack in the south beginning 21 March, the British had sent reinforcements to the Amiens area. After 9 April, the British brought in reinforcements from the south and the French also sent troops to help. There was a substantial loss of railways during the German advance as shown in Figure 8.1. This shows railways lost to the enemy, and railways demolished or taken out of use in the area remaining in Allied hands.

Standard gauge railways
Withdrawal, demolitions and operating losses

These are listed from south to north, the south being most affected by the German advance. Figure 8.1 shows losses in territory captured by the Germans and operations suspended with track left, dismantled, or demolished. A list of standard gauge lines captured by the Germans or taken out of use is shown in Table 8.1.

Hazebrouck to Armentières, Brulooze and Douve Valley lines and branches

The main line to Armentières was lost east of Strazeele, with the Rabot and Romarin branches and the Petit Pont Line. The bridge over the Lys River at Armentières was demolished on 10 April and the track was destroyed from Steenwerck back to Trent Junction (Bailleul) the same day. Strazeele Yard, close to the final front line, and Merris ROD depot were abandoned on 13 April and the line was destroyed back to Carlisle East, junction with the Hazebrouck avoiding line. The Douve Valley line, and the Brulooze (Kemmel) line and branches, were abandoned and demolished on 10 April, as far as Butterfly Farm just south of La Clytte.

Ouderdom and Busseboom lines

The German advance was slower than further south, but by 19 April had reached Wytschaete. The track was demolished from Dickebusch to Elzenwalle on 17 April and salvaged on 19 April despite heavy snowfall. The Germans took Mont Kemmel and were 2 miles (3km) west of this by 27 April. Dickebusch was evacuated on 26 April,

Narrow Gauge in the Ypres Sector

Table 8.1 Standard gauge railways in the Ypres Sector lost or taken out of use 9 April - 30 April 1918 all dates are in 1918

Line	Type	Lost to German Army from	to	date	Taken out of use from	to	date	notes
Hazebrouck - Armentières	CdN	Strazeele	Armentières	13.04	Carlisle East	Strazeele	12.04	demolished east from avoiding line
Rabot, Romarin and Petit Pont Line Depots around Bailleul - Trent and Duke of York	BM	entirely lost		11.04				
	BM	entirely lost						
Brulooze (Kemmel) line	BM	Trent Junction	Butterfly Farm	10.04	Butterfly Farm	Ouderdom	by 27.04	part demolished, part dismantled
Douve Valley line and branches	BM	entirely lost		10.04				
Ouderdom system	BM	Elzenwalle	Oaten Wood & Wytschaete					
Busseboom system	BM			by 30.04	Dickebusch all except Abeele - Ellarsyde	Elzenwalle	17.04 by 30.04	demolished and salvaged used for dismantling & salvage
Ypres - Comines	BSR/PC	Ravine Wood spurs (BM)		by 30.04	Ypres	Ravine Wood spurs junction	15.04	dismantled
Ypres - Roulers	BSR/PC	Hell Fire Corner	Wild Wood	by 30.04	Ypres	Hell Fire Corner	15-28.04	dismantled
Midland Line extension	BM	Wieltje	Spree Farm	by 30.04	Reigersburg	Wieltje	15.04	demolished, yards later salvaged
Midland Line	BM				Brielen	Reigersburg	29-30.04	demolished
Hazebrouck - Poperinghe - Ypres	BSR/PC*				Vlamertinghe	Ypres	30.04	demolished
					Borre ROD depot		by 30.04	later dismantled
Ypres - Boesinghe	BSR/PC				Ypres	Boesinghe	April	dismantled May 1918
Northern Line	BM**				Waanebeke Yards	Boesinghe	by 30.04	in use only for salvage & maint.
					Boesinghe	Pilckem loop 14-16.04		dismantled Boesinghe Bridge demolished
Heidebeeke (Far North) line	FM/BM				Noordhoek	Zuydschoote		dismantled or demolished

*   double track by British Army 1915-1916
** originally BSR/PC from Boesinghe towards Staden
BM   British military
CdN   *Compagnie du Nord* (French)
FM   French military
BSR/PC Belgian State Railways (*Chemins de fer de l'État Belge*) or Private Company

164

Railways of the Ypres battlefields 9 April 1918 to 11 November 1918

and by 30 April the Germans reached their final position in Elzenwalle and Vierstraat. After the end of April, some maps only show the line open from Abeele to Ellarsyde, implying that all operating had ceased. However, standard gauge as well as light railway track was maintained to Reninghelst-Fuzeville and Heksken yards, for dismantling and salvage, so these lines are shown as open in Figure 8.1, even though the depot functions were closed until July.

Ypres to Comines line
The line had been in use from its junction with the Roulers line in south Ypres at least to the Ravine Wood Spurs. On 15 April the line was abandoned, and the rails removed. The final German advance between 19 and 30 April brought the front line to the junction of the main line and Ravine Wood Spurs, just south of Zillebeke Lake.

Ypres to Roulers line
Prior to the German offensive this line had been in operation to the Wild Wood gun spur 1½ miles (2.5km) beyond Hell Fire Corner. On 14 and 15 April, all track north-east of Hell Fire Corner was demolished and salvaged. By 28 April, the rest of the track back to Ypres station had been taken up. On 30 April, the Germans reached their final position, with the front line between Hell Fire Corner and Gordon House Loop. They were only 1½ miles (2.5km) from the Cloth Hall in the centre of Ypres.

Midland Line - St-Jean and Wieltje extension
This was taken out of use by 15 April, and by 30 April all lines east of the Yser Canal had been rendered unusable, but it is not clear how. Later (June and July 1918) lines in the St-Jean yards were salvaged by light railway. By 30 April, the Germans had advanced to the east end of Wieltje yards.

Midland Line – Peselhoek to Reigersburg
The main operating unit, 5(59)Aus BGROC, was withdrawn from the Reigersburg depot and the main Peselhoek depot on 15 April, and moved to Bergues. On 29 April, the track was demolished from near Brielen to Reigersburg.

Hazebrouck to Poperinghe and Ypres line
The track forward from Vlamertinghe to Ypres station was maintained and repaired until 30 April, when it was demolished. The ROD depot at Borre was closed and later dismantled, but the rest of the line remained in use.

Ypres to Boesinghe
This line was taken out of use in April 1918 and was dismantled, probably in May 1918.

Northern line
The line east of Boesinghe to Pilckem loop was dismantled from 14 to 16 April. From 17 April, RCE II decided that the line was only to be maintained for traffic to Waanebeke (Elverdinghe). From there to Boesinghe was for maintenance and salvage only. By 30 April, the Germans had crossed the Steenbeek and advanced onto the eastern slopes of Pilckem Ridge, but had not taken the Ridge itself, and no part of the line which had been in operation had been captured. Boesinghe bridge was demolished by the Belgians after 30 April.

Heidebeek to Noordhoek and Zuydschoote line ('Far North' line)
No part of the line was captured. This was the area into which French reinforcements were moving. On 29 April, the line was handed over to the Belgians, by which time the section from Noordhoek to Zuydschoote had been taken out of use.

Standard gauge control and traffic
On 11 April, due to German advances, the Traffic Office, the control centre for these lines, moved from Hazebrouck to Wormhoud, and on 25 April to St-Omer. Bergues Exchange remained open, operated by 6(60) Aus BGROC. By 30 April, the line from Bergues to Proven had become the main entry point for standard gauge trains to the Ypres area. Traffic on this line more than doubled in April 1918. The total trains handled at Bergues Exchange from 9 to 30 April was 990, and of these 146 were British and French troop trains. As a result, there were many troop trains running especially through Bergues Exchange. Between 18 and 20 April, thirty-seven French troop trains arrived there from the south at hourly intervals, followed by twenty-five trains of British troops at the same intervals. From 19 April, daily civilian refugee trains were provided. Between 24 April and 1 May five trains were loaded at Proven or Roosbrugge. Each train had forty covered wagons and took 1,000 refugees.

Metre gauge railways
No operating metre gauge railways were directly affected by the German advance. Those around Poperinghe and north-west of Ypres came under more direct German observation, especially from Mont Kemmel, with increased shell fire.

Light Railways
Withdrawal, demolitions and operating losses
As for standard gauge railways, these are listed from south to north (the south being most affected by the German advance); losses (territory captured by the Germans); operating suspended (track left, dismantled or demolished) – all as shown in Figure 8.1.

A detailed list of light railway lines lost and taken out of use is shown in Table 8.2. Many tramways were also lost, not shown in Figure 8.1 or Table 8.2.

Table 8.2 Light railway (60cm) lines captured or taken out of use in the Ypres Sector 9 April - 30 April 1918 all dates are in 1918

System	Line	Lost to German Army from	to	date	Taken out of use from	to	date	notes
Romarin	Main line south	La Crèche	1st Army area	by 12.04				'Cape to Cairo' line
	Steenwerck, Romarin, Ploegsteert, and related tramways Inc. R lines and X3 line	entire system		10-19.04				
Douve Valley	S1 line & related tramways	entire system		10-19.04				
Ouderdom system	P lines	entire system		by 30.04				
	V1 V2 V3 V4 V5	entire lines		by 30.04				
	V6	V7 jct	V3 jct	by 30.04				in use to Swan Château
	V8 V9 V10	entire lines		by 30.04				
	W1	entire line		mid April				Kemmel - Wytschaete
	K1	Milky Way jct area		by 30.04	Ouderdom	Willesden	by 30.04	Fuzeville - Ouderdom in use
	K2	"		"	Barbed Wire jct	Milky Way jct	"	
	K3 K4 K8 K10							
	K5	entire lines		19-30.04	Dickebusch W	Willesden jct	30.04	entire line
Busseboom system	F3				Wellington jct	Dickebusch	by 30.06	pr before
	D3				Ottawa (D6)	Dickebusch	by 30.06	pr before
D, C & most	D4				Hull	White House (D12)	by 30.06	pr before
F lines					Frankton	Brisbane (C1)	by 30.06	pr before
	D10				Yale (D1)	Dunedin (F3)	by 30.06	pr before
	D11				Asylum	Ypres (B6)	by 30.06	pr before
	C1				Castle loop	Bedford jct	by 30.04	
	C2	Hellfire jct	Hanebeke	19-30.04	Bedford jct	Gordon House	by 30.04	
	C2C C2D C2F C4 C6 C10 C11	entire lines		by 30.04				
	C3	all but 500 yds (C2 jct)		by 30.04				
	C5	all but 300 yds (C2 jct)		by 30.04				
	C7 C8				short lines nr. Zillebeke Lake		by 30.04	

Railways of the Ypres battlefields 9 April 1918 to 11 November 1918

System	Line	From	To	Date	Notes
Westonhoek system	B4		Yser Canal (B14)	by 30.06	pr before
	B5		just E of Laydon jct (Y1)		
B and Y lines and extensions	B5		Manner's jct	by 30.06	pr before
	B11		Station Yard (B6)	by 30.06	pr before
	Y1		Hellfire jct (C2)	by 30.04	pr before
	Y2		White Château	by 30.04	entire line
	Y3		Ecole jct (Y2)	by 30.04	entire line
	Y5		Vinery jct	by 30.04	entire line
	Bedlington line & Y4	Wooler jct	Devil's Crossing	19–30.04	
	Rotherbury line	Wooler jct	Railway Wood	19–30.04	
	Wilde Wood line	Grey Ruin jct	Wilde Wood	19–30.04	
	Forest Hill line	Cambrai	Pommern jct	19–30.04	
	Pommern line	Pommern jct	Beck House jct	19–30.04	Prowse Farm	by 30.04	entire line
	Y6 Y8 Y9 Y10	entire lines		19–30.04	
		Manner's jct	Argyle Farm	by 30.04	entire line east from Pommern jct
B9 system	B9	Juliet loop	Alouette Farm	19–30.04	Oblong Farm	by 30.04	entire line
	B16 Admiral's Road loop				Wieltje	by 30.04	entire line
	Kitchener's Wood line	Canopus (Juliet Fm)	Beeston	19–30.04	Rudolphe Farm	by 30.04	conn. to Cannock (A system)
	Bochcastle loop and link line		Bourne jct (B9)		Kitchener's Wood line	by 30.04	
	Hindenburgh Farm branch		No Man's Cot		Kempton Park jct	by 30.04	
			Fusiliers loop		5 Chenins	by 30.04	
	B9 south loop	Romeo jct (B9)	Bradford jct (B9)	19–30.04			via Blyth loop
Proven-Boesinghe system	A1 extension	Stafford	Carre Ney	19–30.04	Warwick	by 30.04	
	A6 extension (Hanley line)	Burton	Pheasant Farm	19–30.04	Chantry	by 30.4	Langemarck line
A lines and extensions	Burton – Stafford link	entire line		19–30.04	Steenbeek bridge		
	Cannock link to south		Cannock		Rudolphe Farm	by 30.04	to Kitchener's Wood line

jct junction

La Crèche, Romarin, Ploegsteert, and Douve Valley lines

All these lines were captured by the Germans between 10 and 19 April 1918. The five APBs were sent south to the Somme Sector in late March or early April. After that, no records of light railway construction or operating companies are available for this area, so we have no reports of their withdrawal. The link southwest from La Crèche across the plain of the Lys into the First Army area was lost completely in the first few days of the German offensive. The link north from between Wulverghem and Messines towards the road from Kemmel to Wytschaete was also lost around the middle of April. Because of the early loss of the link south, rolling stock, personnel and materiel would certainly have withdrawn north.

Ouderdom and Busseboom system – the D, F, K, V and P lines

As of 9 April, construction and maintenance were undertaken by 10CRT based at Dickebusch. Operating was by 5NZ LROC based at Poperinghe and 2(16)Aus LROC based at Vauxhall depot, Elzenwalle, on the V1 line from Dickebusch to White Château. They were much nearer the front line than 5NZ LROC.

On 10 April, the 2(16)Aus LROC depot and coal dumps at Vauxhall (Elzenwalle) were heavily shelled. All spare kits and QM store material were packed and returned to Ellarsyde. The next day they began pulling down buildings and dismantling huts. They also maintained close communication with forward batteries, which they kept supplied with ammunition. In the afternoon, an officer and five men took a push truck of water, tea, all spare food, and 3,500 packets of cigarettes, donated by the Company, to the Seaforth Highlanders who were holding the closest part of the front line. At about 7pm, orders were received from the ADLR to evacuate Vauxhall except for control and running staff. The trains leaving carried huts, salvaged coal and as much materiel and personnel as possible. On 12 April, HQ moved into Hitchin depot on the D6 line about a mile north-west of Vauxhall, where the men made themselves secure in old dugouts. Steam locomotives were being withdrawn for back area work, leaving petrol tractors to carry on the work from Hitchin.

By 12 April, part of 10CRT based at Dickebusch had already started construction of the Heidebeek to Watou and Bergues light railway lines (L6 and L7) in the back areas, and 10CRT HQ was moved from Dickebusch to Ellarsyde. By 13 April, 2(16)Aus LROC had sent four steam locomotives with two crews each to the back area. Four remained at Hitchin to run trains from there to further back, and four tractors were busy moving ammunition and all supplies to the forward areas. Men were still removing materials from Vauxhall, which was now being used as a sub-control centre. On 14 April, 5NZ LROC were ordered to take over the district. 2(16)Aus LROC were ordered to go to Swiss Cottage area leaving all tractors and crews at Yale on the D1 line to work under the New Zealanders, and a warrant officer and twenty men were detailed for demolition work at Vauxhall.

On 14 April, 8CRT took over maintenance of all the lines in this area from 10CRT who moved to near Proven to continue back area construction. No. 2 Company of 8CRT was based at Westonhoek and later in April they suffered gas casualties on the D1 and D11 lines. On 25 or 26 April, 5NZ LROC left Poperinghe, where they had been for more than a year, and 'many a sweetheart wept bitter tears at the departure of the boys'. They initially went to Hagle Dump on the B1 line. On 26 April, Zevecoten and Ouderdom Yards were shut down when the Germans took nearby Mont Kemmel. By 30 April, large parts of the D and F systems, almost all the C, V, and K systems, and all the P lines had been captured or taken out of service. Operation of the remaining lines was taken over by 12LROC based on the B lines at Westonhoek.

Y and C systems

As of 9 April, construction and maintenance were undertaken by 10CRT based at Dickebusch. The lines were operated by 3(17)Aus LROC, based at the depot and locomotive yard at Mimico on the D11 line, assisted by a detachment of 14LROC. In the first week of April, 3(17)Aus LROC reported traffic light and the main part of their work was salvage from forward areas. However, on 9 April enemy shelling became very heavy. 'Hell Fire' yard on the C2 line was evacuated to a yard on the Y2 line about 1 mile (1.5km) further in towards Ypres.

On 13 April, Company HQ moved back to Yale on the D1 line. Mimico, previously district control, became forward control, and the limit of operation of steam locomotives. On 14 April, 8CRT took over all the lines in this area from 10CRT and on 15 April 3(17)Aus LROC again moved back, to Pacific on the D1 line. Yale stayed as forward Control. Mimico became 'end of steel' (end of operated line) and a controller and operator stayed there to keep in close touch with the batteries moving back to positions in the new line. During this period, very few staff were required for operations. The balance of the Company were employed salvaging abandoned posts. Old sidings at Mimico were dismantled and huts and coal salvaged. On 22 April, three men were killed and five wounded, all by shelling at Pacific Yard. The same day, 3(17)Aus LROC moved from Pacific to a camp near Proven, for light railway construction work. By 30 April, most of the Y system and almost all of the C system had been captured or taken out of service.

B and B9 systems
As of 9 April, construction and maintenance were undertaken by 8CRT No 2 Company, and operating by 12LROC, both based at Westonhoek. On 11 April, 12LROC were ordered to evacuate Manner's district on the B4 and B5 lines. A forward tractor yard and camp were established at Canadian Siding on the B1 line ½ mile east of Atherley junction, west of the Yser Canal. On 14 April, the forward power depot moved back again to 'Mission', probably Culloden Locomotive Yard another 2 miles (3km) further west. That night a bogie wagon carrying miscellaneous ammunition leaving Hagle caught fire at Triangle, with several casualties. On 15 April, Trois Tours control on the B9 line closed, and the whole area was now operated by Westonhoek control.

On the morning of 27 April there was very heavy shelling in and about the camp and yards. Hagle Ammunition Dump was destroyed with very heavy casualties in the labour party. Triangle control hut was shattered and Mission (Culloden) power depot was withdrawn, presumably to Westonhoek. Cranes and surplus power were transferred to the north. 5NZ LROC were at Hagle Dump from 26 to 28 April and narrowly escaped casualties. All lines east of Potizje and Wieltje were lost. By June, the B5 and B6 lines, with a stub of Y1 at the beginning of the Menin Road, were the only lines operating to just east of Ypres. B4 was only operating to the west bank of the Yser Canal. B9 was operating across the canal but only to Battle, and to Fusilier loop to the north. B15 was operating south from B9 to Zouave sidings and the St-Jean standard gauge yards but all the standard gauge lines were closed. There were no operations beyond the south end of B15.

A system
Construction and maintenance, at least from 15 April, was done by 8CRT. The operating company was 29LROC, for whom there are no War Diaries for this period. In later April, the Germans took Langemarck, crossed the Steenbeek and advanced onto the eastern slopes of Pilckem Ridge. The lines around Langemarck were lost and so by June the loop east of Pilckem village was the only A system line east of the Yser Canal still in service. Lines to the north taken over from the French in January 1918 were also still in use.

Rolling Stock
Light railway rolling stock was evacuated north and west in moderately good order along the existing lines in the Second Army area during the Battle of the Lys, and little was lost. This contrasts with the situation in the south following the attacks starting on 21 March, in which it is likely that Fifth Army lost almost all its light railway stock, and Third Army about one third of its stock.

Recovery and Consolidation 1 May to 27 September 1918
In response to the German advance new defensive lines were built, especially around and to the north of Hazebrouck. There was also a highly secret plan, plan Z, that in the event of a further German offensive and advance the Allies might abandon Belgium and the north of France altogether and retreat to the line of the Somme River. This would have prevented the Germans reaching the sea between the British and French cutting the British off in the north. In May 1918, the Fifth Army was put back into the front line on the plain of the Lys. It replaced the First Army on the south flank of the Second Army around Ypres.

Railways
The plan was to build additional facilities further from the new front line, to replace those lost, and to build new links and lateral lines where necessary. In case of further German advance, some lines were prepared for demolition. On 12 August, the British formed a Directorate of construction, embracing light railway, broad gauge and dock construction.

Standard gauge railways
The main works on standard gauge railways in the area of this book are summarised in Table 8.3.

Salvage
The programme of salvaging standard gauge track and other materials continued from April to August 1918, from lines considered too near the front line to operate. By June, since there was less standard gauge construction, salvage and maintenance were the main activities for most railway companies. Most material was despatched to Audruicq, the rest was used immediately locally.

Calais
The main works were the avoiding line to the port, and the associated 'sand line' to extract sand from the dunes as ballast.

Vendroux
This vast supply depot was enlarged from April 1918 as part of a response to loss of depots further east, by 17 miles (27km) of new sidings. The estimated final track length for the yards was 43 miles (69km).

Table 8.3 Standard gauge railway works in the Ypres Sector April 1918 - 11 November 1918

Area or line	Line or other facility	Construction work start	finish	undertaken by	notes
Calais	Rivière Neuve chord	04.1918	04.1918	2CRT 298RC 14ARE	from Dunkerque to St-Omer line
Vendroux	Calais avoiding line to port	04.1918	07.1918	2CRT 298RC 14ARE	also the 'sand line' to the dunes
Ghyvelde/	extensions - No 4 and No 6 triage	19.04.1918	07.1918	2CRT 298RC 14ARE	see also metre gauge section
Bray-Dunes	sand quarry salvage	26.07.1918	21.08.1918	120RC	quarry shut down 26.07.1918
Bergues - Proven line					
Bergues	Hoymille stores and workshops sidings	16.05.1918	24.05.1918	113RC	
	Les Forts depot dismantled	28.05.1918	29.06.1918	113RC	
	points for Haeghe-Meulen Belgian depot		13.08.1918	113RC	2¼ miles (3.5km) east of Bergues
Haringhe	hospital sidings for French	21.04.1918	24.04.1918	113RC	
Rexpoëde	stone siding	24.06.1918	02.07.1918	113RC	
Rousbrugge	end loading ramps	24.07.1918	09.08.1918	120RC	
Bambecque	ammunition depot	01.03.1918		1CRT 113RC	
	depot	03.04.1918	27.04.1918	113RC	
	No 3 spur RE Park	13.05.1918	18.05.1918	113RC	
	No 2 spur extension (dump line)	27.05.1918	06.06.1918	120RC	
	completion No 1 and No 2 dump spurs		22.07.1918	120RC	
Watou	ammunition depot	26.02.1918	20.04.1918	1CRT 113RC	
	stone sidings	29.04.1918	07.05.1918	120RC	
Poperinghe Avoiding line					
	Zwynland sidings dismantled		06.1918	261RC	
Hazebrouck - Poperinghe - Ypres					
	Borre ROD yard - track salvaging		05.1918	295RC	DT by 23 September
	repair Vlamertinghe to Boesinghe line jct ST	31.08.1918	10.09.1918		station repairs complete 3 October
	line repaired to Ypres station		28.09.1918	261RC	
	restore sidings Borre ROD yard	29.09.1918	14.10.1918	112RC	
Proven - Elverdinghe - Boesinghe line (the Northern Line)					
	dismantling and salvage Wannebeek to Boesinghe	19.06.1918			including reconstruction Boesinghe
	taken over east of International Corner by Belgian Army	16.09.1918			bridge
Crombeke Road - Reigersburg - St-Jean (Great Midland Railway & extension)					
	Salvage Brielen to Reigersburg		06.1918	10RC	
	Salvage St-Jean yards	25.06.1918	19.07.1918	113RC	salvage by light railway
	LR transhipment at Trois Tours		05.07.1918	113RC	for salvaged material from St-Jean
	repair Reigersburg north to Yser Canal bridge	31.08.1918	06.09.1918	113RC	
	repair Yser Canal bridge to Wieltje	06.09.1918	23.09.1918	113RC	made permanent by 2 October
	repair Brielen to Reigersburg, temporary for works and salvage only		27.09.1918	113RC	probably completed earlier
	complete reconstruction of St-Jean and Wieltje yards		by 20.10.1918	120RC	from junction with line from
Ypres - Boesinghe	repairs to Reigersburg (Midland Railway)	01.09.1918	06.09.1918	113RC	Vlamertinghe
	repairs Reigersburg to Boesinghe	06.09.1918	21.09.1918	113RC	

Railways of the Ypres battlefields 9 April 1918 to 11 November 1918

Line	Description	Date	Unit	Notes
Ypres - Roulers	Ypres to Hell Fire Corner reopened	28.09.1918	113RC	Hell Fire stone siding open 29 September
	repaired Hell Fire Corner to Passchendaele station	11.10.1918	113RC 120RC	linked to German line at Passchendaele
	repaired to junction with Roulers - Menin line	17.10.1918	113RC 120RC 261RC	
Ypres - Comines and Menin				
	repaired to Shrapnel Corner	08.10.1918	113RC	
	re-opened to Menin	26.10.1918	113RC	
Roulers - Menin	reconstructed	28.10.1918	113RC	
Ouderdom - Brandhoek	Line dismantled	05.1918		
Ouderdom system	repairs to Fuseville and Heksken lines	26.08.1918	109RC	and yards
	repairs Ouderdom and Zevecoten yards	31.08.1918	109RC	
	line open Ouderdom - St-Hubertshoek	14.09.1918	109RC	
	reconstruction St-Hubertshoek - Dickebusch	29.09.1918	109RC	
	reconstruction Dickebusch - Elzenwalle	08.10.1918	109RC	line abandoned 24 October
Hazebrouck - Dunkerque				
	Bavinchove (Cassel) stone siding	30.04.1918	277RC	
	La Cloche ammunition depot	01.05.1918	5SCFC 277RC	light railway TS line L8
	La Cloche RE Park siding	01.07.1918	277RC	
	La Cloche - Eringhem ammunition extension	29.07.1918		light railway TS line L8
	La Cloche coal siding	27.08.1918	277RC	light railway TS line L8
St-Omer - Hazebrouck				
	Fort Rouge ammunition depot & RE Park	20.04.1918	295RC	
	Malhove ROD depot	16.06.1918	109RC	
	Wallon-Cappel ammunition depot	08.08.1918	295RC	
Watten - Socx (new British line)		01.06.1918	2CRT	dismantled mid September 1918
Hazebrouck - Armentières				
	Strazeele salvage by cable and trollies	22.05.1918	277RC 2CRT	
	reconstruction Strazeele - Bailleul	31.08.1918	295RC	DT 6 October
	reconstruction Bailleul - Steenwerck ST	04.09.1918		
	reconstruction of Merris and Strazeele yards	10.1918	112RC	
	Lys River bridge Armentières ST	06.10.1918	297RC	DT 23 October
	Armentières - St-André (Lille) ST	18.10.1918	109RC	DT 1 November
Douve Valley line	reconstruction Trent jct - Clapham Junction	11.09.1918	295RC	some reconstruction Duke of York and Trent
	reconstruction to de Kennebak ST	29.09.1918	295RC	
	extension to near Warneton	23.10.1918		

DT	double track	RC	Railway Company (RE)
ST	single track	RE	Royal Engineers
TS	transhipment	ROD	Railway Operating Division (RE)
MG	metre gauge	ARE	American Railway Engineers
LR	light railway (60cm gauge)	SCFC	Section des Chemins de Fer de Campagne (Section Field Railway Engineers - French)

Audruicq
This remained the ROD HQ, also the major standard gauge locomotive repair shops in the north, with other yards including a wagon repair yard, and a major ammunition dump. During the German advance of April 1918, many RC trains were sent here for safety but by May most had moved away. 85Can ECC arrived at Audruicq on 12 April and remained until April 1919. Most had come from Bollezeele where they had been operating metre gauge railways. 5(59)Aus BGROC moved to Audruicq on 20 April, and remained until moved to Conchil-le-Temple on 3 September. Both these units were required to train on French lines and operated widely in France from this base.

Dunkerque
The 4(35)Aus BGROC moved from the Somme Sector to Audruicq after the German offensive beginning on 21 March 1918. By 25 April, they had moved on to Dunkerque, taking over from ROD and French operating companies. By the end of May, they were operating from the Les Dunes locomotive depot over the whole range of northern France. They remained in the Dunkerque area until April 1919.

Ghyvelde & Bray-Dunes
On 26 July Ghyvelde sand quarry was shut down and the work moved to Calais Dunes.

Bergues to Proven
This line continued to be the principal British army standard gauge route into the Ypres area. Major works were the Watou depots and yards, the Bambecque depot and the Hoymille and Bergues Canal branches at Bergues. Some of these started before the German offensive of 9 April 1918, but nonetheless they came into their own by the need to have facilities further back from the front. They replaced the St-Jean yard on the Midland Railway extension and the Peselhoek Yards which had been downgraded because of increased shelling. Intermittent heavy shelling of the line continued to 27 September. 6(60)Aus BGROC remained the principal operating company based at Bergues Exchange.

Watou yards
These were built south of the Bergues to Proven main line from a triangular junction between Bandaghem and Mendinghem. When complete they had a standard gauge yard and two further long sidings with light railway transhipments. There was also a metre gauge transhipment with the Belgian vicinal line 115.

Bambecque Depot
This consisted of three sidings south of the main line just east of the Rousbrugge yards. One siding had a light railway transhipment with the L7 line.

Heidebeek to Nordhoek – the 'Far North' line
This was taken over by the Belgian Army on 9 May. The ROD retained running rights.

Proven to Remy (Poperinghe avoiding line)
Heavy shelling of this line, and particularly the Railhoek yard, began on 9 April, and continued at times until August. This yard was a major salvaging centre. On 25 April, seven men of 33LROC were killed and fifteen wounded in Railhoek Yard by shell fire. Two more died the next day.

Northern line (Proven to Boesinghe)
By 19 June, salvage east from the end of the operated line at Waanebeke (Elverdinghe) was complete, including on Pilckem Ridge. By September Boesinghe bridge over the Yser Canal had been demolished by the Belgians.

Midland line and St-Jean extension
The section from Brielen to Reigersburg, taken out of use, was dismantled and salvaged in May and June. Salvage of the St-Jean yards was undertaken by light railway, which is described in the light railway section.

Hazebrouck to Poperinghe and Ypres
The line was operated as far as Vlamertinghe. There was intermittent heavy shelling, especially around Brandhoek and near Poperinghe, and especially in July. There were forty-eight shell breaks between May and August 1918, including sixteen in July. Many new gun spurs were constructed between Hazebrouck and Abeele from May to August.

Ouderdom system
Although maps of June 1918 show only the line from Abeele to Ellarsyde in operation, track was maintained to Ouderdom, Fuzeville and Heksken to allow the salvage from those places. The line from Ouderdom to Brandhoek was dismantled in May. In July, the lines to Fuseville and Heksken were repaired, and by the end of August ammunition was being taken up to new gun spurs in this area.

Hazebrouck to Armentières
In late May 1918, a steam winch and cableway were erected from Carlisle East to salvage materiel from Strazeele, very near the front line. This was too slow, so the line was roughly repaired to allow trollies to be pushed in and then pulled out by the cable, but only night work was possible. The cableway was dismantled by 17 July.

Hazebrouck to Dunkerque
Additional facilities were built at Bavinchove (Cassel station) and yards at La Cloche. The latter provided

transhipments to the L8 light railway (see light railway section).

St-Omer to Hazebrouck
The main additional works on this line were the Fort Rouge ammunition depot and RE park (to replace Strazeele), and the ROD depot at Malhove, to replace Merris and Borre. Later another ammunition depot was built at Wallon-Cappel, but this was dismantled in mid-September.

Watten to Socx line
Built at the request of the British Second Army, this provided a new lateral line north of Hazebrouck, in case of a further German advance threatening or capturing Hazebrouck itself. This line is described in some detail because of its close relationship with the Bergues to St-Momelin, Herzeele to Bollezeele, and Tournehem to St-Momelin metre gauge lines. It was mis-named the Watten to Soex line in many British Army documents up to September 1918. The line is shown in Figure 8.2. Construction started in June, but by the time it was completed for use in late August it was not needed.

Construction started on 1 June by 2CRT, who were already in the area building plank roads. Station names were changed on 12 August to avoid confusion with stations on the Bergues to St-Momelin and Herzeele to Bollezeele metre gauge lines:

old name	new name
St-Momelin	La Ferme Bleue
Lederzeele	Cromestraet
Bollezeele	Smeekaert
Zeggers-Cappel	Langen

8.1 Men of the 297th Railway Company (RE) working on the lifting bridge over the Aa Canal near St-Momelin. This was constructed in June 1918 for the Watten - Socx standard gauge line. It is probable that the bridge for the Tournehem to St-Momelin metre gauge line was similar. (*Imperial War Museum ©IWM Q47347*)

Details of the metre gauge transhipment and other yards at La Ferme Bleue are given in the metre gauge section, and shown in Figure 8.4. Other major facilities were:

- Lederzeele/Cromestraet RE stores siding, stone and supply sidings
- Bollezeele/Smeekaert ambulance and supply sidings
- Peenhof ammunition loops and siding
- Zeggers-Cappel/Langen ammunition transhipment to metre gauge

A lifting bridge with a 20ft (6m) opening had to be constructed over the Aa Canal. 2CRT had already constructed a similar bridge for the Bois de Ham plank road and some smaller bridges over side canals with lifting sections, but other reports state that it was constructed by 297RC, a specialist bridging company. The bridge was open for traffic by 1 July. Work on some station yards was stopped on 13 August. 2CRT left for the Somme on 26 August with 28.5 miles (45km) of track laid, and the line in usable condition, but some ballasting of sidings remained to be done. On 5 September, RCE III ordered that the line be abandoned, but it was still being maintained at the Armistice.

Preparations for demolitions

In June and July 1918, preparations were made for more demolitions on standard gauge lines, because a further German offensive was anticipated. All railway companies were advised of an enemy attack expected on the morning of 18 July, but there was no attack.

Metre gauge railways

On 26 April 1918, an RE major was appointed to be in charge of metre gauge construction, with his HQ at Berck-Plage at the end of the Aire to Berck line. On 12 August, the British Army formed the Directorate of Construction embracing light railway, broad gauge (standard and metre gauge), and dock construction. French sources state that the British laid 47km (29 miles) of new metre gauge track in 1918, of which 20km (12½ miles) was the Tournehem to St-Momelin line. Metre Gauge works from April to September 1918 are shown in Table 8.4.

Table 8.4 Metre gauge railway works in the Ypres Sector April 1918 - September 1918

Line or other facility	Work start	finish	undertaken by	notes
Belgian Vicinal line 115				
Watou depot TS and stone siding	27.04.1918	05.05.1918	113RC	TS to SG and LR
Hondschoote - Bray Dunes line (NF)				
Ghyvelde holding sidings	19.06.1918	11.07.1918	10RC	dismantling (constructed 10RC summer 1916)
Hazebrouck - Hondschoote line (CF)				
Rexpoëde		11.05.1918	113RC	diamond crossings for LR line L7
		02.07.1918	113RC	new crossings with SG - anti-creep frame 15.08.1918
Herzeele (Le Nouveau Monde)	25.05.1918	08.06.1918	113RC	loop and sidings for French ammunition depot
Winnezeele north	08.05.1918	17.05.1918	113RC	loop with 2 sidings, adjacent to LR line L8
south	31.05.1918	09.06.1918	113RC	siding leading to LR line L8
Steenvoorde station	19.07.1918	24.07.1918	113RC	stone siding
Wagenbruge	06.05.1918	13.05.1918	113RC	2 loops between Steenvoorde and St-Sylvestre Cappel
Bergues - Bollezeele & Herzeele - Bollezeele - St-Momelin lines (SE)				
Zeggers-Cappel TS siding	02.07.1918	26.07.1918	2CRT	TS to SG branch from Langen (Zeggers-Cappel) station, Watten - Socx line
Tournehem - St-Momelin (new British line)				
St-Momelin - Watten-Éperlecques station		before 07.06.1918		TS to SG Calais - St-Omer main line (CdN)
Watten-Éperlecques station - Zouafques	by 07.06.1918		BRE	pr completed end-July or early August 1918
St-Momelin (La Ferme Bleue) TS station	22.07.1918	25.08.1918	2CRT	TS to SG Watten - Socx line - on Bollezeele - St-Momelin MG line
St-Momelin Yard (French Yard)	02.08.1918	15.08.1918	2CRT	on Bollezeele - St-Momelin line
Tournehem Junction Yard (Belgian Yard)	02.08.1918	25.08.1918	2CRT	on Bollezeele - St-Momelin line
Anvin - Calais line				
Anvin temporary TS	29.05.1918	07.06.1918	110RC	Étaples - Arras SG line
Anvin Marshalling yard (triage)	26.06.1918	07.08.1918	262RC	
Anvin-Teneur TS	13.06.1918	28.08.1918	262RC	Étaples - Arras SG line
Bergueneuse additional loop	15.06.1918	03.07.1918	296RC	& conversion of siding to loop

Railways of the Ypres battlefields 9 April 1918 to 11 November 1918

Line or other facility	Work start	finish	undertaken by	notes
Verchin additional loop	14.06.1918	06.07.1918	296RC	& conversion of siding to loop
Gourgesson jct chord line Anvin to Aire	27.07.1918	31.08.1918	296RC	loop on Aire line
Fruges 3 sidings and crossovers		22.07.1918	296RC	
Rimeux-Gournay 2 sidings	05.07.1918	02.08.1918	296RC	
Fauquembergues RE workshops	21.07.1918	06.08.1918	296RC	guard house, 2 sidings
Fauquembergues water supply		20.08.1918	296RC	
Wirquin loop line	14.06.1918	27.06.1918	296RC	linked to existing factory siding
Remilly reinstatement of loop on running line	12.07.1918	20.07.1918	296RC	
Lumbres track/sidings to RE park Elnes	20.03.1918	10.04.1918	262RC	
Acquin reinstatement of loop on running line	06.05.1918	18.05.1918	296RC	& conversion of siding to second loop
Journy conversion of siding to loop	05.07.1918	19.07.1918	296RC	siding extension and crossover
Guémy loop line with 3 looped sidings	01.06.1918	03.07.1918	296RC	for Tournehem - St-Momelin line
Tournehem conversion of 3 sidings to loops		11.05.1918	296RC	1 additional siding, for Tournehem - St-Momelin line
Zouafques junction with line to St-Momelin		06.1918	BRE	
Balinghem 2 sidings in a loop		Spring 1918	298RC	
Écluse-Carrée (Vendroux) 2 sidings, and loop line		early 1917	5SCFC	
Vendroux TS depot	05.04.1918	13.04.1918	298RC	
Vendroux supply camp sidings and TS	01.08.1918		298RC	
Pont de Cologne strengthening Guînes canal bridge		11.06.1917	298RC	
Rivière Neuve (Calais) dual gauge to Vendroux		11.06.1917	298RC	
Aire to Berck line				
Mametz siding	27.05.1918	17.06.1918	296RC	
Crecques No 1 & No 2 sidings	10.05.1918	22.06.1918	296RC	5 sidings, 4 looped
Thérouanne loop line and siding	08.05.1918	07.06.1918	296RC	
Thérouanne RE park	21.06.1918	10.07.1918	296RC	
Delette siding	28.05.1918	15.06.1918	296RC	
Coyeques siding			296RC	
Dennebrœucq additional loop line and siding	21.05.1918	12.06.1918	296RC	
Matringhem siding	04.06.1918	17.06.1918	296RC	
Verchocq siding	02.06.1918	18.06.1918	296RC	
Rumilly additional loop line	06.05.1918	23.05.1918	296RC	
Rumilly water supply	30.07.1918	23.08.1918	296RC	
Ergny additional loop line and siding	12.05.1918	07.06.1918	296RC	
Hucqueliers additional loop line and siding		15.05.1918	296RC	
Engoudsent concrete factory sidings	07.09.1918	01.11.1918	296RC	
Beussent additional loop line and siding	14.05.1918	02.06.1918	296RC	
Recques-sur-Course additional loop line	14.05.1918	02.06.1918	296RC	
Attin TS sidings		1917	5SCFC	Étaples - Arras SG line
Wailly additional siding			296RC	
Bahot 2 long loops and spur	28.05.1918	14.06.1918	296RC	
Verton bridge modification for quadrupling SG main line		1918		& 2 additional sidings
Rang-du-Fliers SG/MG workshops and sidings	06.1918	08.1918		
Boulogne to Bonningues				
Ostrohove (Boulogne) siding for camp	26.01.1918	07.03.1918	112RC	

LR	light railway
MG	metre gauge
SG	standard gauge
TS	transhipment
RE	Royal Engineers
RC	Railway Company (RE)
BRE	Belgian Railway Engineers
CdN	Compagnie du Nord
CRT	Battalion, Canadian Railway Troops
CF	*Compagnie des Chemins de fer des Flandres*
NF	*Compagnie des Chemins de Fer d'intérêt local du Nord de la France*
SCFC	*Section de Chemins de Fer de Campagne* (French railway engineers)
SE	*Société Générale des Chemins de Fer Économiques*

Preparations for Demolition

On 21 July 1918, it was agreed that the metre gauge lines Hazebrouck to Herzeele, Herzeele to St-Momelin via Esquelbecq, and Bollezeele to Bergues, would be taken over by RCE III if the *10ème* SCFC left the area. In fact, they did not leave. However, on 26 July RCE III surveyed these lines for demolition. On 14 August, the CRCE ordered that metre gauge lines should be prepared for demolition and the *10ème* SCFC, who were operating the lines, agreed to RCE III making the necessary mines and other preparations. Next day RCE II started to place mines for demolition of metre gauge track in the Poperinghe area. 109RC based at Remy on the Hazebrouck to Poperinghe standard gauge line were responsible for the Hazebrouck to Herzeele and Herzeele to Esquelbecq lines and they made the preparations between 19 and 31 August. 295RC were responsible for demolition of the bridges over the standard gauge lines at Esquelbecq and 2CRT were responsible for the Bergues to St-Momelin and Esquelbecq to Bollezeele lines.

2CRT, while working on the Watten to Socx standard gauge line in the area, received secret instructions at the end of July about demolitions between Esquelbecq and St-Momelin, and between Bergues and Bollezeele. This was 45 miles (72km) in total. However, no action was to be taken until written orders came from RCE III, because the *10ème* SCFC were responsible for maintenance, operation, and demolition on these lines. On 17 August, instructions came from RCE III to 2CRT to prepare charges for bridges and culverts as agreed, and mines for the embankments. *10ème* SCFC were to be responsible for track, station buildings and rolling stock, and for doing the actual demolitions. 2CRT preparations were completed by 24 August.

Hondschoote to Bray Dunes line

On 19 June, a detachment of 10RC moved to Bray Dunes, to dismantle the metre gauge holding sidings at Ghyvelde. This was completed on 17 July and 5 miles (8km) of track was salvaged.

Hazebrouck to Hondschoote line

Developments on this line are shown in Figure 8.4.

Rexpoëde

On 11 May a 60cm and metre gauge diamond crossing was put in by 113RC on the L7 light railway line. On 2 July, a new diamond crossing was put in to cross the Bergues–Proven standard gauge double track main line. This needed an 'anti-creep' device, probably because of the very heavy traffic.

Herzeele (Le Nouveau Monde)

These sidings were built as an ammunition depot for the French at Le Nouveau Monde, about 1 mile (1.5km) south of Herzeele village. On 25 May, 113RC built a 200 feet (60m) clear loop and two dead ends alongside the road. Lack of track caused delays but on 30 May 113RC started transhipment at Esquelbecq of track material and sand ballast by arrangement with the French operating those lines. In early June there was a further delay because of an outbreak of influenza in the detachment camp. Work was completed on 8 June with four sets of points and 450 yards (400m) of track. It was handed over to the *10ème* SCFC on 14 June.

Winnezeele

Two sidings, Winnezeele north and south, were to be built, close to the L8 light railway line. On 8 May, Winnezeele north sidings were started by 113RC as one loop 450 feet (137m) clear, completed on 17 May. On 31 May, more men of 113RC moved from Le Nouveau Monde (Herzeele) to start work on Winnezeele south sidings, completed on 9 June. Both were handed over to the *10ème* SCFC.

Steenvoorde

From 19 to 24 July, 113RC constructed a stone siding at Steenvoorde station, which was handed over to *10ème* SCFC.

Wagenbruge

On 6 May, one platoon of 113RC moved to Wagenbruge for construction of sidings. This was to consist of two loops between Steenevoorde and St-Sylvestre-Cappel, each 450 feet (137m) clear. Work was delayed owing to lack of sand ballast caused by the cancellation of transhipment arrangements at Heidebeek and also by the line being blocked with French ammunition traffic. A further problem was lack of metre gauge trollies and sufficient tools. However, by 13 May it was complete and handed over to *10ème* SCFC.

Esquelbecq to Herzeele, Proven and Poperinghe, Poperinghe to Crombeke

Since January 1918, the Herzeele to Esquelbecq line had been operated by the 85Can ECC with the lines from Herzeele to Proven, Proven to Poperinghe and Poperinghe to Crombeke Road. On 7 April, 85Can ECC had been ordered to go to Merris ROD leaving one officer and sixty men at the base at Bollezeele to assist the new operating company, the 6(60)Aus BGROC, from 9 April. From 8 April, 85Can ECC were involved in considerable movements of troops to and from Poperinghe. On 9 April, the first day of the German offensive, there was considerable shelling of Poperinghe and Proven. One sapper was killed at Poperinghe by a shell fragment. On 6 May, the *10ème* SCFC took over these lines and the 6(60)Aus BGROC men returned to their HQ at Bergues Exchange. The remaining 85Can ECC men were sent straight to Audruicq. When 85Can ECC took over these lines in January 1918, there were forty-two British and fourteen Belgian locomotives. The Belgian locomotives were taken over by the French. Six British locomotives had been sent to Vendroux on 7 April. We do not know the destiny of the remaining locomotives. On 9 May, 113RC put in a diamond crossing at Trois Rois, between Herzeele and Watou, for the L6 light railway line.

Bergues to St-Momelin & Esquelbecq to Bollezeele

These lines were still operated by the French *10ème* SCFC. Two crossings with the new Watten to Socx standard gauge line were built. One was on the Bollezeele to St-Momelin section just north of Volkerinckhove station, put in on 27 June. The other was on the Esquelbecq to Bollezeele section. On 25 July, 2CRT were asked by RCE III to provide guards for the two crossings.

Zeggers-Cappel transhipment siding
(see Figure 8.2)

On 2 July, the Second Army asked RCE III for a metre gauge ammunition siding and transhipment at Zeggers-Cappel. This was to be a branch north between Esquelbecq and Zeggers-Cappel, leading to a transhipment with the standard gauge branch from Zeggers-Cappel (later Langen) station on the Watten to Socx line. By 26 July, 1,500 feet (460m) of track had been laid. By 31 July, 2CRT were transhipping sand from standard to metre gauge using this siding. The sand was sent to St-Momelin.

Tournehem to St-Momelin

In May 1918, the British Second Army made a pressing request for a metre gauge link between a junction at Zouafques, close to Tournehem on the Anvin to Calais system, and another east of St-Momelin, near the end of the line from Herzeele and Bergues. The British had a railway depot at St-Momelin from January 1917, when substantial additions had been made at St-Momelin wharf on the Aa Canal.

The Tournehem to St-Momelin line was built in the spring and summer of 1918, mainly by Belgian railway engineers. The whole link was 20km (12.5 miles) long. This linked the metre gauge systems all the way through from the west coast of the Pas-de-Calais to Belgium. In the event of further German advance this would enable the evacuation of stock westwards. By the summer the tide had turned, and this additional line would have made it possible for the network to carry supply trains all the way from Calais, Rang, Montreuil, and other points on the lines near the coast, to the front which was now moving rapidly back eastwards. However, by that time it was not needed, and it is doubtful if it was ever used except for construction and maintenance.

This line was pencilled in on maps at the planning meeting held in late April 1918. Colonel Henniker in his book states, 'Initially a short line branched off just east of St-Momelin to the standard gauge line from Watten to Bourbourg, with transhipment facilities'. This part of the line probably already existed, with transhipment sidings at Watten station. The route of the final line to Tournehem is shown as 'under construction' on a British Army map of June 1918. This route crossed the main Calais to St-Omer line just south of Watten station, However, this route is known to be incorrect. On the east side of the Aa canal, where it shows the line going over rather than round the hills. Plans of the Watten to Socx line show part of the St-Momelin end of the line, confirming that it did go round the hills, following the edge of the flood plain. Figure 8.2 shows the route of the line. This map also shows named places on the line, which on a civilian line would have been stations, but on this rapidly built military line were probably only passing places, some with other facilities.

We have few details of engineering works associated with this line. The demolition list for the 6 July 1918 version of the withdrawal plan includes the 'bridge over the [*Compagnie du*] *Nord* line for the new Tournehem to St-Momelin line now under construction'. At the eastern end the line crossed the Watten to Socx standard gauge line just before the junction with the Bergues to St-Momelin metre gauge line, close to the yards at La Ferme Bleue. These are described separately at the end of this section. This junction pointed towards Bollezeele. At the western end there was a triangular junction with the Anvin to Calais line at Zouafques, on the Calais side of the *arrêt* there, and ¾ mile (1.2km) north-east of Tournehem station on that line.

Work on the line must have started by 7 June 1918, when 271 Railway Labour Company RE (271RLC) collected timber from Audruicq and delivered it to the 'Belgian section' near Nordausques, at the western end of the line. On 14 June there was a derailment at the transhipment siding at Watten station, reported at 7am, and repaired by 10am, by 271RLC. This was caused by a shunter not properly closing the points. This confirms that the line from St-Momelin must have been built by that time. On 7 July, 2CRT put in the diamond crossing for the Watten to Socx line to cross this line. The timing was determined by the progress of the standard gauge line, since the metre gauge line was already there. The bridge across the Aa canal near Watten was probably similar in construction to that of the Watten to Socx line (see picture 8.1) and would also have needed to be a lifting bridge to allow canal traffic to pass. One would have expected that the 297RC would construct this bridge as they did the other only a few miles away, but we have no confirmation of this.

On 12 September 1918, 298RC based at Calais made a note that 'additional hutting for the St-Momelin–Tournehem metre gauge line put in hand'. Also, on 12 September 271RLC, based at Zeneghem, just north of Watten, received plans from 6 Survey Section for a water supply at Bleue Maison station, and probably another at Le Plouy. The Belgians moved out on 16 September. By 17 September there were detachments of 271RLC at Bleue Maison and Le Plouy stations. The French Area Commandant stopped them moving materiel from the site of the Belgian camp and sent a fatigue party to move it to his quarters. The British Commanding Officer stopped this and placed a guard over the materiel. He spoke to the Area Commandant about the matter in the evening, and we presume that there was no further trouble. The detachments left Le Plouy and Bleue Maison on 24 September. It is not known who took over the maintenance of the line, but it was maintained into 1919.

St-Momelin – later La Ferme Bleue – metre gauge facilities

This station and yards were initially called St-Momelin but the name was changed on 12 August to avoid confusion with the old St-Momelin metre gauge station. In addition to the standard gauge marshalling yard, there were to be, on the metre gauge lines, a metre gauge yard, called St-Momelin yard, or 'French Yard', transhipment sidings to standard gauge and another metre gauge yard. The latter was called Tournehem Junction Yard or 'Belgian Yard'. The layout of the yards at La Ferme Bleue in early September is shown in Figure 8.3. There is some overlap, but the transhipment yard was built first, then the 'French Yard', then the 'Belgian Yard'.

The Transhipment sidings

The plans were agreed and staked out from 19 June. On 22 June, RCE III instructed that priority be given to the Watten end of the line to Socx, and to the St-Momelin transhipment yard, as 'in the event of certain eventualities [i.e., a further German advance] all the stock on the metre gauge railway might have to be evacuated on standard gauge trucks'. The line to Tournehem, which later would allow evacuation of stock, was still being built. On 16 July, 6 Survey Section said that the Belgians could not undertake the metre gauge work at St-Momelin transhipment station, and 2CRT, who were building the Watten to Socx line, would have to do this. On 19 July, 2CRT received the plans, requiring 3.2 miles (5.5km) of track (probably of both gauges). On 22 July, 2CRT started laying track with 85 Chinese labourers.

8.2 Remains of earthworks for the Tournehem to St-Momelin metre gauge line close to the Ferme Bleue yards near St-Momelin, which were on the top of the rise out of sight in the picture. November 2006. (*Authors*)

There were two transhipments, called north and south, each with a single standard gauge siding. Each had a ramp for the metre gauge loading to be level with that of the standard gauge and each had a metre gauge loop with three bays so that trucks could come and go to and from six positions without moving others which were being loaded or unloaded. The south standard gauge siding also had a short end-on metre gauge spur. On 3 August, occupation of the main metre gauge line was secured for installation of the points to the transhipment and the same day fifteen metre gauge trucks of sand came in on the new track. Track laying was complete by 17 August.

St-Momelin Yard (French Yard)

This consisted of two sidings north of the main line and a longer loop south of it. It also included the pre-existing St-Momelin Halt at the east end and the junction to the transhipment yard at the west end of the north loop. Track laying had begun by 2 August and probably finished by 15 August.

Tournehem Junction Yard (Belgian Yard)

The yard occupied the route of the original main line from Bollezeele. A new main line was laid just to the south, and completed on 1 August. The yard consisted of the old main line and two loops, with the junction to Tournehem at the west end. By 24 August, 1,500 feet (450m) of track had been laid, enough for the two loops in the yard. On handing over the Watten to Socx line on 25 August, 2CRT reported that the metre gauge transhipment yard, and the 'French' and 'Belgian' yards were completed with the exception of one set of points. In all, 2CRT had graded, laid and maintained 4.3 miles (7km) of metre gauge track.

Anvin to Calais, Aire to Berck, and Boulogne to Bonningues lines

These lines were in the western Pas-de-Calais and a few works were undertaken along these lines in 1917. Many works were carried out in spring and summer 1918 but an extensive discussion of these is beyond the scope of this book. Readers are referred to the book by the present authors *Tortillards of Artois* (Oakwood Press, 2008). A list of the works is included in Table 8.4 for reference. The works were done in case these lines would be needed if there was another German advance. In addition, some stations on the lines had been identified as supply and ammunition rail heads in case of a need to retreat to the line of the Somme river (Plan Z). The majority of the works were undertaken by 296RC. Works included conversion of three sidings to loops, and one additional siding at Tournehem in May 1918, and an additional loop line with three

8.3 The station at Tournehem on the Anvin to Calais line. In 1918 this was close to the west end of the Tournehem to St-Momelin metre gauge line. Three sidings were converted to loops here, and one additional siding was built, in May 1918, by the 296th Railway Company (RE). (*Authors' collection*)

looped sidings built in June and early July at Guémy. The latter was near Tournehem station in the Anvin direction and there had not been a station or any facility there previously. Both these developments are likely to have been wholly or partly in support of the Tournehem to St-Momelin line.

Vendroux
The standard gauge works at Vendroux are described on page 169. Plans agreed in April 1918 show new transhipment sidings to metre gauge just to the Calais side of the halt on the Anvin to Calais line at Écluse-Carrée. 298RC put in this transhipment, which they completed on 13 April. At the same time, permission was given to construct two sidings connected into a loop at Balinghem, with a connection to the Anvin to Calais line at the Calais end. The plan was to load timber needed for the Vendroux camp at Balinghem, and tranship it at Écluse-Carrée.

On 28 April, a working party of 298RC were 'transhipping' metre gauge rolling stock. This suggests they were moving it from standard gauge trucks onto metre gauge track. On 7 April, 85Can ECC based at Bollezeele had sent six metre gauge 0-6-0 *bicabine* locomotives and 100 wagons to ROD Vendroux from Poperinghe on standard gauge trucks, and on 26 April 6(60)Aus BGROC had despatched forty metre gauge wagons from Heidebeek to Vendroux. On 3 May, 298RC were offloading metre gauge locomotives at Vendroux and they obtained authority from the CRCE for a temporary metre gauge line for 'dumping purposes'. They worked on this on 4 and 5 May. Probably this was for parking metre gauge stock no longer needed, and in June seven ROD 'tram engines' were recorded at Vendroux, as being apparently 'long out of use'. On 10 May 298RC started a new metre gauge transhipment depot and during July and August transhipping materiel became a regular part of their work.

Belgian Vicinal Line 107
In April 1918, this was probably being operated by the SVCFC between Poperinghe and Noordschoote.

Heidebeek transhipment
This was served by a branch from line 107 at Crombeke into the north section of the Heidebeek yards. This closed in April or early May 1918, due to the handover of the standard gauge yard and the line to Noordhoek to the Belgians on 9 May. On 26 April, 6(60)Aus BGROC based in Bergues sent a special rake of empty flat wagons from Calais to Heidebeek to be used for loading metre gauge wagons. A crane loaded forty metre gauge wagons onto standard gauge trucks to be despatched to Vendroux. On 30 April, 6(60)Aus BGROC handled a macadam train from Marquise to Heidebeek for transhipment to metre gauge. The destinations were Winnezeele 150 tons, Hondeghem 150 tons, and Sylvestre-Cappel 100 tons, all for use on roads. All these destinations were on the Hazebrouck to Hondscoote line so the route must have been via Crombeke, Proven, and Herzeele. In later April and in May, 120RC loaded metre gauge material at International Corner on the Northern Line for RCE II. On 2 May 1 mile (1.5km) of metre gauge track and eight sets of points were loaded. However, the loss of the metre gauge transhipment at Heidebeek led to construction delays on the Hazebrouck–Hondschoote line at Wagenbruge because of lack of sand ballast. This suggests that this had been a regular route for materials to the metre gauge lines.

Belgian Vicinal Line 115 and the Proven to Crombeke link
Until 6 May the sections from Poperinghe to Proven, were operated by 85Can ECC and then 6(60)Aus BGROC from Bollezeele. On 6 May the *10ème* SCFC took over these lines. The rest of the line north from Proven to Furnes and La Panne was operated by the SVCFC.

Watou depot
The Watou depot and transhipments are shown on Figure 8.4. Just east of the junction with the link to Herzeele, Line 115 passed the southern part of the Watou depot and standard gauge to light railway transhipments. On 27 April 1918, 113RC started work on a metre gauge stone siding and transhipment siding to standard gauge but work was delayed by a miscalculation which necessitated re-grading. The transhipment was completed on 5 May.

Accidents
On 4 May 1918 a standard gauge train of mine earth approaching Proven 'ran through' a metre gauge train at the crossing of Belgian *Vicinal* line 115, blocking both standard gauge lines. A metre gauge truck of (tar)macadam was smashed under the standard gauge locomotive's cylinders. The Court of Enquiry found that the accident was due to:

1. the crossing attendant leaving his post (shell fire)
2. an unidentified soldier signalling the metre gauge train to proceed
3. the metre gauge locomotive braking heavily and skidding

Light Railways
Most existing lines were maintained and operated within the limits set by the withdrawal in April (see Figure 8.1). The new developments in response to the

withdrawal were mostly the L lines. These are shown in Figure 8.4 and listed in table 8.5. They were built west of the previous light railways with one exception, the L4 line extension towards Bailleul. The L lines had partly been planned before the German April offensive. The LX line from Swiss Cottage to Heidebeek had been constructed in July 1917 and the L4 line had been constructed from Quintin to Boeschepe Quarry in March 1918. On 12 August, it was proposed to form a Directorate of Construction of the British Army to bring light railway construction in with standard and metre gauge railway construction, and dock construction. This was approved on 28 August.

On 10 June, General de Mitry, Commanding Officer French Army of the North, in a letter to the Commanding Officer Second Army, complimented the CRT Battalions for their valuable cooperation on light railways. On 1 July, General de Mitry awarded the Croix de Guerre to Lt. Col. J.K. Cornwall DSO, Commanding Officer of 8CRT. More information is available about 10CRT, who by 10 July were maintaining 30 miles (48km) of track in the back areas and 52 miles (83km) in the forward areas on the A, B, D and related lines.

A lines and associated lines

This includes the A lines themselves, the extensions east of the Yser Canal still in use, the Z1 and Z2 lines to the north of this, the other lines to the north taken over from the French in December 1917, and the lines in the Swiss Cottage depot area. We know that construction and maintenance was by 8CRT based near Lovie Loop, with a detachment at Appledore in June and July 1918. Operating was by 29LROC, probably based at the Pretoria Yard at Swiss Cottage. Later they may have been based wholly or partly at the Yser Locomotive depot on the L7 line.

At the end of May, light railway locomotives and wagons were despatched from International Corner and St-Sixte. The destination may have been south to replace some of the losses in the Somme Sector. On 24 June, 3(17)Aus LROC moved camp from Proven to near Woesten and started work with Canadian troops salvaging light railway track and material from the area in front of Langemarck. These were A line extensions near the front line taken out of use after the German advance in April. 35LROC were reported to be in the Heidebeek area in July 1918, and were at Coppernolle on the A1 line maintaining light railway track with 8CRT from 1 to 13 August. On 14 August, they travelled to St-Sixte where they entrained for Marcelcave in the Somme Sector.

LX line

This had been built in July 1917 from the north end of the Swiss Cottage Yards to the junction to the Heidebeek light railway transhipment. From April 1918 it was extended west as the L6 line. Maintenance and construction were by 2(16)Aus LROC, who arrived from the Ouderdom system on 15 April, leaving all their rolling stock on the D1 line for 5NZ LROC. They camped near Crombeke Ammunition Refill Point (ARP), two looped sidings alongside Belgian metre gauge line 107. Operating was by 29LROC (as for the A lines above). On 14 and 15 April 120RC laid a 600ft (180m) siding for ADLR II on the Heidebeek to Nordhoek line at the west end of Heidebeek Station, for the light railway transhipment, with a spur to Heidebeek junction.

There was a serious incident at Crombeke ARP on 30 April half a mile (750m) from the 2(16)Aus LROC camp, caused by a truck loaded with ammunition catching alight and exploding, blowing up the other truck in the train and 'hurling the engine 150yds' (135m). The explosion destroyed 200yds (180m) of track. From 8 May, C Company of 10CRT moved to near Proven and took over maintenance of the LX line, and 2(16)Aus LROC worked under their supervision. 2(16)Aus LROC were not happy at Heidebeek since they were a skilled unit who had been operating a very busy district, and were not suited to maintenance. They were expected to provide daily parties of fifty men or more to tranship sand and other materials from standard gauge and during April there was much sickness, especially influenza. During May they were also (as were other railway units) armed with rifles, bayonets and the accessories, and given 'infantry and musketry' training. However, some men from 2(16) Aus LROC were able to fill operating roles and on arrival they provided 29LROC with control staff, but these were relieved a few days later. In the longer term, they provided a small number of tractor drivers to 29LROC. By 1 May their Captain was doubling as Second Army Locomotive Superintendent (light railways), a lieutenant was Officer Commanding Locomotive Workshops Second Army (light railways) and another lieutenant was detached for duty as Traffic Controller with 29LROC. On 17 June, a lieutenant was sent to Central Light Railway Workshops at Beaurainville.

By June, 2(16)Aus LROC were maintaining the line between Swiss Cottage and Truro, the first passing loop on the L6 line. Five gangs of sixteen men each were supplied daily. This line was originally laid for emergency purposes, and consequently was lightly ballasted and followed the line of least resistance. Instructions were now to reballast and realign the whole line in expectation of heavy personnel and goods traffic, but there was a shortage of mine earth or other suitable ballast. About this time, 2(16)Aus LROC also made some pointed comments. During the enemy offensive on the Marne Sector, they were unable to obtain reliable daily news.

Table 8.5 Light railway works in the Ypres Sector 9 April 1918 – 11 November 1918

System	line	Section and works	Construction work start	undertaken by finish	notes	
L System lines	LX	new spur at Heidebeek yards	02.08.1918	15.08.1918	10CRT	line built 07.1917
	L4	Boeschepe quarry - Antoinette jct L6		by 31.05.1918	pr 8CRT	L4 Quintin - Boeschepe 03.1918
	L4	Hopoutre ARP		14.08.1918	10CRT	
	L4	Antoinette jct (L6) - near Flêtre		by 30.06.1918	pr 8CRT	L4 extension
	L4	Caestre TS sidings	26.08.1918		295RC	"
	L4	near Flêtre - Bailleul		09.1918		"
	L6	Heidebeek LX - Watou jct L7		by 31.05.1918	10CRT	
	L6	Watou jct L7 - Antoinette jct L6		by 30.06.1918	pr 8CRT	
	L6	Watou yard sidings and Palatine ARP		by 31.05.1918	10CRT	
	L6	Rattekat ARP		29.07.1918	10CRT	
	L7	Watou jct L6 - Rexpoëde	14.04.1918	26.05.1918	10CRT 3(17)Aus LROC	
	L7	Rexpoëde TS, RE park sidings		13.06.1918	10CRT	
	L7	Yser Locomotive depot		26.05.1918	10CRT	main depot for Second Army LR
	L7	Croix Rouge ARP (near Bambecque), Rousbrugge ration and station spurs, and Bambecque junction spur				
	L8	Beauvoorde Junction to La Cloche	08.05.1918	by 31.07.1918	10CRT	Beauvoorde jct - Esqulbecq taken out of use 24 September 1918
	L8	La Cloche - Esquelbecq		by 30.06.1918	10CRT	
	L8	Ledringhem locomotive depot		by 31.07.1918	10CRT	
	L8	La Cloche ammunition depot	01.05.1918	by 31.07.1918	pr 10CRT	
	L8	La Cloche - Eringhem ammunition ext	29.07.1918	15.06.1918	5SCFC 277RC	SGTS Hazebrouck - Dunkerque line
	L8	La Cloche coal siding	27.08.1918	15.08.1918	277RC	SGTS Hazebrouck - Dunkerque line
	L8	Winnezeele north & south MG sidings	08.05.1918	31.08.1918		SGTS Hazebrouck - Dunkerque line
	L9	Steenvoorde jct - Wagenbruge		09.06.1918	113RC	possible TS with L8 line
	L9	Wagenbruge - Hondeghem		by 31.05.1918	pr 8CRT	MGTS (113RC May 1918)
	L9	Hondeghem TS to SG		by 30.06.1918	pr 8CRT	
	L9	Hondeghem - Wallon-Cappel		06.1918		pr not used, bridge dismantled 18 September
	L10	Abeele (L4) - Loye		by 31.08.1918		
	L10	Loye - Heksken yards (F4)	12.08.1918	06.1918		ballasting stopped 3 September - pr not used
	L12	Peselhoek (AB lines) to Penshurst (L6)	22.07.1918	15.08.1918	10CRT	
	L12	Watou ammunition spur No. 3		15.08.1918	10CRT	SGTS
	L14	Hopoutre (L4) - St-Jan-ter-Biezen jct (L12)	12.08.1918	13.09.1918	10CRT	
Westonhoek system	B9 & B15 repairs		by 23.06.1918			
B lines	Y2	Repairs Hexham (Y1) - Austral loop		by 23.07.1918	113RC	for St-Jean standard gauge yard salvage
and	B4	reconstruction Yser Canal - Manner's jct		by 28.09.1918	10CRT	used for 2nd special gas attack 23 July
extensions	B5	reconstruction Laydon jct (Y1) - Manner's jct		by 28.09.1918	pr 8CRT	
		Reconstruct Forest Hill line and Pommern lines		28.09.1918	pr 8CRT	
		New northern Ypres Salient lines - Pommern line to Passchendaele Station		c. 10.10.1918	pr 8CRT	

Railways of the Ypres battlefields 9 April 1918 to 11 November 1918

System	Line	Description	Date	Unit	Notes
		Passchendaele Station to Hulste	by 31.10.1918	pr 8CRT	pr earlier in October – using German lines
		branch Moorslede – Ledeghem	by 31.10.1918	pr 8CRT	pr earlier in October – using German lines
		north – south link Ledeghem to near Duizendzinnen	by 31.10.1918	pr 8CRT	pr earlier in October – using German lines
Busseboom system	C1	repaired Castle Loop to Trois Rois (Three Kings)	19.06.1918	10CRT	
	D6	repair including Harrow Yard D6D	by 21.07.1918	10CRT	used for 1st special gas attack 22 June
D, C & most F lines	D12	reconstruction Dawson (D6) to White House	by 29.07.1918	10CRT	
	F3	Napier jct – Dickebusch	08.1918	10CRT	
	C2	reconstructed Bedford to near Hell Fire Corner	by 28.09.1918	10CRT	possibly further
	C2	Hell Fire Corner jct – Birr Cross Roads	by 04.10.1918	10CRT	
	C2B	reconstruct Birr Cross Roads – Yeomanry jct (C4)	by 04.10.1918	10CRT	to Sanctuary Wood
	C3 & C4	reconstruct Zillebeke jct – Sanctuary	by 11.10.1918	10CRT	jct with C2B, tractor feeder lines
		New southern Ypres Salient lines –			
		C2B extension Yeomanry jct (C4, Sanctuary) – Gheluvelt	by 04.10.1918	10CRT	reconstructing German lines
		Gheluvelt – Ledeghem (H4 line)	by 18.10.1918	10CRT	reconstructing German lines
		branch to Menin	by 18.10.1918	10CRT	reconstructing German lines
		Menin to Ledeghem	by 18.10.1918	10CRT	reconstructing German lines
		Menin to Wevelghem	by 22.10.1918	10CRT	partly using Vicinal line 41 path
		branch to Moorseele & Gulleghem	11.1918		extended to Becelaere December 1918
		Vijtwegen (H4) – Terhand			
Ouderdom system	P2	Ouderdom – Heksken TS yard part re-opened	by 26.07.1917	10CRT	
		Re-open Fuzeville TS, Ouderdom Yard, Heksken ARP	by 11.09.1918	10CRT	
	K1	reconstruct Ouderdom – Hedgerow	13.09.1918	10CRT	
	K1	reconstruct Hedgerow – Willesden jct	by 04.10.1918	10CRT	
	V6	reconstruct D4 jct – Swan Château (V7)	25.09.1918	10CRT	
Messines front		Oostaverne – south-west of Gheluwe	by 30.11 1918		joined H4 branch to Menin of October 1918
		reconstruct Hyde Park Corner – near Deûlémont	by 31.10.1918	pr 8CRT	

c.	circa (about)		RE	Royal Engineers
LR	light railway(s)		RC	Railway Company (RE)
MG	metre gauge		jct	junction
SG	standard gauge		TS	transhipment
pr	probably		ARP	Ammunition Refill Point
CRT	(Bttn) Canadian Railway Troops			

The comment was that people 'at home' knew more about the general progress of the War. Another complaint was that a tank for drinking water was set up and had to be sterilised by boiling as the area was badly contaminated, with a comment that Australia was a 'new Country but considerably ahead of France and Belgium in this regard'. Also in June, influenza returned and two Armstrong huts were converted into temporary wards in the camp, On 7 June, twenty-four patients were admitted, and sixty-one were sent to hospital during the month.

As well as transhipping sand and other materials inwards, light railway materials were transferred onto the standard gauge for shipping out. On 20 May, Traffic St-Omer instructed 6(60)Aus BGROC to provide stock for thirty truck-loads of light railway sleepers to load at Heidebeek for Zeneghem. On 2 June, eighty-one men were working in Heidebeek Yard loading 20,000 steel sleepers into standard gauge trucks, and on 22 June seventy men were loading steel sleepers, rails and other material. In July the unit were transhipping mine earth at St-Sixte (A1 line) and from 1 July to 6 August a detachment of four men was working in the evacuated area of light railways around Ellarsyde and Ouderdom, salvaging water pipe and other materials.

In June, there was a visit by General Sir H.R. Birdwood KCB ADC, Commander of the Australian Corps from November 1917, and then of the Fifth Army from 31 May 1918. 150 men paraded and the General congratulated the unit on its work and appearance. While the General was addressing the men, a train of personnel proceeding towards Heidebeek was about 30 yards from the parade when a truck door dropped open, catching on a sleeper. The truck completely turned over, derailing three other trucks. Fortunately, only minor injuries resulted. The sleepers and three lengths of rail were replaced, and the line was open again by 4pm. Birdwood was impressed with the rapidity of the work.

By July, 2(16)Aus LROC were cultivating vegetable plots, but they also used their own funds to buy vegetables locally. On 26 July, a German two-seater plane was hit by bullets from the 2(16)Aus LROC Lewis gun. It descended in flames. One aviator was killed, the other was wounded. Others claimed credit, but the Area Commandant Crombeke attributed it to 2(16)Aus LROC. Also in July, the section to be maintained was extended past Swiss Cottage to the main Poperinghe Road in the east, and in the west along the L6 line to the crossing of the Proven to Bergues standard gauge line.

On 2 August, 10CRT started a 'new' Heidebeek Spur, completed on 15 August. From 10 August to 21 September a party from 2(16)Aus LROC were building control shelters on other lines with only twenty-six NCOs and men (see separate section page 188).

By later in the month three officers and seventy-seven men were detached and it became necessary to reduce maintenance gangs from sixteen to eleven men. We have some examples of daily orders at the time, which included:

- not thieving potatoes from communal plots
- not machine gunning 'our own' planes
- observing the blackout
- not carrying cameras without permission (which was rarely granted)

Other news in August was that parcels mailed to Australia had been lost at sea through enemy action.

B lines and extensions, and D lines and associated lines

What remained of the D system, and the few lines of the F, C, V and Y systems, were operated and maintained during this period with the B lines north of the main Poperinghe to Ypres standard gauge line. Maintenance and construction were undertaken mainly by 10CRT, based initially near Proven. The B lines had been maintained by 8CRT but on 8 May B Company of 10CRT moved to Steentje Mill (near Westonhoek) to take them over. 8CRT continued to maintain the D area lines until the end of May, when 10CRT took these over too. Operating was mainly by 12LROC based initially at Westonhoek. This whole area was heavily shelled, most of it being under observation from the new German positions on the hills around Mont Kemmel. In May 1918 B Company repaired 214 shell breaks in these light railway systems.

Although the main part of 5NZ LROC had left Poperinghe by the end of April, some must have remained. The 'Toronto' district, formerly the Quintin lines, that is the D and related lines, was not taken over by Westonhoek district and 12 LROC until 7 May. On 2 May, Westonhoek yard and 12LROC camp area were heavily shelled and Baldwin 4-6-0 locomotive No. 968 was struck by shell fire. Continuing heavy shelling led to the removal of all motive power from the yard on 28 May. On the next day, intermittent shell fire damaged two tractors. On 30 May, there was incessant shelling. Motive power was shifted to West Spur but had to be moved again that evening to a new camp site at Byng, 1½ miles (2.5km) north of Westonhoek. By 31 May, it was impossible to maintain the yard and control office at Westonhoek any longer. A tractor depot was established at East Spur loop. The control office moved to a farmhouse half a mile east of Westonhoek on the B12 double track.

In June, 10CRT rotated companies through this 'hot seat', but by 29 June they had to move north to near St-Sixte. 12LROC also had to clear their camp frequently between 20 and 25 June. There were 223 shell breaks on light railways repaired by 10CRT during June. This included a record fifty-one on 6 June, forty-one of these on the B1 line, especially in the area of Hagle dump and Triangle junction. On 19 June the use of the C1 line was extended from Castle loop to Three Kings (Trois Rois), about 500yds (450m), under machine gun fire and gas attacks. This was about 1,000yds (900m) from the front line.

In July 1918, 10CRT continued the rotation of companies into the heavily shelled areas. By now there were markedly fewer shell breaks in the lines, about sixty. Dismantling and salvaging continued. In later July, they salvaged twenty-eight trucks of light railway material, rebuilt D6 yard and bridges (probably D6D Harrow yard), reopened part of the P2 line in the forward artillery area and began constructing the D12 line. On 24 July, the entire crew of three sappers of 12LROC was killed near Culloden locomotive yard at 8pm when their PE tractor was struck by a shell.

From 1 July to 6 August a detachment of four men from 2(16)Aus LROC were working in the evacuated area of light railways around Ellarsyde, Ouderdom and Reningshelst, salvaging 4 inch (10cm) water pipe, and other materials. By the end of July, over 10,000 feet (3km) of water pipe had been salvaged. They also worked in some areas on the D lines still in use. On 19 August, bombing close to 12LROC camp killed one man of 14LROC and three artillery horses. In heavy shelling at Quintin yards, three men were killed, and one man of 12LROC died soon after. On 23 August, two tractor drivers and four guards of 2(16)Aus LROC (Heidebeek) were sent on detachment to 12LROC at Westonhoek.

St-Jean salvage
On 23 June, RCE II ordered that sleepers, fastenings and if possible rails to the east of the Yser Canal must be salvaged by light railway. The bridge and standard gauge track had been rendered unusable. Work had to be at night, with St-Jean yards just over 1 mile (1.6km) from the front line. Before the work could start, 113RC had to replace part of the approach via the B9 and B15 lines and the light railway drivers did not 'know the road', indicating that part of the route was not in regular use. Initially, the salvaged material was offloaded at Waanebeke (Northern line), but by 5 July there was a transhipment for this at Trois Tours (Midland Railway). Work continued nightly to 17 July, hampered by poor labour, derailments, late running due to drivers not knowing the route, and bad tractors. By 17 July, about 3,400 standard gauge sleepers had been salvaged, with spikes and rail fastenings. From 13 July, rails were also salvaged. A rail carrying truck was fitted up by using 60cm bogie trucks with a distancing link piece of 2½ inch pipe 28 feet long. Labour was withdrawn on 16 July, and salvage stopped after that night. This was said to be temporary, but in fact salvage never restarted. By the stoppage 138 36ft rails had been salvaged, with 4 sets of points.

The L lines
These lines are shown in Figure 8.4.

L4 line – Quintin to Antoinette junction
This line had been built from Quintin to Boeschepe Quarry in March 1918, with a branch into the quarry. Sand from Boeschepe quarry was approved as ballast in spring 1918. On 25 May, 109RC were salvaging cable and rollers from the gravity funicular at the quarry. The main line was extended to Antoinette, junction with the L6 line south of Godewaersvelde, in May and the Wippenhoek spur was added in July. The line was operated from Antoinette junction north to Quintin by 5NZ LROC from May 1918. By 14 August, 10CRT were constructing an Ammunition Refill Point at Hopoutre.

L4 line – Antoinette Junction to Bailleul
This line was built from Antoinette junction in June 1918 as far as 1½ miles (2.5km) south-east of Flêtre, only ¾ mile (1.2km) from the front line. We do not know the construction company, but it was most likely a company of 8CRT who were also probably the maintenance unit. On 26 August, 295RC started construction of transhipment sidings at Caestre. The line was operated by 5NZ LROC from their base at Ledringhem on the L8 line. After the withdrawal of the Germans from the Bailleul area, the line was extended in September 1918 to join the lines being reconstructed around there. There is a suggestion that there might at least have been a line on the Bailleul end of this route before the German offensive of April 1918.

L6 line – Heidebeek to Antoinette junction - This was constructed from Heidebeek to Watou depot in April 1918 and the rest of line in May 1918. Construction began at Heidebeek on the 14 April by A Company 10CRT. A double diamond crossing was laid over the double track Bergues to Proven line at the west end of Mendinghem Yard. Crossings were also laid over the double track line to Waayenberg and over the metre gauge line from Proven to Rousbrugge (*Vicinal* 115). Special instructions were issued with regard to the control of the two double track standard gauge crossings:

1. The Crossing will be protected by a red flag by day and a red light by night, placed 100ft each side of the crossing on the light railway

8.4 Wounded men of the 9th Division awaiting evacuation in 4 wheel box wagons near Meteren during the battle for Outtersteene Ridge, 8 August 1918. The line is the L4 extension between Flêtre and Meteren, or more likely a tramway leading to it. Note the very light rails and metal sleepers (see Figure 8.4, locations 97-98). (*Imperial War Museum* ©*IWM Q6959*)

2. Before a light railway train is allowed to cross, the light railway blockman must obtain permission from the standard gauge blockman
3. After the light railway train has cleared the crossing advice will be sent to the standard gauge blockman accordingly
4. All light railway trains will come to a stand before crossing the standard gauge

By 5 May, ADLR II confirmed that telephones had been installed, and that the agreed regulations were now in operation.

By the end of April, 4 miles (6.5km) of track had been laid, with 50 culverts, a storage yard, and spur sidings near the Heidebeek end to accommodate 150 wagons. 113RC also put in a 600ft (180m) siding at Watou for a light railway repair train, completed on 24 April, and a 60ft (18m) bridge had been constructed over Heidebeek Creek. South from the Watou junction with L7, construction and maintenance was undertaken by 8CRT. During May, diamond crossings were put in for the metre gauge Herzeele–Watou link line at Trois Rois by 113RC, and for the standard gauge Hazebrouck to Poperinghe double track line just west of Antoinette junction by 109RC. During May, 10CRT completed the sidings at the Watou marshalling yards, and the spur to Palatine ARP. The northern part to Beauvoorde junction with the L8 line was probably operated by 29LROC whereas the southern part was operated by 5NZ LROC. In July, 2(16)Aus LROC took over the maintenance of the line from Heidebeek to Proven and erected water tanks at the Dawpool loop. On 22 July, 10CRT took over maintenance from 8CRT from Watou Junction as far as Kouter. 10CRT completed Rattekat ARP on 29 July. A new spur at Heidebeek and the ammunition spur no. 3 at Watou were completed by 15 August.

L7 line – Watou junction to Rexpoëde

10CRT began construction of this line on 14 April. From 22 April, 3(17)Aus LROC joined 10CRT and by the end of April they had constructed to the west of the Yser River. This included a 60ft (18m) bridge over the river. Construction continued in May to Rexpoëde station on the Bergues–Proven standard gauge line, more than 1 mile (1.6km) from Rexpoëde village and the metre gauge station. On 11 May 113RC laid in a diamond crossing for the Hazebrouck–Hondschoote metre gauge line and by 26 May 10CRT had completed the 6 miles (9.5km) from Rexpoëde to Watou junction. At Rexpoëde there were terminal sidings and a standard gauge transhipment. The line had been planned on to Bergues, with another south to Esqulebecq, but these were never constructed.

Also on 26 May, 10CRT finished the Yser locomotive yard just east of the river. This became the main locomotive depot for Second Army light railways. Captain Gahan, Commanding Officer 2(16)Aus LROC (Heidebeek), became Second Army Locomotive Superintendent, based here. Lieutenant Anderson, also 2(16)Aus LROC, became Commanding Officer of the depot. The detachments there included steam locomotive and petrol fitters, boilermakers, blacksmiths, and wagon repairers and by August twenty-two men of that unit were there.

In June, 10CRT were heavily affected by the influenza epidemic, however, by 13 June, the RE Yards at Rexpoëde had been completed. In July 10CRT completed Croix Rouge ARP (near Bambecque), Rousbrugge ration and station spurs, and Bambecque junction spur. That same month 120RC were loading salvaged sleepers brought in by light railway to Bambecque transhipment. They also continued loading sleepers at Waanebeke on the Northern Line.

L8 line – Beauvoorde Junction to La Cloche and Esquelbecq

- This line was constructed in May 1918 by 10CRT. By the end of May, D Company were laying track east of Winnezeele, while 113RC were putting in the new metre gauge sidings at Winnezeele. On 2 May, the *19ème Génie* started a new standard gauge ammunition depot at La Cloche, just south of Esquelbecq on the main line from Hazebrouck to Bergues and Dunkerque, and asked for 100 skilled men to assist. This was agreed provided that they did track laying only, and forty-one men of 277RC started work on 7 May. On 18 May, the La Cloche construction ran short of sand. The French traffic authorities had refused paths for the sand trains! On 15 June, *5ème Génie* left La Cloche, and later in June 277RC were given the go ahead for an RE Park siding at La Cloche.

10CRT started track laying west of Winnezeele in early June. The line was operated by 5NZ LROC, who moved into the new Ledringhem Locomotive Yard, near the La Cloche end, as soon as it was available. From this base they operated the L8 line, the L6 line south from Beauvoorde Junction, and the L4, L9 and extension, and L10 lines, called the Ouderzeele District. By July, the light railway had been extended north to Esquelbecq and transhipments at La Cloche and at Esquelbecq were proposed. La Cloche included an ammunition depot and light railway coal sidings. On 24 July, the Second Army Transportation Meeting approved work to go ahead for the ammunition depot. It was completed on 15 August without French approval. On 25 August, the La Cloche coal siding was approved and completed by 109RC on 4 September. On 17 September, 12LROC transferred to Ledringhem to take over operating from 5NZ LROC, who moved back to Quintin (Poperinghe). However, on 24 September 12LROC moved to Beauvoorde on the L6 line and operating on the L8 line was discontinued. It was one of the lines no longer needed.

L9 line and extension – Steenvoorde junction to Wallon-Cappel

This line was built in May 1918 from Steenvoorde junction on the L6 line to the Wagenbruge metre gauge transhipment, just south of Steenvoorde. Sidings were built in June immediately north of Terdeghem Halt. An extension to Hondeghem was built in June 1918, with a branch leading to transhipment sidings to standard gauge on the Hazebrouck Avoiding Line. This line and branch crossed the metre gauge line three times around Hondeghem.

A final extension was built in August 1918 from Hondeghem to near Wallon-Cappel. This needed to cross three main lines. The first crossing was a double diamond crossing over the standard gauge double track junction link between the Hazebrouck Avoiding Line and the Hazebrouck–Bergues line. The second was a bridge under the double track main line from Hazebrouck to Bergues at Les Ciseaux. The third caused some problems since CRCE disapproved the first design for the trestle over-bridge for the main line between St-Omer and Hazebrouck, but plans were finally approved on 24 August. These plans also had to be approved by the agent of the *Nord* railway and the French military Commandant. During building, on 27 August, CRCE warned that a central bent (support) for the trestle bridge was being put between the two *Nord* tracks, leaving only 11ft 9in (3.58m) openings. On 23 September, 112RC met the French *Chef de Section* about dismantling the centre bent of this bridge and it was removed.

The line was probably built by 8CRT. The line was operated, at least as far as the Hondeghem transhipment, by 5NZ LROC, and from 17 September by 12LROC.

On 18 September, RCE III inspected the light railway undercrossing at Les Ciseaux, on the Hazebrouck to Dunkerque line. He agreed that 112RC should dismantle the bridge and fill in the undercrossing with mine earth leaving the piles in place, so it is clear that by late September the line was out of use at least from Hondeghem to Wallon-Cappel.

L10 line – Abeele to Heksken

This line was constructed by 10CRT and operated from Abeele to Loye by 5NZ LROC. When complete it linked the L4 line near Abeele with the former yards at Heksken, at the end of the F4 line. The first half from Abeele to near Loye was constructed in June 1918. The extension from Loye to Heksken was started on 12 August. However, ballasting was discontinued on 3 September. This probably means that the line was no longer needed.

L12 line – Peselhoek to Penhurst (Watou)

This line was constructed by 10CRT and probably operated by 12LROC. Work started on 22 July and track laying was complete by 15 August. It started from the end of the AB1 line at Peselhoek Yards and ended on the L6 line at the Penhurst loop, just south of the Watou yards. It was 10 miles (16km) long. L12 had a spur and standard gauge transhipment at the south end of Watou Yard, Watou ammunition spur No 3.

L14 line – Hopoutre to St-Jan-ter-Biezen Junction

Construction was started by 10CRT on 12 August 1918, from the junction with the L12 line north of the village of St-Jan-ter-Biezen, to link to the L4 line. The L4 end was completed in September with a junction at Hopoutre. There were also diamond crossings at Hopoutre, two across the Hazebrouck–Poperinghe line and one across the Poperinghe Avoiding Line, put in on 13 September.

Lines planned but not built
St-Jan-ter-Biezen to Crombeke

In July 1918 it was announced that this line was going to be started, with the L12 line, but it was not built. It would have linked the L12 line near St-Jan-ter-Biezen with the LX line at Crombeke or Heidebeek Junction.

Caestre to Renescure and Campagne (Arques)

From 22 May to 3 June, two small parties of 2CRT surveyed lines around Hazebrouck and west to near Arques. The new line would have formed a complete ring around Hazebrouck. It would then have gone west to the Campagne standard gauge depot near Arques. Its total length would have been 23½ miles (38km). None of these lines were built.

Control shelters

In August, a party from 2(16) Aus LROC began the construction of control shelters, probably standardised replacements rather than new control points. Shelters were constructed at Hagle, Peselhoek, Pacific and Culloden. Further shelters were constructed in September at Yale on the D1 line, and at Barbed Wire junction (K1 and K6 near Ouderdom). They continued to build shelters until 21 September.

Visit by the King and Queen of the Belgians 8 August 1918

Although the light railways in the Ypres Sector were visited on 6 August by George V, starting from Watou, much more is known about the visit on 8 August by the King and Queen of the Belgians (Albert I and Elisabeth). No doubt the arrangements for George V would have been similar.

The itinerary and programme were:

Leave Vox Vrie 9.05am (AB line).

Peselhoek Farm 9.10am (AB line) – see Poperinghe system defences on the boundary between the Belgian Army and II (British) Corps, and see Americans at work.

Leave Peselhoek 9.35am – Hagle (Triangle, B1/B12 junction) – see Brandhoek defences where light railway crosses the system.

Leave Hagle 9.55 – arrive where the line crosses the road and Kortebeek standard gauge branch (Midland Line) 10.05am (B1 line). Inspect a Company of the Reserve Brigade – see Vlamertinghe system of defences.

Then train or walk to Mission Junction (about 1 mile – 1.5km) (B1/B9 junction) where a Company of Americans will be inspected.

Leave Mission junction 10.35am – Reading Junction (Dawson's Corner) 10.45am (A4/B9 junction) – see 30 siege battery in action.

Leave Reading 11.00am.

Arrive near Welsh Camp (A8 line south-west of Elverdinghe) 11.15am – see 12in Howitzer in action.

Leave 11.25 via A8 and AI lines – arrive Ring Dump, Swiss Cottage 1200noon.
End of tour.

The royal 'special' train consisted of a Simplex 40 tractor, a D class wagon, and a special inspection coach with a covered section. Two further trains were provided, with petrol tractors. One of these was a 'pilot' 5 minutes in advance of the 'special'. This was flying a white flag. The second was the 'breakdown' to follow the 'special' on sight. The driver for the 'special' was detailed by the SLR. 29LROC was to provide experienced tractor drivers for 'pilot' and 'breakdown', and guards for all three trains who knew the route. 29LROC and 12LROC were to man control posts, all road crossings and all facing points. Facing points were to be blocked in position with a wooden wedge. Points men were to have red and green flags. The line was to be cleared and all traffic shunted off 30 minutes before the 'special' was due. The 'special' train was to leave the Yser Locomotive Yard at 5.30am and to be at Vox Vrie by 8am. It should be noted that the royal party travelled through an area still subject to significant shell fire, and also that Albert I had led the Belgian Army in resistance to the German invasion in 1914 and remained in the small unoccupied part of Belgium. In his letter of appreciation, the King remarked that the train had been more punctual than the Paris-Boulogne express!

Operating units

As is apparent from the history of the lines, the major light railway operating units in the Ypres Sector after the retreat of April 1918 were 29LROC, 12LROC and 5NZ LROC but there were a number of operating units in the Ypres Sector of which less is known because there are no detailed war diaries for this period. One of these is 29LROC, but enough is known from other war diaries about this unit (see sections above). For the others, brief notes of what is known of their contribution in 1917 and 1918 follow.

10th Light Railway Operating Company

The company was formed at Longmoor on 10 March 1917 and from 10 July 1917 they were at International Corner on the A lines. The company remained in the Ypres area at least until June 1918. Ten men from the Company were buried in the area between July 1917 and June 1918.

14th Light Railway Operating Company

The Company was formed at Longmoor on 16 May 1917 and arrived with the Second Army on 27 May. They operated from Ouderdom before and during the Battle of Messines in June 1917. In March 1918, one sapper of 14LROC was awarded the Military Medal for his gallant conduct at Hell Fire Corner Yard when 'during heavy shelling of the yard they cleared all ammunition and rolling stock out of the danger zone'. On 16 March, 12LROC were heavily shelled at Manner's Yard on the St-Jean B lines and five men of 14LROC were injured. One of these died on 17 March and was buried at Nine Elms Cemetery. In April 1918, six men of 2(16)Aus LROC were attached to 14LROC, who remained with the Second Army at least until August 1918.

33rd Light Railway Operating Company

In April 1918, five steam locomotive men and two petrol tractor men of 2(16)Aus LROC were attached to 33LROC. Between 17 and 30 April they were working at Railhoek on the Poperinghe Avoiding Line. We know they were not operating. On 25 April, RCE II reported that six men of 33LROC had been killed and fifteen wounded in Railhoek Yard by a shell. The Company may have left the Ypres Sector in May or June 1918 and by September 1918 they were certainly with the Fifth Army south of the Lys River.

21st Light Railway Train Crews Company

The Company arrived on the Western Front on 27 March 1917 and were with 5NZ LROC based at Quintin during that spring. On 16 March 1918 one man of 21TCC, who were at that time attached to 12LROC, was killed. On 28 August 1918 one man of 21TCC was slightly wounded while guarding the 10CRT ration train. From Commonwealth War Graves Commission (CWGC) records, we see that they remained with the Second Army throughout, suffering twelve deaths between May 1917 and August 1918.

22nd Light Railway Train Crews Company

They were formed at Longmoor on 8 May 1917, arrived in the Somme Sector on 13 May and moved to the Ypres Sector in June 1917. CWGC burials show that they stayed in the Ypres sector until December 1917 with five deaths. The distribution of burials suggests that they were operating on the B system lines. They were back in the Somme sector from Jan 1918, attached to the Fifth Army.

Special gas attacks

Between June and August 1918 three special gas attacks were mounted. These were combined operations of British Army Armourers, an RE specialist unit, light railways, Royal Artillery and front-line infantry. For routes and locations refer to Figures 7.4 and 7.6.

First attack – 22 and 23 June 1918

This involved 12LROC and 10 CRT. This was labelled by 12LROC as a 'special gas stunt'. 10CRT's job was to keep the C1 line fit for use to Bedford junction, about 500yds (460m) from the front line. At 5.30pm, three CSMs, eight tractor drivers and eight guards left the 2(16)Aus LROC base at Heidebeek junction, with one tractor and truck, and picked up five further tractors at Vox Vrie, and two at Peselhoek. At Peselhoek rail head, seven trains of seven box wagons each were formed. Each wagon carried 40 gas cylinders (1,960 cylinders in all).

They left Peselhoek for Brandhoek on the D1 line at 7pm followed by a breakdown train. The first train left Brandhoek at 9.30pm, with the other six and the breakdown train following at five minute intervals. The route was via D1 to Pioneer Junction, then D2 and D4 to the C1 Junction. There was a slight delay due to three derailments, but all arrived safely. The success of the operation depended on secrecy, so quietness was essential. It was considered advisable to detach the tractors of the 1st, 2nd and 7th trains at Castle Siding and to hand shunt the trucks about 800 yards (730m) into position on a double track between Bedford junction and Tunnel junction on account of these tractors rattling. The 3rd, 4th, 5th and 6th tractors were detached at Trois Rois and stood by in the siding. The trucks were hand shunted about 500 yards (460m) into position, so there were 49 trucks in a line. The last was in place at 12.15am. The position was on a siding just in No Man's Land. The cylinders were connected, and the gas discharged in one dense volume at 12.54am. We do not know which gas was used. The trucks were hand shunted back to Trois Rois where they were attached to the waiting tractors. The last train departed at 2.10am. It is likely that 2(16)Aus LROC took a large part in this operation because up to April this had been their operating area, from Vauxhall (Voormezeele), not far away. One lieutenant of 12LROC was affected by the gas and was in hospital until 25 June. One 2(16)Aus LROC guard was also gassed. This is thought to be the first such use of light railways during this war.

Second attack – 23 and 24 July 1918
We have more details of this than of the first attack because a special report was drawn up. This is not attributed but it may have been by 49th Division on whose front the attack took place. It is similar to the first attack but at a different location. It was originally scheduled for 9 and 10 July but the weather was unfavourable. As previously, 10CRT kept the line open to reach the location which was the stone siding at Hell Fire Corner about 150 yards (135m) in front of the British front line. Nine trains of seven trucks each were used. These were three-ton box wagons and each carried forty cylinders. 2(16)Aus LROC supplied three NCOs and four tractor crews. The rest came from 12LROC assisted by 4th Foreway Company.

The trains were picked up already marshalled at Pretoria (Swiss Cottage) and moved to Austral Dump Loop on the Y2 line. The route there is not stated but was probably D1 and D2 to Frankton, then D11 past Mimico into south Ypres to join the B6 line. This ran through the ruins of Ypres and out through the Menin Gate to join the Y1 line. The Y2 line began at Hexham junction a short way along the Menin Road. Five trucks were lost on the way, three to shell fire at Peselhoek, one was delayed by an over-heated axle box, and one derailed. They arrived at Austral Loop from 10pm to 11.05 pm. To drown the noise of tractors and truck movements, arrangements were made for Divisional Artillery and machine guns in the area to keep persistent harassing fire, to increase if tractors or trucks could be heard. Nothing was heard at machine gun emplacements 500yds (460m) from the light railway.

At Austral Loop, the tractors were detached and parties of the 4th Battalion Duke of Wellington's Regiment were waiting. A party of five was allocated to each truck with ropes to enable troops to keep windward after discharge. The trucks were moved into position, about 700 yards (640m). One burst of machine gun fire swept Austral Dump shortly after arrival of the first train, wounding two men. A portion of the British defence line was evacuated but a covering party was in position in front of the trucks during preparations. This was withdrawn when the trucks were in position. Since they did not hear any truck movement they wondered if the operation was postponed. 'P' Special Company RE wired the electrical detonators. 12.15am was set for Zero by the Commanding Officer, but a burst of machine gun fire, shortly after the trucks had been pushed into position, punctured a few of the cylinders. Since the escaping gas might have alarmed the enemy, and the covering party had been withdrawn, the plan to synchronise the discharge with a burst of artillery fire was abandoned and discharge took place at once.

It is estimated that 2,274 cylinders were discharged. In addition to those lost on the way, the wires of five batteries were cut by machine gun fire. There were a few other failures, mostly due to faulty detonators. The infantry pulled the trucks back 15 minutes after discharge. With the smell of gas, respirators were worn. The removal was the most difficult part of the operation. At the request of the light railway authorities, the trucks were not uncoupled but were pushed forward as complete rakes of seven. They had very great difficulty pulling the trucks up the incline to 10yds from the front line. Heavy rain made it difficult to stand and the ground sloped away from the track. Pulling one side only did not work as men fell over each other. The return of the trucks to Austral Loop was completed at 2.45am when it was getting dangerously light. The last train left Austral Loop at 3.25am. The operation was considered successful since the enemy front was very quiet after the attack. In September, a lieutenant of 10CRT was awarded the Military Cross, for making the track passable for the gas attack under machine gun fire.

Third attack – 26 and 27 August 1918

This was essentially a repeat of the first attack, at the same location but with more trucks and different starting arrangements. 29LROC were to remove trucks from XG No 1 ammunition depot at Swiss Cottage in three trains of twenty-one by 6pm and place each train on a separate siding in Pretoria (Swiss Cottage south). All crews and tractors were to arrive at Pretoria at 6pm. The first train was to leave at 7pm and was to run straight through to Brandhoek making use of the new triangle at Toronto, the B1/D1 junction. From Brandhoek the first train was to depart at 8.30pm then the rest at intervals of five minutes. The last train was to leave Trois Rois at 4am. As a result, sixty-three wagons of gas cylinders were discharged at 1.30am. There was one fatal and several minor gas casualties to US Troops.

Later attacks

There was at least one more gas attack on 19 September. We do not know where, but we do know it was reported that 10CRT kept the track passable under extremely dangerous conditions. One party found a switch (points) blown out by shelling and spent half an hour repairing it under machine gun fire. The gas attack was successful.

Late August to 27 September

Although the breakout from the Ypres Salient and along the Belgian coast did not begin until 28 September, the German retreat had begun on 8 August at the Battle of Amiens, on the Somme. From the middle of August, the Fifth Army started advancing on the plain of the Lys, south of the Hazebrouck to Armentières railway. There were also some actions in the Second Army area, such as the attack on Outtersteene Ridge, west of Ballieul, in August. As a result, towards the end of August the Germans began to withdraw south of Ypres, between the Ypres to Comines railway and Armentières. However, they did not withdraw north from there to the coast. On 30 August, the Germans evacuated Bailleul and next day they retired from Mont Kemmel. By mid-September they had retreated to Ploegsteert, north of Armentières, and had left Wulverghem and Vierstraat, but they still held Messines Ridge, with the villages of Messines and Wytschaete.

Railways

The withdrawal west of Messines Ridge, and from Mont Kemmel, enabled some reconstruction to be started towards Armentières. To the north of that, large areas were now no longer under direct observation from Mont Kemmel and although shelling continued, reconstruction in some areas was started. Also, there was some preparation for the break-out on 28 September.

Standard gauge railways

All demolition charges were ordered to be removed from bridges on 4 September, and on 17 September the HQ of RCE II moved forward from Rexpoëde to Dosinghem on the Northern line. The main works during this period are shown in Table 8.3. As can be seen, the priority after 31 August, on receipt of the information that the Germans had retired from Mont Kemmel, was the restoration of the lines from Vlamertinghe to Wieltje via Reigersburg and St-Jean. The Midland Railway was only restored from Brielen to Reigersburg later, temporarily for works trains and salvage. The Belgians took over the Northern line east from International Corner on 16 September, including the reconstruction of Boesinghe bridge. The term repair includes reconstruction of dismantled sections. To the south, reconstruction was started towards Wulverghem and Armentières.

Light Railways

On 5 September, No 1 (A) Company of 8CRT moved from Hondeghem to Bailleul and on 9 September No 4 (D) Company of 8CRT moved from Ledringhem to La Crèche. This reflected developments between Hazebrouck and Armentière, but by 23 September the whole company was at Elverdinghe (A lines). For operating 29LROC remained in the Heidebeek area. On 5 September 12LROC moved from Byng on the AB line to Quintin on the D1 line to take over all the lines south of the Poperinghe to Ypres road. 5NZ LROC were still at Ledringhem on the L8 line. On 17 September, 12LROC moved to Ledringhem and 5NZ LROC to Quintin but on 24 September 12LROC transferred to Beauvoorde on L6 as operations ceased on the L8 line.

By 28 September, the policy was that traffic in back areas was to be reduced as much as possible to secure additional operating personnel for forward areas. It was therefore expected that some light railway units would transfer to standard gauge work. 2(16)Aus LROC remained at Heidebeek maintaining the LX Line and north end of the L6 lines and loading and unloading stores in Heidebeek Yard. However, they continued to have large numbers on detachment, and on leave. By 1 September they had eleven PE drivers, guards and brakemen on detachment with 12LROC, transferred on 4 September to 29LROC, who also took over another seventeen tractor drivers and guards. On 6 September they were ordered to send one officer and fifty men to Savy-Berlette near Arras, to join the First Army at the light railway training school. Eight steam drivers, eight firemen, eight guards, three petrol tractor drivers, ten mechanics, four district control staff, and nine station masters were required. Some of these had to be recalled from other attachments.

The men left by train on 10 September. This left only fifty men in camp. The company were relieved of maintenance on the LX line and loading in the Heidebeek yard. In late September they sent another fifty to Savy-Berlette, and in early October the rest of the unit joined them there.

10CRT worked on reconstruction on forward lines, especially on the D lines and extensions, F and K lines, the Ouderdom and Busseboom systems. These were lines taken out of use, back towards Kemmel and the Messines Ridge. In late August they were working on the F3 line. In early September they re-opened the Fuzeville transhipment, the Ouderdom Yard, and the Heksken ammunition refill point. The K1 track had been completed to Hedgerow, about halfway from Ouderdom to Willesden. Heavy shelling continued at times, interfering with work. They also received extra locomotives to maintain and reconstruct in the vicinity of White House on the D12 line and to Swan Château on the V6 line, leading into forward areas. 10CRT also worked on the A and B systems north of Ypres. Meanwhile 8CRT were working on reconstruction towards Armentières and along the Douve Valley, but we have no details of this work.

The breakout from Ypres to the Armistice 28 September to 11 November 1918

In September 1918, Albert I, King of the Belgians, led the Army Group Flanders (Belgian, British and Dominion, and French) in the breakout along the front from the Ypres salient to the North Sea. This was the Fifth Battle of Ypres. By the time of the attack on 28 September, the British Second Army were responsible for the front from immediately south of Armentières to just north of Ypres. The Belgians were responsible from immediately north of Ypres, at about the end of the Yser Canal, to the sea, except for a short stretch north of Boesinghe, where the French took part.

The Fifth Battle of Ypres

The attack on the morning of 28 September took place from Dixmude in the north to the Douve Valley area in the south. On 30 September, the Germans retired from Armentières and by 1 October the Allies held all the high ground around Ypres, had reached the road north from Menin to Roulers north of Ledeghem, and were on the north bank of the Lys River from Comines to Wervicq. However, the advance stopped on 2 October, with rain and mud, the arrival of German reinforcements, and supply difficulties. The advance was resumed later with attacks by the Belgians along the coast and by the British Second Army from 14 October. By the Armistice on 11 November, the Germans had retreated from Ypres to Oudenarde 30 miles (48km), and 38 miles (60km) east from Nieuport, almost to Ghent. Further south, the British were back in Mons, where the BEF had first met the Germans in 1914.

Railways

The development of standard gauge and light railways around Ypres and east of Ypres up to 11 November 1918 is shown in Figure 8.5.

Standard Gauge Railways

Developments for this period are included in Table 8.3. As the front line moved east, reconstruction and operations increasingly took place on lines further east, beyond the scope of this book. On 18 October, RCE II moved from Dosinghem (Northern line) to Den Aap halt, south of Roulers on the line to Menin. The most important lines in the Ypres area were those to Staden, Roulers, and to Comines and Menin. Reconstruction of the Staden line was handed over to the Belgians. The Roulers line connected with German track at Passchendaele Station and was open to Roulers by late October, as was also the line to Menin. The north to south line between Roulers and Menin was crossed in four places by the eastern light railway extensions in October.

RCE III moved from Renescure to Borre (Hazebrouck–Poperinghe line) on 30 September, and on to La Madeleine (Lille) on 26 October. On 3 October, the Germans left Armentières, and on 17 October they left Lille. For the official entry of the British into Lille on 23 October a special train was run from Hazebrouck to Lille St-André with *Nord* officials, the CRCE and RCE III. The same day, the Douve Valley line extension reached the former line from Armentières to Menin in the Warneton area. By 30 October, the ROD were operating to Menin via Comines, probably on the line from Ypres.

Metre gauge railways

The lines in operation in French and Belgian Flanders continued to operate as before. There was no reinstatement of lines east of Nieuport and Ypres before the end of the war because of damage, and the Germans did more damage, including blowing bridges, as they retreated.

Light Railways

Following the advance from 28 September, reconstruction of former lines in newly captured areas was very selective. Work was concentrated on reaching and reconstructing German light railways which would lead east quickly to keep up with the advance and with standard gauge construction. By late October, both the advance and the standard gauge reconstruction outran the light railway developments, and many units were transferred to standard gauge work. However, light railways were needed in the early stages of the attack, although rain and the ensuing mud made maintaining supplies difficult. By 2 October, the initial attack had out-run supplies. 15,000 rations were dropped by parachute from 80 Belgian and British aircraft.

The northern part of the area which had been British until early September had passed into Belgian hands for the breakout, but there were three main areas of development of light railways in the British area:

- northern Ypres salient - starting from lines around Zonnebeke
- southern Ypres salient - lines south of and then further east around the Menin Road
- the Douve Valley area, east to Menin

These lines are shown on Figure 8.5, and listed in Table 8.5.

Northern Ypres salient lines

We have no details of the construction of these lines but assume they must have been by 8CRT, given the disposition of the companies. 8CRT War Diaries at the time consist almost entirely of personnel information. On 28 September, 8CRT HQ was near Lovie Loop on the standard gauge Poperinghe Avoiding Line and a detachment of No 1 (A) Company 8CRT moved to near Elverdinghe. On 29 September, No 1 (A) Company 8CRT moved from near Elverdinghe to the B1 line between Hagle and Tavistock and No 3 (C) Company 8CRT moved from Ondank on the Northern line to just north of Manner's Junction on the B4, B5 and Forest Hill lines at St-Jean village. By 25 October, No 2 (B) Company 8CRT was at Moorslede and No 3 (C) Company 8CRT was at Boschmolens. Both locations are by the northern light railway extension to Hulste. On 28 October, 8CRT was moved from light railway work to work on the Lille-Courtrai standard gauge line under RCE III.

After 28 September, their first task would have been the reconstruction of the Forest Hill and Pommern lines. The B4 and B5 lines to Manner's Junction to St-Jean must have been reconstructed before the 28 September advance. From the Pommern line, the 'new' line probably followed the path of the former Y8 line to just north of the Ypres to Roulers standard gauge line near Zonnebeke. The former Y9 line was restored to near Zonnebeke standard gauge station, and at Seine Junction it was linked back to the former Y6 line and Kansas Cross tramway, but we do not know the state of repair of these lines. From there it would have followed the path of existing German lines as much as possible. There was a spur to Passchendaele station, but the main line crossed the standard gauge west of the station. This crossing was laid in by 113RC on 10 October. The main line continued east, crossing the standard gauge line from Roulers to Menin and passing Boschmolens and Lederlende. It ended at Hulste, close to the Lys River. The total 'straight line' length was 14 miles (22km) but it was considerably longer as it was quite sinuous. South-east of Moorslede a branch linked with a line north from Ledeghem. All these developments were completed in October 1918. We are uncertain who operated these lines. It is known that 5NZ LROC based at Quintin (Poperinghe) operated on some lines in the area held by the Germans before the attack, but this may have been partly or wholly on the southern lines. 29LROC based around Heidebeek would also have been available but there are no records.

Southern Ypres salient lines

Construction was undertaken by 10CRT and 8CRT. Operating was undertaken by 5NZ LROC based at Quintin and from 3 October by a detachment of 12LROC based at Birr Cross Roads, the C2 and C2B junction. 10CRT spent the week before the offensive of 28 September in preparation. A major effort was made to finish the lines as close as possible to the front line. These were the C2 and C2B lines that almost ran to Hell Fire Corner. 10CRT also studied the nearest German light railways on photographs, and information on the German supply system, so that connections to German lines could be made. On 28 September, an officer of 10CRT followed the infantry 'over the top' to study the German light railway system.

Construction into the captured area was started immediately and during the following few days lines were pushed forward, especially the C2B line extension through Sanctuary Wood. The work was done initially under constant shelling and machine gun fire. 10CRT also worked on the Y1 line to give an alternative approach to Hell Fire Corner. The 29 and 30 September were showery and on 1 October there was heavy rain during the night which caused many parts of the track to become flooded. 10CRT found the existing lines in 'terrible shape', and grading the C2B extension was 'the worst ever experienced by this battalion'. Track laying was undertaken by 8CRT and by 4 October the line was completed to Gheluvelt. By the next day, it had been connected up to the German lines.

During the week ending 4 October, 10CRT also completed two loops, two spurs, one ARP on the B1 Line, Cross Roads yards on the K5 line, the K1 line, and Zevecoten dump. Hell Fire stone siding and Zillebeke RE spur were finished by 11 October when the weather was better and there was less shelling. The C3 and C4 lines were reconstructed and opened up for tractors as an auxiliary line to feed the main forward line to Gheluvelt. This was felt to be most important on account of the advance. The K1 line was opened for traffic, the D15 line was completed and the main C2B extension was finished off with two spurs. In all, 13 miles of line were opened for traffic in two weeks. On 12 October B Company moved to Jagenhof near Gheluvelt and were followed by A Company next day.

14 October was a fine day, and the offensive was resumed with a barrage at 5.30am. A and B Companies of 10CRT moved up to the front line to repair breaks and extend the line as fast as possible. By 8.30am, trains were picking up the wounded on the former front line. From then lines were rapidly extended by reconstruction of German lines mainly using German materials from dumps. One problem was that the Germans had left some mines under the tracks to be triggered by trains passing over, but these were found and dealt with. On 16 October, A Company of 10CRT moved to near Ledeghem and on 18 October B to Dadizeele.

During the week ending 18 October, the H4 line was extended to Ledeghem and completely reconstructed. Between Ledegham and Moorseele, the metre gauge line (Vicinal 41) was converted to 60cm gauge and connected with the H4 line at Ledeghem. Sixteen miles were constructed and opened for traffic during the week. By 22 October they had completed and opened for traffic a further 18 miles of light railway. The object had been to reach Courtrai but this was not achieved, and the line only reached Gulleghem. Almost certainly, 10CRT also constructed the branches south to Menin and to Wevelghem. On 23 October, they moved to near Lille for standard gauge work. In November, a line was started west from the H4 line near Dadizeele, and completed to Becelaere in December, but we do not know who built this.

5NZ LROC had replaced 12LROC at their old base at Quintin on 17 September, and operated these lines, including later those taken over from the Germans, until 2 November. Here they found that their services were increasingly employed running personnel to the firing line and back, and they were supplied with additional motive power and rolling stock for this. They list twenty units of troops conveyed during this period including twelve infantry units. These included the New Zealand Rifle Brigade, and the 1st and 2nd New Zealand Infantry Brigades. They also carried two batteries of New Zealand Field Artillery.

The detachment of 12LROC based at Birr Cross Roads (C2 and C2B junction) from 3 October also noted that from 18 October there was very heavy traffic. Evacuating wounded was especially mentioned. 12LROC were disbanded on 31 October and on 2 November 5NZ LROC left Quintin and were replaced by 29LROC. Presumably they took over the operating, but we have no details. By this time traffic was winding down.

Messines Ridge, the Douve Valley, and the Armentières area

In early September two companies of 8CRT had been in this area, but by 23 September the whole company was at Elverdinghe (A lines). However, by 2 October two companies were back at Bailleul and La Crèche, and they remained in the area until late October. We have no details of the work they did, but it was on light railways. They were replaced west of Bailleul on 25 October by 239 Light Railway Forward Company (LRFC). 12LROC moved to Caestre on 1 October for operations around Bailleul. This must have been using the L4 line from Caestre to Bailleul. On 17 October the Company HQ moved to Bailleul, and opened Duke of York light railway yards. They noted a big decrease in light railway traffic due to rapid advances on the front and by 30 October transfers had reduced the Company to eighty-seven men. On 31 October they moved to Birr Cross Roads to join their detachment there, and the whole company was disbanded the same day. What remained of Bailleul District was closed on 31 October.

Maps show a new line constructed in October 1918 from Ploegsteert Wood, near Hyde Park Corner, to the standard gauge line from Armentières to Menin opposite Deûlémont. This was most likely constructed by 8CRT. In November 1918 another line was constructed from the end of the 1917 line from Wytschaete to Oosttaverne. This line had branches, but the main line went east. It joined the October line from Gheluwe, a branch of H4. Both these lines mean that the lines leading to them must have been reconstructed or repaired but there is no information about this. By December 1918 there were two links south across the Lys River towards Lille, one from Wervicq and one from just west of Menin.

Units changed or stood down

By late October, the usefulness of light railways had largely come to an end. The change from mainly stationary to more mobile warfare, and the rapid reconstruction of the standard gauge networks, made them mostly redundant. Many light railway units were stood down or changed to standard gauge work but a few returned to light railway work for clearing up in 1919 (see Chapter Nine).

Construction and maintenance units

On 23 October, 10CRT moved to Breucq (Lille) for standard gauge work for RCE III. On 31 October, 8CRT moved from Proven to Mouscron for standard gauge work for RCE III on the Lille-Courtrai line but one company remained on light railways for a few more days.

Operating units

On 23 October, the Commanding Officer of Australian Railway Troops was informed by the DRT that all light railway operating work, excepting clearing back areas, would be dropped, and the light railway operating companies would be temporarily transferred to broad gauge operating duties. 1(15)Aus LROC went to Bergues for standard gauge operating. They moved to Roulers on 3 November. However, they did go back to light railways on 3 February 1919 when they took over from 13Can LROC at Tincourt (Somme Sector). 2(16) Aus LROC had already moved to Savy-Berlette on 9 October but found very little light railway operating to do there, and what there was consisted of clearing up ammunition, ration and stores dumps. On 22 October they moved to Conchil-le-Temple for standard gauge work. 5NZ LROC marched out of Poperinghe on the afternoon of 2 November and entrained at Quintin for Audruicq Broad Gauge Depot, and 'so brought to a conclusion their history as a Light Railway Unit'.

12LROC HQ received word at Bailleul on 25 October that they were to be disbanded and the men transferred to Forward Companies. There were loud complaints from their Commanding Officer – 'even steam drivers and firemen to be transferred, even though Forward Companies have no steam locomotives!' By 30 October, they were reduced to eighty-seven men. On 31 October they joined their detachment at Birr Cross Roads, where they were disbanded and transferred the same day. 'Thus the 12th LROC RE, the finest Company in Second Army, passes into oblivion' ends their War Diary.

239LRFC was one of four Light Railway Forward Companies (LRFCs) allotted to the Director of Construction for standard gauge work. On 25 October they moved to Hazebrouck to take over maintenance of the standard gauge lines from there to Bailleul for RCE III. The other Companies (231, 238 and 240) were allocated to other RCEs/Armies further south. 234 LRFC were at Beythem on the Roulers–Menin line until 30 October, where they may have been assisting in the operation of light railways. They were then placed at the disposal of the Director of Construction, in addition to the four companies already allotted, and moved to Cambrai for work under RCE IV (Third Army).

Chapter Nine

Light railways and some standard gauge railways after 11 November 1918

This Chapter tells the story of 60cm gauge light railways in the Ypres Sector after the Armistice. We have included a few relevant standard gauge lines.

Standard Gauge Railways

The only standard gauge railway built for military purposes to continue in long term use was the Belgian/British line from Adinkerke to Poperinghe. All the rest were eventually dismantled, and all the lines doubled during the war reverted to single track operation.

Military railways

Adinkerke to Poperinghe

This line had been built by Belgian Army Engineers from Adinkerke to Proven by April 1917, and by British Army Engineers from Poperinghe to Proven by April 1916. The British section was still being maintained by Royal Engineers in July 1919. The Belgians decided to retain this single track line for civilian use. The southern junction, with the Hazebrouck–Poperinghe line, was turned round to point towards Poperinghe. The line ran for a short distance, about 1km, in France, just west of Beveren (see Figure 7.4). Since there were no stops or other facilities in France, the French turned a 'blind eye' to this, and there were no border controls for the trains.

The line, 34km (21 miles) long, opened to passengers on 30 September 1920, and to freight later. It was given Belgian State Railways (later SNCB) line number 76. The line closed to passengers in 1934, when it was decided not to repair the Yser river bridge between Beveren and Rousbrugge. The two parts closed for freight in 1939, and the German Army dismantled the track in 1942.

Other lines in Belgium

A map of 1921 shows some remaining British military lines in the Ypres area (shown in Figure 9.1). Parts of the Northern line, the Great Midland Railway, and the Elverdinghe loop remain, but not connected in the east to the Ypres to Staden line. There is no information about the western ends of these lines, but they probably still connected to the Adinkerke to Poperinghe line, opened by then to civilian traffic. Further south, parts of the Ouderdom system remain as far as Dickebusch. These lines probably still connected at Abeele with the line from Hazebrouck to Poperinghe. There is no information concerning the use and operation of any of these lines.

Watten to Socx line

Work on this British Army line was abandoned in September 1918 when it seemed unlikely that the line would be needed. The line was maintained but by the end of December at least part had been dismantled by 298RC. However, from 30 January 1919 track was relaid, with 259RC laying from the Socx end of the line, and 120RC from the St-Momelin end. Track laying was completed by 1 March, but there was no explanation as to why the line was rebuilt. On 13 March 1919, 259RC began dismantling the track again from the St-Momelin end. On 4 April, 120RC left the area, but 259RC remained at Langen (Zeggers-Cappel). The last train of dismantled material left Langen on 18 July, and 259RC moved to Hondeghem on 25 August 1919. There they worked on dismantling the Hazebrouck Avoiding Line.

Vendroux

Half of 114RC worked at the Vendroux yards in December 1918 and January 1919. We do not know when these yards handed over, and in what state, but at least part was still being operated by the French in 1921. There were two crossings of the metre gauge Anvin-Calais line in the yards, and in 1921 one was the site of a collision between an Anvin-Calais metre gauge train and a standard gauge locomotive of CdN.

Hoymille (Bergues)

An ordnance depot was built from February 1919, presumably for the collection of ammunition from the battlefields, with light railway (60cm) lines in the depot. An extension was agreed in April, and the whole depot was finished by 17 June 1919.

Civilian railways (in place before 1914)

Many British and Dominion units worked on the reconstruction and re-opening of French and Belgian lines. By the end of April 1919, all lines had been handed over to the relevant national authorities. Operating companies continued to work on civilian and military lines until spring 1919. Early 1919 was very busy with the opening of evacuated territory, supplies for the Belgian and French civil populations, and heavy troop and demobilisation trains. Standard gauge operating units worked from Bergues, Dunkerque and Courtrai as far as Abbeville, Doullens, Liège, Tourcoing and Cologne (Köln). The French and Belgians gradually took over, and by April 1919 this was complete.

Demobilisation

British standard gauge railway units were demobilised from January to October 1919, most by June, and a few in August, with only one remaining until October. Canadian Railway Troops were gone by February 1919, Canadian broad (standard) gauge operating companies by April, and the Australian equivalents by May. The office of RCE II closed on 23 February 1919, and that of RCE III closed at Montreuil on 19 April 1919. The office of RCE for the lines of communication (RCE Comms) remained open, possibly at Abbeville, until 31 August 1919.

Light Railways
Salvage and clearing up

At the end of November 1918, the British Fifth Army, now in charge of the Ypres Sector, decided that salvage should be organised on a Corps basis. II Corps were positioned in the north of the Ypres Sector. We do not have much detail of the organisation in this area, but it was divided into a western and eastern sector, possibly demarcated by the Yser Canal. It included most of the Y System, and the A, AB and B9 Systems, and the October 1918 northern extensions eastwards.

XIX Corps were ordered to work in the Ypres Sector and on the plain of the Lys south of the main line from Bailleul via Armentières to Lille. ADLR V was still in charge for Fifth Army Light Railways. The lines known to have been reconstructed with their dates are listed in Table 9.1, but there were probably more. The lines reconstructed, constructed, or taken over from the German Army in October and November 1918, including the new lines eastwards, did not need further work. Probably many of those west of the Yser Canal also did not need more than maintenance.

The XIX Corps area was divided into seven sectors, A–G. Of these B, E, F and G were south of the River Lys and out of the area for this book, and only sectors A, C and D are relevant. XIX Corps area ran to just north of the standard gauge line from Poperinghe to Ypres, and areas to the east and west of this. It included the Westonhoek Yards and the B1 line, part of the B System, the D, C, F, and K Systems, the Messines front lines, and the October 1918 southern extensions eastwards.

Within each salvage area there would be a labour group HQ responsible for the work of collection. The materiel would be assembled by the light railway and carried to dumps in each area. There was at least one dump for RE and general salvage in each section, and another for ammunition. Each HQ would requisition any necessary light railway repair or reconstruction

Table 9.1 Light railway works in the Ypres Sector December 1918 - July 1919.

Corps	Sector	line		Construction work start	finish	undertaken by	notes
II	East	B15	Sidings St-Jean Yard	19.01.1919	06.02.1919	240LRFC	
			new sidings St-Jean Yard	07.04.1919	16.04.1919	240LRFC	
			new sidings St-Jean Yard	24.05.1919	09.06.1919	240LRFC	
			new sidings St-Jean salvage dump	17.07.1919	ongoing 31.07.1919	240LRFC	
II	East	Y9	Pillbox Corner (Y8) to Zonnebeke	23.01.1919	25.02.1919	240LRFC	reconstruction
II	East		Wild Wood line - Grey Ruin jct to C2F	01.02.1919	25.02.1919	240LRFC	reconstruction
II	East		Pommern jct - Fortuin jct	07.07.1919	23.07.1919	240LRFC	reconstruction
II	East		Fortuin jct - Gravenstafel & Corner Cut	23.07.1919	ongoing 31.07.1919	240LRFC	reconstruction
II	East		A extension, B9 Weedon to Langemarck and Keerselaare	23.07.1919	ongoing 31.07.1919	240LRFC	reconstruction
II	East		Zuidschoote to Kruisdoorn via Luna Park	23.07.1919	ongoing 31.07.1919	240LRFC	upgrade for steam
XIX	C	C2D	C2 Lake Farm Spur to C2 nr Calomel	bef 07.04.1919	30.04.1919	239LRFC 240LRFC	reconstruction
XIX	C	C3	Zillebeke jct (C2) - Essex (C5)	07.04.1919	17.06.1919	240LRFC	reconstruction
XIX	C	C5	Tunnel jct (C2) - Essex (C3)	07.04.1919	27.05.1919	240LRFC	reconstruction
XIX	C	C2	Bedford (C1,C5) to Westhoek	01.05.1919	12.05.1919	240LRFC	reconstruction
XIX	C	Y1	Menin Road diversion, Ypres	06.05.1919	22.05.1919	240LRFC	
XIX	C		Greenjacket - Klein Zillebeke	01.06.1919	17.07.1919	240LRFC	new construction
XIX	C		Menin Road diversion for MG line (Ypres)	08.07.1919	18.07.1919	240LRFC	Y1 line using MG path

jct   junction
LRFC   Light Railway Forward Company
MG   metre gauge

equipment. The dumps were, if possible, to be on an existing light railway but if necessary a spur would be constructed. Care was taken that the dumps were always sited in places where there was ample room for extension.

ADLR V decided that maintenance and any necessary repairs and reconstruction of light railways would be undertaken by LRFCs, with two labour companies. Construction traffic was to have preference on lines so they could carry on work during the short daylight hours of the season. Rations to all troops on salvage work would probably need to be distributed by light railway. Details of the organisation of salvage sectors are incomplete, but those available are given in Table 9.2. There is, however, some doubt about the use of the St-Jean Salvage Yards, where many new sidings were built in spring 1919. This is thought to have been on the site of the existing St-Jean standard gauge yards rather than at Manner's Junction. This would put it into II Corps East Sector rather than XIX Corps C Sector. Group HQs were ordered to move in as quickly as possible.

II Corps West

This area was probably entirely on the west bank of the Yser Canal.

Maintenance, repair and construction

This was allocated to 240LRFC, formed from 10 AT&FC RE. They were working in November and December 1918 on standard gauge maintenance and construction south of Lille but on 31 December they left for Busseboom and on 7 January moved on to take over the light railways in II Corps area. Their HQ and mounted section were based at Byng junction, just south of the Vox Vrie yards. Two sections of the Company were based at Elverdinghe from 17 January, and a third were at St-Jean (II Corps East). II Corps West were initially responsible for the old French lines taken over in January 1918, the Z lines, the A lines (probably those west of the Yser Canal), and the AB lines. From March they took on additional responsibility for lines in the north-west of the area and the northern part of the L System, Proven area and Watou depots and yards.

Table 9.2 Organisation of salvage in the Ypres Sector December 1918 to July 1919

Sector	II West	II East	A	C	D
Army Corps	II	II	XIX	XIX	XIX
Labour HQ				St Dunstan's Farm	Bailleul
LR location of HQ				B6 (Harrow Yard)	
RE and General dumps		St-Jean Yards	Menin	Bedford Yd C1/C2/C5 (Zillebeeke)	La Crèche (Steenwerck)
				nr Zealand jct D5/F1 (Busseboom)	
				?St-Jean Yards	
Ammunition dump		?St-Jean Yards	Menin	Machine Gun Farm B8	Trent depot
LR					
Maintenance & Construction	240LRFC	240LRFC	235LRFC	235LRFC	3Aus LRFC
Base	Vox Vrie	St-Jean (1 section)	Dadizielhoek	Dadizielhoek	Steenwerck (1 section)
	Elverdinghe (2 sections)				
from	08.01.1919		12.1918	12.1918	05.12.1918
to	at least 31.07.1919			pr 28.02.1919	19.03.1919
				240LRFC	
Base				Vox Vrie	
				Elverdinghe (2 sections)	
				St-Jean (1 section)	
from				1 March 1919	
to				at least 31.07.1919	
Operating	34LROC	pr 34LROC	pr 235LRFC	2(16)Aus LROC	pr 3Aus LRFC
Base	Vox Vrie	Vox Vrie		Manner's Yard (St-Jean village)	
from	07.01.1919	07.01.1919		21.02.1919	
to	at least 31.05.1919	at least 31.05.1919		21.04.1919	
		(possibly 2(16)Aus LROC)		21LROC detachment	
Base		Manner's Yard (St-Jean village)		Manner's Yard (St-Jean village)	
from		21.02.1919		21.04.1919	

pr	probably
jct	junction
LR	light railway
RE	Royal Engineers
LROC	Light Railway Operating Company
LRFC	Light Railway Forward Company

Light railways and some standard gauge railways after 11 November 1918

Operating

Until 9 November 1918, 34LROC were working light railways in the Somme Sector. They moved to Cambrai to assist 58Can BGROC ROD with standard gauge operations. On 7 January 1919, they moved to Vox Vrie Farm under the direction of ADLR V to operate. They continued there at least until 31 May 1919 when the War Diary ends with no mention of leaving or demobilisation.

II Corps East

This area was probably entirely on the east bank of the Yser Canal.

Maintenance, repair and construction

This was also allocated to 240LRFC. The HQ was at Vox Vrie with II Corps West. One section moved to St-Jean on 15 January. They were initially responsible for the A lines and the B9 System east of the Yser Canal, the Y System east from Wieltje, and the October 1918 northern extension eastwards as far as Hulste. Reconstruction work known to have been undertaken by 240LRFC is shown in Table 9.1. They worked on additions to the salvage yards at St-Jean, continuing to July 1919. Other work included the former French line from Zuydschoote to Kruisdoorn via Luna Park being made fit for steam locomotives.

Operating

This area was probably operated by 34LROC, based at Vox Vrie from 7 January 1919, but it may have been partly or wholly operated by 2(16)Aus LROC. They were based at Manner's Yard in St-Jean Village from 21 February to 21 April 1919.

A Sector XIX Corps

This is the sector about which we know least. The area did not include any of the 'old' (pre-April 1918) light railways, but it had all the autumn 1918 lines in the southern part of the eastern extensions. These would not have required much repair. This included territory north of the Lys river to the northern boundary of XIX Corps and the RE, general and ammunition dumps in the Menin area.

Maintenance, repair and construction

These were allocated to 235LRFC, with areas B (south of Lys River) and C. We do not know the origins of 235LRFC. They were probably undertaking standard gauge work in the Ledeghem/Dadizielhoek area on the Courtrai–Roulers line in November 1918. We know that 235LRFC were still at Dadizielhoek in January 1919, but there are no records after that, nor any record of any other unit undertaking work in the A Sector. 235LRFC may have been disbanded on 14 June.

Operating

Nothing is known, but it is likely that 235LRFC also undertook the operating in this section.

C Sector XIX Corps

This included the main areas of light railway development in 1917 and early 1918, south from the B1 line down to Kemmel, and as far east as Gheluvelt.

Maintenance, repair and construction

This was allocated to 235LRFC, with areas B (south of Lys River) and A. We know that 235LRFC were still at Dadizielhoek in January 1919, but there are no records after that. From March 1919, most of the C Sector was taken over by 240LRFC, who remained responsible for II Corps West and East and were still based at Byng near Vox Vrie. By 7 April, 240LRFC were responsible for the whole of XIX Corps C Sector. They took over reconstruction of C2D line from 239LRFC on 7 April, but we do not know anything else about the activities of 239LRFC in this area. From April 1919, 240LRFC undertook considerable reconstruction in the former C System lines. One new line was constructed, from Greenjacket to Klein Zillebeke. It is likely that this was the line along the north side of Zillebeke lake (see Figure 9.1), and that Greenjacket was at the west end of the lake, or near Shrapnel corner. Two lines are identified as 'Menin Road diversion' (Table 9.1), one constructed in May and the other in July 1919 (both shown in Figure 9.1). One was probably the line to White Château, between Ypres and Hell Fire Corner. The other was the line following the standard gauge line from near the Lille Gate to near Hell Fire Corner, then connecting south to Zillebeke. Both helped replace the Y1 line along the Menin Road, allowing the SNCV metre gauge line 121 near Ypres to re-open from September 1919 (see Chapter Eleven).

Operating

This was done by 2(16)Aus LROC who had been at Courtrai from 22 November operating standard gauge railways. This work was taken over by the Belgians in February 1919. 2(16) Aus LROC moved to Manner's Junction in St-Jean village, arriving on 21 February 1919, and their standard gauge train was parked on a siding at the dump in St-Jean Yard salvage depot. One comment was 'nothing but mud, shell holes and broken-down dugouts being visible as far as the eye can see'. Manner's Yard was on B4 line, at the beginning of the Forest Hill line, at its junction with B5. It was also close to B15 line to St-Jean standard gauge yards and the B9 System. St-Jean standard gauge yard clearly became a major salvage dump with additional light railway sidings. War Diaries make it clear that 2(16)Aus LROC considered this part of XIX Corps area. By spring 1919, some aspects of 'sectorisation' were breaking down, with units leaving. Salvage work was performed with Chinese and POW labour, under ADLR V North. There were eighty-two men actively operating, with the remainder erecting a locomotive depot building at Manner's. They were officially allocated five locomotives, two petrol electrics, one 40hp and two 20hp Simplex tractors. In practice, their actual power was two locomotives, two petrol electrics and two 20hp tractors. The locomotives were in bad order.

At the beginning of March 1919 the ration strength of 2(16)Aus LROC was 195 men but only one officer and 104 men were operating. In March, operating was variable but at times slack. The weather was very cold with snowfalls. Two Armstrong huts were erected to use as a canteen and hall. The War Diary commented that these were essential since 'on this Park civilisation is almost unknown'. They were opened on 10 March with a concert by a Labour Company. Two lectures were given by members of XIX Corps Education Department based in Ypres, one on the battlefields, and one on the League of Nations. The Company ran a train to Poperinghe and Voormezeele on 2 March. Voormezeele (Elzenwalle) had been their first base in 1917. They found it 'knocked beyond recognition'. On 9 and 30 March, a Sunday special train ran to Kemmel Hill for the benefit of Company members and on 23 March a party was driven round old battlefields, probably by train.

Traffic remained variable in April, but power could be short, requiring long working days. On 21 April, they handed over all power and supplies to a detachment of 21LROC, formerly the 21TCC, who took over operating. On 23 April, 2(16)Aus LROC left by standard gauge train for demobilisation.

D Sector XIX Corps

This sector consisted of the Messines front from Armentières north to Kemmel, as far east as the Lys River at Warneton then north through Messines and Wytschaete. The Labour HQ was at Bailleul. The RE and general salvage dump was at La Crèche (Steenwerck), and the ammunition dump at Trent depot.

Maintenance, repair and construction

This was allocated to 3Aus LRFC. They were also allocated sectors E, F and G, outside the area of this book. They were formed from 3(17)Aus LROC on 7 September 1918 and moved HQ to Lomme (Lille) on 5 December. One section, probably a platoon, was based there (F Section), others were at Annaples (G Section), La Gorgue (E Sector) and Steenwerck (D Sector). The whole Company were allocated 12 Simplex 20 tractors, 16 F class and 8 D class wagons. F class was a well bogie wagon with detachable stanchions. D class was an open bogie with side doors. Through January 1919, reconstruction was carried out with difficulty due to weather conditions and falling off of labour due to demobilisation. In February, their own demobilisation started. The Steenwerck section (D Sector) was withdrawn on 19 March. On 21 April they completed work in their last area, F Sector, and handed over to ADLR V before demobilisation.

Operating

Nothing is known, but it is most likely that 235LRFC also undertook the operating in this sector.

The end of the story?

This provides a very unsatisfactory end to the story of British Army light railways in the Ypres Sector. By June 1919 there were only British units, and all Australian and Canadian units had gone home. At the end of July active reconstruction and operation for salvage purposes was still under way. Then, suddenly, there are no more light railway records. It may be that with the signing of the Treaty of Versailles on 28 June 1919, the war officially came to an end, ending the requirement for British armed forces units to keep war diaries. Some standard gauge railway companies continued their diaries a little longer. 240LRFC, based at Peselhoek, ended their diary on 31 July 1919, but were still actively at work. 34LROC were operating from Vox Vrie at least until 31 May 1919. 21LROC (formerly 21TCC) were at St-Jean, and 239LRFC were in the Ypres area, at least until April 1919, and 235LRFC at least until 15 June 1919. Any of these may still have been at work after July.

Belgium

For civilian passenger services, the Belgian authorities concentrated on rebuilding and re-opening the metre gauge *Vicinal* lines in the battlefield areas (see Chapter Eleven).

The map of the Ypres area in 1921, already referred to in the standard gauge section above, shows a considerable number of light railways remaining (shown in Figure 9.1). North of Ypres are two systems, one north and west of Boesinghe, and one south of this. These are mostly on new alignments, with few sections based on, or intersections with, former British military lines. The lines mostly end at farms, or at main roads or railways, and have the characteristics of farm support systems. They would have been constructed using materials bought from or left behind by the British Army.

South of the standard gauge line from Ypres to Poperinghe, more lines remain, and although there are some new alignments, the majority follow the paths of former British military lines and are labelled as such on Figure 9.1. Some of the new lines, notably the two Menin Road diversions, and the line to Zillebeke, are known to have been of British Army construction (see Table 9.1). The fragmentation around Voormezeele may reflect demolitions.

We have no information on the operation of any of these lines in 1921. By the 1930s, all these light railways have gone. There only remains a line west from Warneton along the Douve Valley, as far as the road from Neuve Église to Kemmel, near or at the site of the former de Kennebak yards. It is in an area off the 1921 map to the south. It is shown using the symbol for 'secondary networks, industrial lines and tramways', which is used elsewhere on our 1930s map for standard gauge, metre gauge, and 60cm gauge railways in these categories, so the gauge of this line is in doubt.

The Tramway from Adinkerke to La Panne

This 60cm gauge pre-war horse tramway is described in Chapter Three. It was operated with mechanical traction by the British in 1917, and probably by the French and Belgians at other times during the war. It reopened after the war with mechanical traction, but because it was later incorporated into the metre gauge network, its later history is described in Chapter Eleven.

France

On 6 October 1919, the majority of the British light railways in France were sold to the French government for 85 million francs, about £2.4 million. The deal included all laid track east of the N16. This road ran from Dunkerque to Paris. In the area of this book, it ran from Dunkerque south to Hazebrouck via Bergues and Wormhoudt. As such this would have included almost all the light railways which were not separate systems forming part of yards, and probably, in practical terms, all of them. Some rolling stock including steam locomotives and petrol tractors were included in the deal. The lines were taken over by the *Ministère des Régions Libérées* (Ministry for the Liberated Regions). In the Nord *Département*, which includes the area of this book, no public passenger service was offered on the old military lines, which were taken up from 1921 onwards.

Chapter Ten

The metre gauge railways of French Flanders and related lines and tramways 1919 to 1954

After the war these lines gradually began to close. The two main networks lasted the longest, but even they succumbed by the early 1950s to under investment and road competition. Buses and road freight were more flexible, but often the replacement bus services were less frequent or less convenient than the trains.

The Coordination Committees and SNCF

In 1934, the National Economic Council, who had been asked to take evidence on the problems of local transport, presented their report. They recommended that each Region (in a railway sense) should have a 'coordination committee', to advise which transport mode or modes should provide local and long-distance passenger and goods services. The Coordination Committees represented local and national railway and road operators, but did not include users, as had been recommended. Closures of *Intérêt Local* lines, usually of metre gauge, were widespread in the 1930s and by 1937 France had lost nearly half of these lines. Many companies running them began to take an interest in owning and running bus services, often through a separate but related company. In 1938, the major standard gauge railway companies were joined together to form the *Société Nationale des Chemins de Fer Français* (SNCF) which was, and still is, effectively a nationalised railway company.

10.1 Railcar No. 402, one of the two constructed by VFIL at the Lumbres workshops on the Anvin to Calais line in 1936, at Warhem station. Postcard postmarked October 1939. (*Collection Bernard Rozé*)

Hazebrouck to Hondschoote, and Rexpoëde to Bergues

In 1919, these lines of the CF were integrated into the new *Compagnie Générale des Voie Ferrées d'Intérêt Local* (VFIL), formed by the neighbouring Anvin to Calais Company. This Company is usually abbreviated VFIL in British literature but CGL in French books. It was started in 1919 by grouping the Anvin to Calais line (proprietor Émile Level), the Aire to Berck line (proprietor Alfred Lambert), the Tramway from Ardres to Pont d'Ardres, and these lines (Alfred Lambert). Between 1920 and 1930 it acquired another 15 metre and standard gauge lines of *Intérêt Local* in the Nord, Pas-de-Calais, Oise and Aisne *départements*, but no more in the area of interest of this book.

Most sources say the wartime line from Herzeele to Watou in Belgium was taken up soon after the armistice. However, a Michelin map of the 1930s shows the track still in place from Herzeele to Houtkerque and only the last 2km into Belgium missing. There was no passenger service, and we assume that the line was kept for goods.

Stations and track layouts

The layout in 1933 of the station shared with the CdN at Hazebrouck is shown in Figure 10.1, and that at Bergues shared with SNCF in 1948 in Figure 10.2.

Rolling stock

In 1932 the closure of the *Chemins de fer de l'Artois* (Béthune to Estaires) allowed the purchase of four 2-6-0T SACM locomotives of 19 tonnes. These had been delivered to the Artois company in 1924, with the numbers 1, 4, 5 and 6. They were 7.29 metres long (without buffers), 2.30 metres wide, 3.20 metres high, 19 tonnes empty and 24.35 tonnes ready for service.

Two bogie diesel railcars, numbers 401 and 402, were put into service in 1936. These were constructed by VFIL in the workshops at Lumbres on the Anvin-Calais line. Their bodies were constructed by Million-Guiet. They had a Berliet engine of 115hp, were 14.65 metres long, and weighed 18 tonnes, with a maximum speed of 75 kilometres per hour. They provided forty-eight places seated and twenty standing.

Finally two 0-6-0 three axle diesel tractors, numbered 351 and 352, were delivered in 1951. These were also constructed by VFIL at Lumbres. They were capable of hauling 130 tonnes at 15 kilometres per hour up a gradient of 1.5%. Their specification was -

10.2 Diesel locomotive No 351, one of the two constructed by VFIL at the Lumbres workshops on the Anvin to Calais line in 1951, at Hazebrouck station in the early 1950s. The long *passerelle* (footbridge) over the station and yards is in the background (see Figure 10.1). (*Collection Bernard Rozé*)

Operations

For a list of the stations see Table 2.1. Post-war, the service restarted with three trains each way on each section, as opposed to the four each way in 1914. The timetable for 1929 is shown in Table 10.1. As prior to 1914, the basic services were between Hondschoote and Hazebrouck, and between Hondschoote and Bergues, but some of the latter started or finished at Rexpoëde, which was the junction. The timetable varied somewhat on Mondays, which was market day in Bergues. From 1936, two trains out of the three each way on each section were provided by diesel railcar. A timetable (*Chaix*) from October 1936 is shown in Table 10.2. In the last years of the line, there were again four services each way per day between Hazebrouck and Hondschoote, but one was broken at Rexpoëde. There were three each way between Bergues and Hondschoote. The timetable (*Chaix*) for April 1952 is shown in Table 10.3.

Closures

Despite the line's modernisation the *département* decided to close the line from Rexpoëde to Bergues on 31 January 1954, and from Hazebrouck to Hondschoote on 31 December the same year.

Disposal of assets

One railcar, No 401, and the two tractors, Nos 351 and 352, were sold to SE for the Baie de Somme network.

Diesel engine	Willeme	power	180hp
Cylinders	8	Size	130 x 170 millimetres
Length (without buffers)	8 metres	Width	2.00 metres
Height	3.15 metres	Weight (empty)	19.5 tonnes
Transmission	De Dion		
Maximum speed	45 kilometres per hour		

Narrow Gauge in the Ypres Sector

Table 10.1 Hazebrouck - Hondschoote & Hondschoote - Bergues
Timetable from 1929 (main stops only, excludes *arrêts*)
Compagnie Générale de Voies Ferrées d'Intérêt Local **(VFIL-CGL)**

Hazebrouck (CdN)			08.55	12.25		19.45
SG lines Lille - Calais, Paris - Arras - Béthune - Dunkerque, Hazebrouck - Merville (IL), Hazebrouck - Poperinghe (Belgium)						
Hondeghem			09.07	12.37		19.57
St-Sylvestre-Cappel			09.18	12.48		20.08
Steenvoorde			09.33	13.03		20.23
Winnezeele			09.45	13.16		20.34
Herzeele			10.00	13.40		20.48
junction with line to Esquelbecq (from 1910) & St-Momelin (from 1912) (SE)						
Bambecque			10.08	13.51		20.54
Rexpoëde (1)			10.22	14.11		21.05
Killem			10.30	14.19		
Hondschoote			10.37	14.26		

	(3)	(2)				
Hondschoote		07.05		11.15	16.40	
Killem		07.13		11.23	16.48	
Rexpoëde (1)	06.20	07.24		11.35	17.01	
Warhem	06.35	07.39		11.50	17.19	
Bergues (CdN)	06.50	07.54		12.07	17.35	
SG line Hazebrouck - Dunkerque, origin of MG line to St-Momelin (from August 1914) (SE)						

	(2)	(3)	(2)		(4)	(5)
Bergues (CdN)			11.00	13.08	18.45	18.55
Warhem			11.16	13.26	19.03	19.15
Rexpoëde (1)	06.30	08.12	11.34	13.41	19.20	19.32
Killem	06.38	08.20	11.42	13.49	19.28	19.40
Hondschoote	06.45	08.27	11.49	13.56	19.35	19.47

Hondschoote	05.55		09.15		15.50
Killem	06.03		09.23		15.58
Rexpoëde (1)	06.15		09.35		16.12
Bambecque	06.27		09.47		16.24
Herzeele	06.34		10.01		16.41
junction with line to Esquelbecq (from 1910) & St-Momelin (from 1912) (SE)					
Winnezeele	06.49		10.14		17.14
Steenvoorde	07.05		10.26		17.20
St-Sylvestre-Cappel	07.19		10.39		17.38
Hondeghem	07.30		10.56		17.56
Hazebrouck (CdN)	07.41		11.03		18.05
SG lines Lille - Calais, Paris - Arras - Béthune - Dunkerque, Hazebrouck - Merville (IL), Hazebrouck - Poperinghe (Belgium)					

CdN *Compagnie du Nord*
SG   Standard Gauge
NF   *Compagnie des Chemins de fer d'intérêt local du Nord de la France*
SE   *Société générale des Chemins de fer Économiques*

(1) Junction of lines to Hazebrouck - Hondschoote and Hondschoote - Bergues
(2) Mondays only (Monday market day in Bergues)
(3) Except Mondays (Monday market day in Bergues)
(4) daily except Wednesdays and Saturdays from 1 September to 29 June
(5) daily from 30 June to 31 August

(departure times shown, except at end of journey, sometimes arrivals a few minutes earlier)

The railcar, now known as M-41, arrived there in 1955, the tractors in 1957. The tractors are still at work, preserved on the *Chemin de fer de la Baie de Somme*. However, railcar 402 was cut up for scrap.

Tournehem to St-Momelin

This line was built in 1918 for the British Army from St-Momelin to Tournehem to join the Anvin to Calais line. It linked up the metre gauge networks of the Pas-de-Calais and Nord *départements*, and, with the line from Herzeele to Watou, with Belgian Flanders. The line was being maintained by the British Army at least until 4 April 1919, but it was later dismantled. This is perhaps surprising since in 1919 the Anvin to Calais, Aire to Berck and Hazebrouck-Hondschoote-Bergues lines had all came under the management of the new joint company

Table 10.2 Hazebrouck - Hondschoote & Hondschoote - Bergues
Timetable from October 1936 (main stops only, excludes *arrêts*)
Compagnie Générale de Voies Ferrées d'Intérêt Local **(VFIL-CGL)**

Hazebrouck (CdN)	08.10	12.00	19.30
SG lines Lille - Calais, Paris - Arras - Béthune - Dunkerque, Hazebrouck - Poperinghe (Belgium)			
Hondeghem	08.17	12.07	19.39
St-Sylvestre-Cappel	08.24	12.14	19.47
Steenvoorde	08.32	12.22	19.56
Winnezeele	08.38	12.28	20.02
Herzeele	08.47	12.37	20.11
junction with line to Esquelbecq & Bollezeele (SE)			
Bambecque	08.51	12.41	20.15
Rexpoëde (1)	09.00	12.52	20.23
Killem	09.05	12.57	
Hondschoote	09.10	13.02	

	(3)	(2)			
Hondschoote	05.43	07.06		14.10	16.56
Killem	05.51	07.14		14.15	17.01
Rexpoëde (1)	06.16	07.21		14.21	17.09
Warhem	06.29	07.34		14.33	17.22
Bergues (CdN)	06.44	07.49		14.48	17.37

SG line Hazebrouck - Dunkerque, origin of MG line to St-Momelin (SE)

		(2)	(3)	(2)		(4)	(5)
Bergues (CdN)			07.40	11.00	15.22	18.45	18.55
Warhem			07.56	11.16	15.37	19.02	19.11
Rexpoëde (1)		06.20	08.12	11.34	15.52	19.20	19.29
Killem		06.26	08.17	11.41	15.57	19.27	19.36
Hondschoote		06.30	08.27	11.48	16.02	19.34	19.43

	(6)	(7)		
Hondschoote		06.37	09.46	
Killem		06.47	09.51	
Rexpoëde (1)	06.00	06.50	09.59	17.08
Bambecque	06.08	06.58	10.07	17.16
Herzeele	06.12	07.02	10.11	17.20
junction with line to Esquelbecq & Bollezeele (SE)				
Winnezeele	06.21	07.11	10.20	17.14
Steenvoorde	06.27	07.18	10.25	17.29
St-Sylvestre-Cappel	06.35	07.26	10.34	17.35
Hondeghem	06.42	07.34	10.44	17.52
Hazebrouck (CdN)	06.49	07.41	10.48	18.01

SG lines Lille - Calais, Paris - Arras - Béthune - Dunkerque, Hazebrouck - Poperinghe (Belgium)

CdN   *Compagnie du Nord*
SG   Standard Gauge
NF   *Compagnie des Chemins de fer d'intérêt local du Nord de la France*
SE   *Société générale des Chemins de fer Économiques*

(1) Junction of lines to Hazebrouck - Hondschoote and Hondschoote - Bergues
(2) Mondays only (Monday market day in Bergues); service put on for the market, liable to be advanced or delayed
(3) Except Mondays (Monday market day in Bergues)
(4) Sundays from 2 October to 14 May, except Easter Sunday
(5) daily from 15 May to 1 October, and Easter Sunday and Monday
(6) Mondays (market day in Hazebrouck)
(7) Mondays (market day in Hazebrouck) or alternative market day

(departure times shown, except at end of journey, sometimes arrivals a few minutes earlier)

VFIL, and the ex-military Tournehem to St-Momelin line would have seemed a useful link between the lines. Perhaps the problem was that the only way to link the whole network was through this intermediate line belonging to SE, and maybe they were not interested. An additional disincentive for VFIL may have been that the resulting line would have been aimed at longer distance travel and the metre gauge railways were satisfying a more leisurely and local style of transport. A further problem could have been that this ex-military line was without logic in terms of civilian services, avoiding rather than serving communities, and had no civilian infrastructure.

Since the SE lines were suffering from the failure to extend from St-Momelin into St-Omer before 1914, it is perhaps surprising that they did not at least keep the section from Ferme Bleue (St-Momelin) to Watten

Table 10.3 Hazebrouck - Hondschoote & Hondschoote - Bergues
Summary Timetable April 1952
Compagnie Générale de Voies Ferrées d'Intérêt Local (VFIL-CGL)

Hazebrouck (SNCF)			08.50	12.30	17.00	19.35
SG lines Lille - Calais, Paris - Arras - Béthune - Dunkerque, Hazebrouck - Poperinghe (Belgium)					
Steenvoorde			09.16	12.56	17.26	20.01
Herzeele			09.33	13.13	17.43	20.18
Rexpoëde (1)			09.47	13.26	17.55	20.31
Hondschoote			09.57	13.36		20.41
Hondschoote	06.18	08.26		13.52		
Rexpoëde (1)	06.30	08.37		14.04		
Bergues (SNCF)	07.00	09.06		14.33		
SG line Hazebrouck - Dunkerque						
Bergues (SNCF)		07.43		12.03	16.57	
SG line Hazebrouck - Dunkerque						
Rexpoëde (1)		08.13		12.34	17.28	
Hondschoote		08.23		12.45	17.39	
Hondschoote			10.09	14.46	17.50	
Rexpoëde (1)	06.54		10.20	14.57	18.00	18.05
Herzeele	07.07		10.33	15.10		18.18
Steenvoorde	07.24		10.50	15.27		18.35
Hazebrouck (SNCF)	07.49		11.15	15.52		19.00
SG lines Lille - Calais, Paris - Arras - Béthune - Dunkerque, Hazebrouck - Poperinghe (Belgium)						

SNCF *Société Nationale des Chemins de Fer Français*
SG   Standard Gauge

(1) Junction of lines to Hazebrouck - Hondschoote and Hondschoote - Bergues

standard gauge station, on the main Calais to St-Omer and Lille line. Possibly it was built in a hurry, and perhaps badly engineered. There was probably only a temporary trestle bridge across the Aa Canal which might have presented problems. However, on a map of 1929 a standard gauge branch is shown as leaving the main line and taking the route of the former metre gauge line to the Aa canal, with a bridge across the canal, so by that time the bridge had become more permanent. On the other side is just a stub, turning north to the tile factory. By 1934, when detailed plans were produced to widen the Aa canal from St-Momelin to Watten, there is no sign of this bridge or branch.

Herzeele to St-Momelin, and Bergues to Bollezeele

These lines continued to be operated by the SE, who also ran the Somme network. Between the wars this part of SE suffered greatly from not completing the line from St-Momelin to St-Omer.

Stations and track layouts
The layout in 1948 of the station shared with the SNCF at Bergues is shown in Figure 10.2.

Rolling stock
After the war the lines acquired an additional locomotive, 2-6-0T No 3.857 constructed by Haine-St-Pierre, which came from the Somme network in 1921. The reduced service after the war allowed the sale in 1936 of two mixed first and second class (ABf type) passenger carriages, and four all second class (Bf) passenger carriages to the Paris-Orléans-Midi company for the Cerdagne line. After two renovations by SNCF these are still in service on the '*petit train jaune*' of Cerdagne.

In 1932, eight '*truck porteurs*' were delivered and numbered 1 to 8. These trucks allowed the carriage of standard gauge wagons on metre gauge ones, and this was the model used by the *Chemins de fer de la Banlieu de Reims* and the Paris-Lyons-Marseille Company line from St-Gervais-Le Fayet to Chedde. They were put into service on these lines in 1934 and used up to the end of the Second World War to transport standard gauge wagons between Esquelbecq and the *huilerie* at St-Momelin.

Operations
After the armistice, each line, Herzeele to Bollezeele and Bergues to St-Momelin, ran three trains each way per day plus two partials, Herzeele to Esquelbecq and Bollezeele to St-Momelin. The timetable from the 4 October 1936 is shown in Table 10.4. By this time most of the passenger service was provided by buses, and these are not shown in Table 10.4. There was no passenger train service between Bollezeele

The metre gauge railways of French Flanders and related lines and tramways 1919 to 1954

Table 10.4 Bergues - St-Momelin & Herzeele - Bollezeele
Summary timetable from October 1936
Société générale des Chemins de fer Economiques (SE)

	(1)	(2)			(3)	(4)	(5)
Bergues (CdN)	08.39	10.15					
SG line Hazebrouck - Dunkerque, origin of MG line to Rexpoëde, Hondschoote & Hazebrouck (CF)			St-Omer		bus service only		
			St-Momelin		bus service only		
Bollezeele	09.41	11.21	Bollezeele		11.50		
junction with line to Herzeele & St-Momelin			Junction with line to Bergues				
			Esquelbecq CdN	arr	12.18		
				dep	15.20		
			SG line Hazebrouck - Dunkerque				
			Wormhoudt		15.36		
			Herzeele		15.53		
			junction with line Hazebrouck - Rexpoêde, Hondschoote & Bergues (CF)				
			Houtkerque		bus service only		
			Houtkerque			bus service only	
			Herzeele			17.03	17.55
			junction with line Hazebrouck - Rexpoêde, Hondschoote & Bergues (CF)				
			Wormhoudt			not stopping	not stopping
			Esquelbecq CdN	arr		17.40	18.30
				dep		18.53	18.53
			SG line Hazebrouck - Dunkerque				
	(3)		Bollezeele			19.20	19.20
Bollezeele	06.45		Junction with line to Bergues				
junction with line to Herzeele & St-Momelin			St-Momelin			bus service only	
			St-Omer			bus service only	
Bergues (CdN)	07.49						
SG line Hazebrouck - Dunkerque, origin of MG line to Rexpoëde, Hondschoote & Hazebrouck (CF)							

CdN *Compagnie du Nord*
SG   Standard Gauge
CF   *Compagnie des Chemins de fer des Flandres*

(1) Except Sundays, public holidays, & Mondays
(2) Mondays only (market day in Bergues)
(3) Except Sundays and public holidays
(4) Except Sundays and public holidays, and only from 1 October to 31 December
(5) Except Sundays and public holidays, and except from 1 October to 31 December

and St-Momelin, but the bus service did extend the service from St-Momelin to St-Omer *Compagnie du Nord* station, and from Herzeele to Houtkerque. The main bus route was St-Omer, St-Momelin, Bollezeele, Esquelbecq, Wormhoudt, Herzeele, finishing at Houtkerque near the Belgian border. Between Houtkerque and St-Omer, one bus service ran all the way every day. This left Houtkerque at 7.27am, arriving in St-Omer at 9.24am. There was a wait of 32 minutes at Esquelbecq for standard gauge connections. In the evenings the return bus left St-Omer at 5.43pm and reached Houtkerque at 7.23pm, with 16 minutes at Esquelbecq. The rest of the services were partial and not daily. This partly depended on market days, which were Saturday in St-Omer and Wednesday in Wormhoudt. However, one passenger train service ran daily each way between Bergues and Bollezeele, and between Bollezeele and Herzeele, except Sundays and public holidays. There were buses in addition between Bollezeele and Bergues, three times each way per day, but the times varied with market days, Monday in Bergues, and on Sundays and public holidays. Freight services continued between Bollezeele and St-Momelin.

In 1947 the train service was very similar, but with a less frequent bus service. This is shown in Table 10.5, which does include the bus times. There were no services on Sundays and public holidays. On other days one train ran from Bollezeele to Bergues and back in the early evening. This was supplemented daily by one morning bus service starting from Bollezeele and returning at lunchtime. The only exception to this was on Mondays, market day in Bergues, when there was a second train, starting from Bollezeele in the morning and returning there after lunch. A daily train service was provided from Bollezeele to Esqulbecq and back, with another

Narrow Gauge in the Ypres Sector

Table 10.5 Bergues - Bollezeele & Herzeele - St-Momelin (St-Omer)
Summary timetable from October 1947 - bus times in italics
Société générale des Chemins de fer Economiques (SE)

	(1)	(2)	(3)			(3)	(2)	(1)	(4)	(5)
Bergues (CdN)	*11.30*	13.05	18.45							
SG line Hazebrouck - Dunkerque, origin of MG line to Rexpoëde, Hondschoote & Hazebrouck (CF)				St-Omer*		16.20				
				St-Momelin**		16.35				
Bollezeele	*12.30*	14.07	19.45	Bollezeele		06.30	*17.10*			
junction with line to Herzeele & St-Momelin				Junction with line to Bergues						
				Esquelbecq CdN	arr	06.52				
					dep		*07.25*	07.25	*18.50*	20.14
				SG line Hazebrouck - Dunkerque						
				Wormhoudt			*07.35*	07.42	*19.00*	20.20
				Herzeele			*07.46*	07.59	*19.11*	20.31
				junction with line Hazebrouck - Hondschoote & Bergues (CF)						
				Houtkerque*					*19.21*	20.41

						(3)	(1)	(2)	(2)	(1)	(3)	(5)
				Houtkerque*		06.15						
				Herzeele		06.27			08.25	08.25		
				junction with line Hazebrouck - Rexpoëde, Hondschoote & Bergues (CF)								
				Wormhoudt		06.39			08.37	08.38		*19.38*
				Esquelbecq CdN	arr	06.51						*19.50*
					dep		*07.20*	07.20	*09.45*	09.45		
	(1)	(2)	(3)	SG line Hazebrouck - Dunkerque								
Bollezeele	*07.45*	08.00	17.10	Bollezeele			07.36	07.41	*10.01*	10.06	12.45	
junction with line to Herzeele & St-Momelin				Junction with line to Bergues								
				St-Momelin**							13.10	
Bergues (CdN)	*08.45*	09.08	18.14	St-Omer*							13.25	
SG line Hazebrouck - Dunkerque, origin of MG line to Rexpoëde, Hondschoote & Hazebrouck (CF)												

*   Bus service only
** For passengers, bus service only
CdN *Compagnie du Nord*
SG Standard Gauge
CF *Compagnie des Chemins de fer des Flandres*

(1) Except Sundays, public holidays, & Mondays
(2) Mondays only (market day in Bergues)
(3) Except Sundays and public holidays
(4) Except Sundays and public holidays between St-Omer and Wormhoudt; except Wednesdays and Saturdays when not public holidays between Wormhoudt and Houtkerque
(5) Only on Wednesdays and Saturdays when not public holidays

from Bollzeele to Herzeele, except on Mondays. The bus service from Bollezeele to St-Omer was provided once daily in each direction, except on Sundays and public holidays. The bus started from Bollezeele at lunchtime. The return from St-Omer in late afternoon ran through to Wormhoudt daily, except on Sundays and public holidays. Again the bus had a long lay-over at Esquelbecq for standard gauge connections. The times from Esquelbecq to Houtkerque depended on the day of the week. Goods services between Bollezeele and St-Momelin continued. There were probably other bus services not regarded as replacing former railway services.

Closure

Passenger traffic was abandoned on 1 August 1950. A goods service remained for the sugar beet season up to the 31 December 1951. After the closure locomotive No 3.661, a 2-6-0T Blanc-Misseron of 1909, was sent to the Somme network. We do not know about the disposal of other assets.

Hondschoote to Ghyvelde-Bray-Dunes and Bray-Dunes-Plage

For a list of the stations see Table 2.4. The line from Ghyvelde-Bray-Dunes standard gauge station to Bray-Dunes-Plage was probably taken up during the war. The wartime link from Pont-aux-Cerfs to Houthem in Belgium was dismantled soon after the war, when the crossing of the *Nord* line at Ghyvelde-Bray-Dunes was removed. The concessionaire M. Michon had not by this time handed over the line to the NF, the Company formed to run it. M. Michon was not in a position to restart operations, so this was entrusted to the SE,

Table 10.6 Bray-Dunes - Hondschoote
Timetable June 1921
Société Générale des Chemins de Fer Économiques (SE)

	(1)	(2)	
Bray-Dunes (Plage)	(no service)		
Ghyvelde (Nord)	06.08	07.00	12.10
SG line Dunkerque - Dixmude (Belgium)			
Ghyvelde (Ville) (3)	06.16	07.08	12.18
Les Moëres	06.38	07.34	12.45
Hondschoote	06.52	07.48	12.59
origin of MG line to Rexpoëde, Hazebrouck and Bergues (CF)			
	(1)	(2)	
Hondschoote	07.04	11.04	15.00
origin of MG line to Rexpoëde, Hazebrouck and Bergues (CF)			
Les Moëres	07.20	11.18	15.17
Ghyvelde (Ville) (3)	07.42	11.41	15.39
Ghyvelde (Nord)	07.49	11.46	15.46
SG line Dunkerque - Dixmude (Belgium)			
Bray-Dunes (Plage)	(no service)		

SG   Standard Gauge
CF   *Compagnie des Chemins de fer des Flandres*

(1) Except Fridays
(2) Fridays only
(3) On Sundays service provided by the *chef du train*

(Friday market day in Hondschoote)

who already operated the Herzeele to St-Momelin and Bergues to Bollezeele lines.

Services restarted on the 4 November 1919 but only between Hondschoote and Ghyvelde-Bray-Dunes standard gauge station. From 1919 to 1921, SE ran two trains daily in each direction. The timetable for June 1921 is shown in Table 10.6. The passenger service was suspended on 1 September 1921. The service to the *cartonnerie* [cardboard factory] at Pont-aux-Cerfs continued until 1929. The line was then finally handed over officially to NF, a necessary formality to allow the *déclassement* in 1936.

Armentières to Halluin

This line continued to be operated by the CEN, part of the conglomeration run by Baron Empain. It was an isolated part of the large group of lines centred on Valaenciennes. Although electrification was a priority for CEN lines, Armentières to Halluin remained worked by steam up to closure.

In 1928 the line was entrusted to *La Société Générale des Transports Départementaux*. At that time three trains ran daily the whole length of the line. In addition, there were six shuttles to Houplines, four to Frelinghien, one to Deûlemont, and four from Comines to Halluin. After 1928 SGTD introduced a mixed service, preponderantly buses, but retaining one train service from Armentières to Halluin and one from Comines to Halluin. These remained until closure in 1935. Before 1914, this tramway had twenty-one carriages. Two more were delivered in the 1920s to replace those which disappeared in the war. After 1918 the tramway needed eleven new covered wagons, thirty-seven open wagons, and eight flat wagons, to replace those damaged by the war. The line closed in 1935 and the whole service was replaced by buses.

Cassel Tramway

This tramway had continued to operate during the war. However, the service was suspended several times afterwards because of lack of raw materials. In 1922 the production of town gas, another function of the business, was separated from operating the tramway, but the financial equilibrium was never regained. By the beginning of the 1930s there were sixteen passenger trams each way per day, ensuring the connection with all the *Nord* trains. Ultimately the town of Cassel decided to buy back the concession and to replace the tramway with a bus service. The tramway closed on the 30 November 1934.

Tramways of Dunkerque

During heavy bombardments, the tram network suffered serious damage, particularly to the overhead lines. This led to the interruption of services on some lines. However, the network continued to function on almost all its lines. In 1923 the short *ligne des Darses* which ran west to the dock area was closed, with a replacement bus service. In July 1934 the seasonal line from Malo-Terminus (Line 3) closed. In February 1935 Line D to the Basse-Ville and Coudekerque-Branche closed, and in August 1937 the line to

St-Pol-sur-Mer (line E). Therefore by 1938 only the following lines remained:

Line A from the station to Malo-Kursaal
Line B and No 1 from rue de l'Église to Rosendaël Gare
Line No 2 from the Place de la Mairie de Rosendaël to the Place des Kursaal at Malo-les-Bains

In May 1940, the tramways were much damaged during the Battle of Dunkerque and the associated evacuation of British and French troops from the nearby beaches. The depot in the Avenue de la République was burned down. Services came to a complete stop. From 9 January 1941 Line A was re-opened from the station to the Place Turenne (Malo-les-Bains) on the orders of the occupying Germans. This was achieved by recovery of rails from recently closed lines, and rough and ready repairs to the overhead lines and the rolling stock. The rest of the rails and overhead pylons were used for the construction of the Atlantic Wall as beach anti-tank traps known as *pieux Rommel*.

In 1945, the town remained in German occupation until the capitulation on 8 May, and subject to Allied bombing. The tram network was again totally destroyed. In 1946 a minimal bus service was established using three buses which had not been requisitioned. Only line A from the station to Malo-les-Bains was roughly repaired and put back into service from 1 March 1947. It ran until 3 November 1952. After that the service was provided entirely by buses.

Rolling Stock

In 1923, 20 years after its introduction, the rolling stock was renovated, with the removal of the small ventilation windows, and reconstruction of the end platforms with side walls, thereby making vestibules. This assured best comfort for the passenger and a good visibility for the driver.

Steenwerck to Ypres

The section from the Belgian border at Le Seau (De Seule) to Steenwerck station on the main standard gauge line from Armentières to Hazebrouck reopened on 14 January 1923. The rest of the line in Belgium had reopened progressively from September 1919 to reach the border in December 1922. From 1919 the line was operated by SNCV. The section in France closed to passengers and goods on 20 May 1932. The section in Belgium from Le Seau to Ypres remained open for passengers until 25 April 1949 and for goods until 20 December 1951 (see Chapter Eleven).

Chapter Eleven

Metre gauge tramways of Belgian West Flanders 1919 to 2022

The Belgian metre gauge lines fared only slightly better than those in France. By 1955, all the steam operated rural lines had closed. Only parts of the electrified coastal lines remained which still operate as the *Kusttram* (Coastal Tramway). Parts of the electrified line 85 west from Courtrai survived until 1963. Most of the tramways in West Flanders were operated by SNCV, the *Société Nationale des Chemins de Fer Vicinaux*, NMVB, the *Nationale Maatschappij van Buurtspoorwegen* in Dutch.

La Panne–Adinkerke (60cm Gauge)
This line was commonly known as the 'Adele', the name of a locomotive that ran during the 1920s. It was taken over by the Allies in the war. The line reopened soon after the war with a WD Simplex tractor. This wore out and was replaced in 1920 with 2 0-4-0T steam locomotives, *Adele* and *Laura*. We do not know their origin. These wore out in 1928 and were replaced with a Fordson engined tractor, said to have been built on one of the old locomotive frames. A photograph shows the tractor looking like a garden hut. The trailers were re-bodied without clerestories and with open sides. They were entered by end platforms. On 4 October 1931, there was agreement with SNCV to replace this tramway with a metre gauge electric line, and it closed in August 1932. The SNCV line 201 opened on 25 June 1932. It was electrified from its opening.

Metre Gauge
From 1918 to 1939
Military metre gauge lines
Allied
The link lines from the Belgian to the French systems were not as far as we know used after the war and were dismantled, at least in Belgium (see Chapter Ten). The military internal link lines, from Proven (line 115) to Crombeke (line 107), and on line 29 north from Linde, were also not used, and were quickly dismantled. The military branches to Heidebeek and Swiss Cottage standard gauge yards, both from line 107, were also probably not used except for clearing up and salvage and were dismantled. An additional line linking line 115 round to the west of Furnes, and some branches on line 115 between Furnes and La Panne, also disappeared quickly.

German
The proposed branch of line 137 from Couckelaere to Dixmude, via Bovekerke and Vladsloo, had not been built by 1914. It is reported that during the conflict, the Germans built a line from Couckelaere to the Vladsloo area. If so, the line was dismantled after the war and not brought into civilian use.

SNCV (Belgian Vicinal) lines
The non-electric network peaked in 1925 at 3,938km (2,447 miles). However, soon parts of this network started to close as the usage of buses, lorries and electric trams increased. SNCV had started to use owned or hired buses in 1924. That year the electric tramway network ran to 523km (325 miles). By 1939 SNCV operated 161 bus lines amounting to 279km (173 miles).

Rails
After the war, the development of heavier motor traffic and trains led to a need for higher trackwork standards. Off road sections were laid with 32kg/m rails in 18m lengths. Rails in the roadway increased to 48kg/m and to 51kg/m if the curve was less than 50m radius. It had become more difficult to relay track with the new tarmac surfaces. Electrified street lines had now changed to welded rail. On other lines, long-welding was employed for some rails.

Steam locomotives
The First World War greatly reduced SNCV motive power units. There are few details about German use of SNCV steam locomotives, but we do know that they were used on many existing lines on their side of the battle zone. For example, at Verdun in 1916 they were used on the SE *Réseau de la Woeuvre* from Montmédy southwards. SE had removed all the stock, but the Germans restocked, partly with SNCV materiel. At the end of the war SNCV reclaimed 14 locomotives, 53 coaches and over 700 goods vehicles from that area. Various items were also reclaimed from the Somme

and Arras fronts. In addition, new Type 18 locomotives were manufactured for SNCV up to 1920.

At the end of the war, SNCV were able to obtain forty-eight locomotives from the British War Department. Of these, twenty-eight had been built by Robert Stephenson and twenty by Hawthorn Leslie in 1916 and 1917 (see Table 6.4). These had been based on the SNCV Type 18 design, and became their Type 19 with SNCV numbers 950 to 997. They operated until the early 1950s. Of the original fifty locomotives, only two of those produced by Robert Stephenson were not acquired by SNCV. There is a photograph of one of these, destroyed at Lestrem in 1918. We assume it had been working on the Bethune-Estaire line. We do not know the fate of the other.

SNCV also acquired some American locomotives after the war. These had been built by the American Locomotive Company (Alco) for the British and worked for the Belgian Army north of Ypres from November 1915. They were heavy locomotives of 26.5 tons. After the war they became SNCV Type 21 Numbers 1001–1020. However, they were out of gauge for tramway work, and unreliable, and thirteen were soon sold to industrial owners. The remainder were used between Antwerp and Turnhout. The last were working hauling sand trains close to the Dutch frontier in 1954. In 1925, 800 steam locomotives were in use on the network.

Passenger carriages
These were four-wheel two axle carriages. These were in use from 1912 and eventually 453 were built, 215 of these after the war. Initially they had wooden bodies and Charpentier underfloor hot air heating. Paraffin lamps supplied lighting and the carriages had open end balconies protected by glazed screens. The only change to the final 215 was that they had metal panelling over a wood framework. The last delivery of two-axle carriages for the steam service (1920-21) comprised three types; second class, mixed first and second class, and some with baggage space.

Goods wagons
The number of these increased after the War, from a total of 7,766 in 1912 to 12,617 by about 1925.

Autorails
Faced with the threat of the rise of bus use, and the deficit facing many small lines, in 1925 SNCV constructed light railcars. The first were transformed buses. The first autorails were put into service in Luxembourg Province. In 1930, SNCV began to investigate the possibility of railcars that would perform much better. From 1933 to 1936, 280 vehicles were constructed, some with petrol and some with diesel engines. Putting these into service brought better comfort, and they were also much faster than steam.

Electric Trams
The earlier style of electric tram had been in use from 1907. These had a panelled body with a short clerestory over the passenger saloon. They had six unequal, usually arched, windows. The end platforms were partially enclosed with small side panels with three-window bow end screens and teak panels below. Another style was introduced just before 1914. These had five unequal windows to allow for division into two unequal saloons for first and second class. These were in use from 1914 to the mid-1920s. From the 1920s most new cars were bogie vehicles designed to tow trailers, especially as intercity routes were being electrified. A few 'Bogota' trams, 9623–9628, appeared on the coast lines. They were constructed between 1914 and 1918. This little series was intended for the network of the capital of Colombia, with short bogies. The 'Titanic' series, 9499–9510, had been evaluated in 1912 but only delivered in 1921. Towards 1925 SNCV ordered a series of much heavier power cars with 2 axles. This type appeared on all the electrified networks.

In 1930, SNCV equipped itself with large modern bogie power cars of 'Standard' type. By 1933, 152 with wooden bodies were in service or production, including 35 for the coast network. Electrification of the networks was by now far advanced. They were a great success, smart and technically efficient. With modification they could reach 65kph (40mph) on the level with a trailer of the same size. By 1935-36 wooden bodies had been succeeded by metal, riveted or welded. Eventually more than 400 'Standard' trams were produced. Some were still running in 1957 on the coast lines. They were replaced by modern trams of Type SO, which were refurbished 'Standards' with reuse of the bogies and electrical equipment.

Lines in the Ypres Sector
Details of lines in the Ypres Sector from the end of the war are shown in Table 11.1 Many lines re-opened in 1919 and almost all those open prior to 1914 in the Ypres Sector were open again by 1922. The exception was the part of line 75 (Ypres to Steenwerck) which opened from the border to Steenwerck station in France in January 1923.

Lines in the Ypres Sector 1918 - 1939
Littoral Group
Although some of the lines west of Nieuport had been in military use during the war, they were close to the front line and suffered much damage, and extensive repairs were required. Line 193 Coxyde–La Panne

had to be completely reconstructed. This group saw most development in the Ypres Sector between the wars, since tourism to the coastal resorts took off. Another factor in their success was that they were the only electrified lines in the Sector, except those near Courtrai.

Line 2 Ostende to Furnes

This was re-opened during 1919 from Ostende to Furnes via Nieuport. The line was electrified from Ostende to Mariakerke in two stages in 1923 and 1932, but never electrified through to Nieuport. With the electrification of line 142 along the coast, and its extension to rejoin line 2 at Lombartzyde in July 1927, line 2 from Mariakerke to Lombartzyde village was almost redundant. Nieuport-Ville to Groenen-Dijk-Bains via Nieuport-Bains re-opened September 1922. Groenen Dijk Bains to Groenen Dijk did not re-open. In September 1926 the line was extended west for 4.4km along the coast from Groenen-Dijk-Bains to Coxyde-Bains, where it linked up with line 193 from Coxyde to La Panne. This re-opened in 1920, completing the metre gauge line along the coast. During the 1930s, the original part of line 2 which had not been electrified, running slightly inland through the old communities from Mariakerke to Lombartzyde, became largely a goods line, with passenger trains running only three times a week on market days.

Line 142 Ostende to Westende-Bains and Lombartzyde village

This line ran from Ostende to Westende Bains along the coast before 1914 and was electrified from the time of opening. This re-opened during 1921. However, as before the war the through link from Ostende to Nieuport was line 2 which ran slightly inland. Line 142 was extended from Westende-Bains to Lombartzyde in July 1927, electrified from the start.

Line 193 Coxyde to La Panne

This line re-opened in October 1920.

Consolidation and development 1927–39

The lines continued to be run by the pre-war concessionaires until 1927. Then they were taken over by a new operating company, the *Société pour l'Exploitation des Lignes Vicinales d'Ostende et des Plages Belges* (SELVOP). This company was formed in Brussels in February 1927. It was wholly owned by the 'Electrorail' group of Baron Empain. From 9 March 1927, this company took over all the coast lines.

Electrification

A programme of further electrification followed the takeover. With the electrification of line 2 by July 1928, from Lombartzyde to Nieuport Ville, and on along the coast to Coxyde Bains and La Panne, in July 1929, the coast electrification west of Ostende was completed. Line 193 from Coxyde-Bains to Furnes was also electrified in July 1929. A branch inland of 2km from Oost-Dunkerke-Bains to Oost-Dunkerke village, where it linked with line 2 from Nieuport to Furnes, was opened with electrification in June 1930, with electrification from Oost-Dunkerke village to Coxyde village. Thus, the main route from Nieuport to Furnes became the electrified line along the coast. The inland section of line 2 directly from Nieuport-Ville to Oost-Dunkerke village was never electrified and became little used. The line between Westende and Nieuport and the whole line along the coast from Nieuport to la Panne was doubled. On 16 September 1931, through trains running all the way along the coast from Knokke to La Panne were introduced. Under SELVOP the line numbering was rationalised from 1 January 1930. Line 2 now incorporated lines 142, and 193 from Coxyde Bains to la Panne. This possibly also included the inland line to Coxyde.

Line 201 Adinkerke to La Panne

This line was opened, with electrification, in June 1932. It took over the route from the 60cm gauge line from Adinkerke to La Panne (the 'Adele'), but only from the crossing of that line with the existing metre gauge line 115 from La Panne to Furnes. It ran inland from La Panne to Adinkerke station, linking the west end of the coast lines to the standard gauge line from Dunkerque to Furnes. Although originally given the number 201, this line also was annexed to line 2 in 1935.

Depot and workshops

The main depot and workshops were in Ostende. They repaired their own rolling stock and also carried out work for other parts of the 'Electrorail' organisation.

Dixmude Group

After the war, ODI accepted only a short term lease of the re-opened lines, and from 1 January 1921 SNCV took over control.

Line 29 Ypres–Furnes

This was re-opened from Furnes to Alveringhem very quickly, on 19 December 1918. This was probably because it had been in military use and therefore maintained. The remainder of the line to Ypres restarted in March and April 1919.

Line 75 Ypres to Steenwerck (France) and Warneton

This line re-opened from Ypres to the French border between 1919 and 1922. It reached Steenwerck station

Table 11.1 Metre Gauge Tramways in Belgium in the Ypres Sector 1919 to 2022

Line no.	(Capital) From	to	length (total) km	length (section) km	section	before WW1 opened	after WW1 opened/ reopened	electrified	closed passengers	freight
Littoral Group										
2	Ostende	Furnes	29.3	2.2	Ostende - Mariakerke Ruslandstraat	13.7.1885	24.2.1919	24.7.1923	13.3.1955	1.2.1952
				0.7	Mariakerke Ruslandstraat - Mariakerke	"	"	"	"	"
				5.4	Mariakerke - Middelkerke	"	8.3.1919	9.11.1932	18.11.1944	18.11.1944
				2.2	Middelkerke - Krokodil	15.7.1885	10.5.1919		"	"
				2.4	Krokodil - Westende	"	28.5.1919		"	"
				1.5	Westende - Lombartzyde	"	21.6.1919		"	"
				1.6	Lombartzyde - Nieuport Palingbrug	"	26.7.1919	1.7.1928	17.5.2019	1.2.1952
				2.1	Nieuport Palingbrug - Nieuport Ville	"	23.8.1919		(1)	
		(Nieuport-Ville - Furnes)	11.2	5	Nieuport Ville - Oost-Dunkerke	22.7.1886	"		16.11.1940	16.11.1940
				3	Oost-Dunkerke - Coxyde (dorp)	"	"	8.6.1930	9.1944	9.1944
				3	Coxyde (dorp) - Furnes	"	"	16.7.1929	12.5.1941	12.5.1941
142	Ostende	Lombartzyde village (4)	19	4	Ostende - Mariakerke digue	19.7.1897	24.3.1921	19.7.1897	(1)	1.2.1952
				6	Mariakerke digue - Middelkerke Bains	31.7.1897	"	31.7.1897	(1)	
				5	Middelkerke Bains - Westende Bains	29.6.1903	8.9.1921	29.6.1903	(1)	
				3	Westende Bains - Lombartzyde Bains	1.7.1927		1.7.1927	(1)	
				1	Lombartzyde Bains - Lombartzyde	1.7.1927		1.7.1927	(1)	
				2.5	Lombardsijde bad - Nieuwpoort Palingbrug	17.5.2019		17.5.2019	17.5.2019	
2	Nieuport Ville	Groenen Dijk	3.6	1.1	Nieuport Ville - Nieuport Bains	5.1889	19.9.1922	7.7.1929	(1)	
				1.4	Nieuport Bains - Groenen Dijk Bains	31.5.1903	"	"	(1)	
				1.1	Groenen Dijk Bains - Groenen Dijk	"	(2)		(2)	(2)
2	Groenen Dijk Bains	Coxyde Bains		4.4			11.7.1926	7.7.1929	(1)	1.2.1952
2	Oost-Dunkerke Bains	Oost-Dunkerke		2.0			8.6.1930	8.6.1930	9.1944	9.1944
193	Coxyde	Coxyde Bains	7.5	3.0 (3)		1.7.1909	1.10.1920	16.7.1929	1.2.1952	1.2.1952
	Coxyde Bains	La Panne		4.5		1.7.1914		23.7.1929	(1)	
201	Adinkerke	La Panne	(5)	3.7 (3)			25.6.1932	25.6.1932	5.9.1954	5.9.1954
				4			1998	(new trace)	(1)	
Dixmude Group										
29	Furnes	Ypres	36.9	10	Furnes - Alveringhem	15.7.1889	19.12.1918		4.10.1952	30.1.1953
				14	Alveringhem - Oostvleteren	"	22.3.1919		"	"
				13	Oostvleteren - Ypres	"	24.4.1919		"	"
75	Ypres	Steenwerck (F)	18.2	5	Ypres - Voormezeele	22.12.1897	17.9.1919		25.4.1949	29.12.1951
				5	Voormezeele - Kemmel	"	15.9.1920		"	"
				4.5	Kemmel - Neuve Église	"	25.12.1920		"	"
				3	Neuve Église - le Seau (border)	1.5.1909	7.12.1922		"	"
				1.5	Le Seau - Steenwerck (France)	"	14.1.1923		"	"
75	Kemmel	Warneton	12.0	7	Kemmel - Messines	22.12.1897	1.1.1920		20.5.1932	20.5.1932
				5	Messines - Warneton	"	1.2.1922		18.5.1936	5.12.1938
115	Poperinghe	La Panne	46.2	8	Poperinghe - Watou	24.10.1905	4.12.1918		10.4.1949	12.7.1954
				7	Watou - Rousbrugge	1.7.1906	"		8.9.1949	"
				24	Rousbrugge - Furnes		"		10.8.1952	15.10.1953
				7.2	Furnes - La Panne (6)	25.7.1901		1.8.30	5.9.1954	1.2.1952

Metre gauge tramways of Belgian West Flanders 1919 to 2022

Line	From	To	km	Route	Opened				
107	Dixmude	Poperinghe	41.8	Dixmude - Oostvleteren	25.9.1906		26.6.1919	27.6.1953	20.7.1953
			19	Oostvleteren - Poperinghe	"		22.3.1919	7.9.1949	17.7.1953
	Merckem	Elverdinghe	*12*	*10*	8.3.1922			20.1.1941	
132	Ostende	Dixmude	26.1	Ostende Petit Paris - Elisabethlaan	29.6.1907		1.8.1919	24.7.1923	29.12.1951
			1	Ostend Elisabethlaan - Steene	"		"	31.3.1955	
			2	Steene - Dixmude	"		"	13.3.1955	
			23					9.11.1932	
150	Roulers	Woumen	*21*	(to 107 Ypres - Dixmude)	15.2.1911		7.9.1919	15.7.1951	29.12.1951
150	Roulers	Langemarck	*14*	Roulers - Westroosebeke	1.3.1913		15.6.1920		22.3.1943
			8	Westroosebeke - Langemarck			15.8.1920	22.3.1943	
			6						

Courtrai Group

Line	From	To	km	Route	Opened				
41	Courtrai	Wervicq	29.1	Courtrai station - Bisseghem	13.2.1893		19.3.1919	1933	14.10 or 15.11 1957
			3	Bisseghem - Moorsele	"		"		21.5.1955	29.8.1955
			3	Moorsele - Ledeghem	"		"		16.9.1949	16.9.1949
			6	Ledeghem - Dadizeele	8.12.1892		21.3.1919			
			3	Dadizeele - Gheluwe	"		1.4.1919			
			5	Gheluwe - Wervicq	13.2.1893		1.7.1919		16.12.1942	16.12.1942
			4.0	Gheluwe - Menin	8.12.1892		25.5.1919	2.10.1932	22.5.1954	22.5.1954
85	Courtrai	Menin	5	Courtrai - Mouscron Station	15.6.1902		1.7.1919	25.5.1939	25.3.1963	17.3.1958
			12	Mouscron Station - Mouscron Risquon Tout	8.8.1900		"	3.7.1932	16.10.1954	22.11.1954
			1.5	Mouscron Risquons Tout - Menin Barraken	"		"	19.6.1932	"	"
			7.5	Menin Barraken - Menin Grote Markt	"		"	10.7.1932		
			1	Mouscron - Mont-à-Leux	4.5.1906		9.3.1919	3.7.1932	29.5.1961	
			2	Mouscron La Marlière - La Planche			1932	24.12.1932	30.9.1954	17.3.1958
			1							
121	Ypres	Gheluwe	17.7	Ypres station - Porte de Menin	14.7.1905		5.10.1919		7.9.1949	7.9.1949
			1.5	Porte de Menin - Zillebeke	"		1.9.1919		"	"
			3	Zillebeke - Gheluvelt	"		14.8.1919		"	
			5	Gheluvelt - Gheluwe	"		25.5.1919		"	
			8.5							
153	Iseghem	Wevelghem	13.9	Iseghem - Gullieghem	11.4.1911		24.2.1919		6.4.1952	6.4.1952
			9.5	Gullieghem - Wevelghem	11.4.1911		14.4.1922		9.11.1937	9.11.1937
	Courtrai	Menin (direct)	4.5	Courtrai - Bisseghem (7)	13.2.1893		19.3.1919	1933	21.5.1955	29.8.1955
			10.5	Bisseghem - Menin			1933	1933	14.10 or 15.11.1957	
			7.6							

Swevezeele Group

Line	From	To	km	Route	Opened				
33	Houglede	Thielt	32.8	Houglede - Roulers	24.12.1899		19.12.1919		12.4.1936	27.11.1936
			5	Roulers - Thielt	(not relevant)					
					(not relevant)					
137	Bruges	Dixmude	25	Bruges - Couckelaere	22.3.1910		6.12.1918		18.3.1951	29.12.1951
			7	Couckelaere - Leke (132 Ostende - Dixmude)						

Distances in italics estimated from maps and other information

(1) in use as part of coast tramway Adinkerke - Knokke-Heist
(2) did not re-open after First World War
(3) Reported length of previous horse tramway
(4) Incorporated into line (Capital) 2 1930
(5) Incorporated into line (Capital) 2 1935
(6) Taken over by Littoral Group under SELVOP from 9 March 1927
(7) using existing Line 41 Courtrai - Bisseghem

11.1 The main road to Elverdinghe in north Ypres in 1919, with the metre gauge line 29 to Furnes in the roadway. (*Authors' collection*)

in France in 1923. Line 75 closed from Le Seau on the French border to Steenwerck in 1932. The branch from Kemmel to Messines and Warneton closed to passengers in May 1936 and to freight in December 1938. The track was taken up in November 1942.

Line 115 Poperinghe to La Panne via Watou and Furnes

This line also re-opened very quickly, in its entirety on 4 December 1918, probably because it also had been in military use. The section from Furnes to La Panne was taken over by the Littoral Group (SELVOP) and electrified in August 1930.

Line 107 Dixmude to Poperinghe and Ypres

The section Poperinghe to Oostvleteren, where it linked with line 29 (Ypres to Furnes), re-opened in March 1919. This was followed by Oostvleteren to Dixmude in June. The branch from Merckem to Elverdinghe re-opened in March 1922. It is surprising that this was ever reconstructed. There was virtually no traffic by the 1930s, just one train each way per day with extra on market days.

Line 132 Dixmude to Ostende

This line was re-opened in its entirety in August 1919. In July 1923 it was electrified from the centre of Ostende to Elisabethlaan in the suburbs, and from November 1932 the electrification was extended to Steene, about 3km from Ostende centre. This effectively incorporated this part into the Ostende urban tramways. The rest remained steam hauled.

Line 150 Roulers to Dixmude & Roulers to Langemarck

The line from Roulers to Woumen, where it joined line 107 just south of Dixmude, re-opened in September 1919. The line branching off west of Roulers, to Langemarck, was re-opened in two stages in June and August 1920. The extension to Bixschoote, on line 107, planned before 1914, was never built.

Courtrai Group

Only those lines and part lines relevant to this book are mentioned here. The area is partly French speaking and partly Dutch speaking. Prior to 1914, all these tramways were steam powered. Some were electrified after the war, with the development of a Courtrai local tram network.

Line 41 Courtrai to Menin and Wervicq via Gheluwe

The section from Courtrai to Ledeghem is not relevant to this book. The section from Ledeghem to Wervicq was re-opened to Gheluwe in March and April 1919, and on to Wervicq in July 1919. The section from Gheluwe to Menin, along the main Ypres - Menin road, was re-opened in May 1919 and electrified in October 1932.

Line 85 section Courtrai to Mouscron and Menin
This line was in the area of German Occupation from 1914 to late 1918 and may have been used and maintained by them. It did not close until 1 July 1917 and was re-opened on the 1 July 1919, except for the spur from Mouscron to Mont-à-Leux on the French frontier which was already open from 9 March 1919. A further very short branch of about 1km in Mouscron to la Planche, on the French frontier, was opened by 1932. The line was electrified in 1932 from Mouscron to Menin, with the two short branches at Mouscron. It was electrified from Courtrai to Mouscron in 1939.

Line 121 Ypres to Gheluwe
Long sections of this line, which ran mostly along the Ypres to Menin road, with a deviation north from Gheluvelt to Becelaere, were destroyed through enemy action. No use was made of it during the war, and parts of the track-bed were used for Light Railways. The line re-opened from Gheluvelt to Gheluwe, the part least damaged, in May 1919. The rest followed from August to October 1919. The last part to reopen ran around the north of Ypres from the standard gauge station to the road junction a short distance outside the Menin Gate. Here the stop was called *Porte de Menin*.

Line 153 Iseghem to Wevelghem
This line re-opened from Iseghem to Gulleghem in February 1919, and on to Wevelghem in April 1922. By 1930 it was finally linked to line 152 (Iseghem to Ardoye) with a 1.20km link across the standard gauge railway and the canal. Services were now provided from Courtrai through to Ardoye. On 9 November 1937, Gulleghem to Wevelghem was closed and abandoned.

Courtrai to Menin direct
Part of the Courtrai local development was a direct route from Courtrai to Menin. From Courtrai the new line shared line 41 to Bisseghem, then went straight on to Wevelghem at the end of line 153, and then to Menin Grote Markt, where it met line 41 from Gheluwe and line 85 from Courtrai via Mouscron. The line from Bissenghem to Menin ran for 7.6km. This was authorised in July 1933, and it opened near the end of that year.

Swevezeele Group
As for the Courtrai Group, only those lines of relevance to our area are mentioned here.

Line 33 Houglede to Thielt
The section from Houglede to Roulers re-opened in December 1919. The section closed to passengers on the 12 April 1936 and to goods on the 27 November 1936.

Line 137 Bruges to Leke and Dixmude
The section from Couckelaere to Leke reopened on 6 December 1918. The rapid re-opening reflects that it was in German hands from 1914 to Autumn 1918, and that the German Army made use of it as far as Leke. At Leke it met line 132 (Ostende - Dixmude), which did not re-open until August 1919. The proposed branch from Couckelaere to Dixmude, via Bovekerke and Vladsloo, was never opened.

Second World War
Littoral Group
Some lines near the coast were closed during part of the Second World War but re-opened before or shortly after the end of hostilities. This may have been associated with the German construction of sea defences, the so-called Atlantic Wall. The closures were:

- **Line 2** Mariakerke to Lombartzyde closed from 18 November 1944.
  Nieuport-Ville to Oost-Dunkerke village closed from 16 October 1940.
  Oost-Dunkerke-Bains to Oost-Dunkerke village closed from September 1944.
  Oost-Dunkerke village to Coxyde village closed from September 1944.
  Coxyde village to Furnes closed from 12 May 1941.
  The direct line inland from Nieuport-Ville to Oost-Dunkerke village was dismantled in autumn 1940. It had never been electrified and was now little used.

- **Line 142** Middelkerke-Bains - Westende-Bains closed from 1942 to December 1944.
  The line 142 closures were because the overhead wires had to be taken down because they were in the firing line of German coastal batteries. This section was run with steam locomotives for a while but later all the steam trains had to go along the inland line from Ostende to Lombartzyde village. The line reopened on liberation in December 1944.

Dixmude Group

- **Line 107** the section from Merckem to Elverdinghe was closed permanently to passengers and freight in January 1941. The track was dismantled in May 1944.

Line 150 line 150 from Roulers to Langemarck was closed permanently to passengers and freight in February 1943.

Courtrai Group

Line 41 the section from Gheluwe to Wervicq was closed permanently to passengers and freight in December 1942.

After the Second World War
Closures

Closure dates for passengers and freight are shown in Table 11.1. Concessionary, non-SNCV, buses replaced the closed lines around Diksmude and Ypres. By the end of 1949, Ypres had lost all *Vicinal* lines except that north to Furnes, line 29, and by the end of 1955 the area of this book was without metre gauge railways except for some electric lines along the coast, and two electric lines west and south-west of Courtrai. The Courtrai lines had all gone by 1963. For the whole of Belgium there were in 1945 4,811km (2,989 miles) of electric and non-electric lines. By 1950 there were 4,236km (2,632 miles) of which 1,528km (949 miles) were electric. By 1960 the whole network was only 977km (607 miles).

The coast line

Line 201 from De Panne (la Panne) to Adinkerke closed for passengers in September 1954. This again cut the coast line off from its connection with the standard gauge service at Adinkerke. The remainder of the line along the coast, incorporating parts of the original line 142 and line 193, remained open. From 1963, this was the only *Vicinal* metre gauge tramway still open in the area of this book. In 1955, SELVOP returned the operation of the coast concessions to SNCV. New unidirectional tramcars were delivered in the late 1950s. They were type SO for the coast routes, with a wider loading gauge (2.4 metres wide versus other lines at 2.2m). Turning circles were provided at the line ends and at selected points in between.

Lines around Courtrai

Mouscron station did not close until 25 March 1963 for passengers, because roads were unsuitable for buses, but it closed on the 17 March 1958 for goods. The local service from Mouscron to Mont-à-Leux ceased on 1 October 1954 but through services from Courtrai continued until 29 May 1960 or 1961.

Rolling stock
Steam locomotives

SNCV continued to use some steam locomotives until well after the Second World War.

Autorails

In 1946, metal bodied Autorails with bogies, equipped with Deutz 180hp engines, were constructed by SNCV

11.2 SNCV Type 19 locomotive, formerly ROD manufactured 1916 or 1917, operating a passenger train in Namur (Belgium) in June 1945. (*Collection John Scott-Morgan*)

for certain lines suitable for high speeds. After the Second World War some autorails built in 1933–34 were converted into tractors in SNCV workshops and replaced steam locomotives for hauling goods trains. These ran mainly in the east of Belgium, the Province of Luxembourg, and the Meuse Region.

Electric trams
Type 'Braine-le-Comte'
10374-394 appeared in 1948 and they broke with the *Vicinal* traditions since they were unidirectional. They had a short life because of the lightness of their construction.

Type N 10
Eighty-one of these were produced between 1949 and 1958. After the closing of some Brussels lines in the 1960s about twenty became trailer cars for the Ostend group.

Type S
These were refurbished 'Standards' with reuse of the bogies and electrical equipment. Power was increased from 160 to 268hp and 200 were delivered up to 1959. Among these, twenty-eight were made for the coast lines. They were unidirectional with greater interior comfort (type SO). They were 14.3m long, 2.35m wide and 3.28m high. Their power was 144.2 kilowatts and they weighed 19.5 tonnes. They accommodated forty-two seated and thirty-three standing.

Type A
In 1977, SNCV ordered articulated power cars with three bogies, including fifty for the coast, made by PN Constructions Ferroviaire et Métallurgiques. Series 60 started with No. 6000 delivered to Ostend in June 1980. They were in service from 1 September 1980 on the coast. They were 22.8m long, 2.50m wide, and 3.26m high. Their power was 432.6 kilowatts in continuous use, 457 kilowatts intermittent use. These were unidirectional, with four doors, on the right only. Maximum speed was 75kph (47mph), and they accommodated 59 seated and 88 standing.

11.3 A two vehicle electric tram from Oostende to De Panne on the coast line (line 2) in the Route Royale at Koksijde-Bad (Coxyde-Bains). The lead car is No. 10052 of Type SO, the unidirectional type of rebuilt 'Standard' power cars introduced between 1953 and 1959, which remained in service until the 1980s. The second vehicle is almost certainly of the same type. (*Authors' Collection*)

Recent history

In 1977, the buses of Belgian railways (SNCB) were transferred to SNCV. From 1980 onwards, political federalism saw the splitting of many national institutions into separate bodies for Flanders (Dutch speaking), Wallonia (French speaking) and the Brussels-Capital Region (bilingual). In 1991 SNCV/NMVB was broken up into De Lijn (Flanders) and TEC (Wallonia). Both are now primarily operating buses. De Lijn inherited the tram systems of Ghent and Antwerp (including the pre-metro), and the coastal tramway.

The Coast tramway

In the early 1970s this was under threat; however, in 1977 SNCV decided to upgrade it and made systematic improvements. All the track from Knokke to De Panne (La Panne) was relaid with heavy rail. The yard at Oostende (Ostende) Kaai was refurbished as a running depot. Electrification was improved with heavy duty catenary and six new sub-stations. The new Type A two section unidirectional articulated trams were introduced. These were mostly converted to three section trams by 1991. CAF Urbos multi-section low floor trams were introduced in 2020–21.

Line 201 De Panne (La Panne) to Adinkerke had been closed on the 5 September 1954, but 'incidental traffic' continued to the 2 June 1956. It was reopened in a new trace in 1998 and forms part of the coastal tramway. The Adinkerke terminus is at Adinkerke standard gauge station. Just east of Nieuwpoort (Nieuport), the line from Lombardsijde-Bad to Nieuwpoort-Palingbrug via Lombardsijde-Dorp (village) was closed on 17 May 2019 and a new more direct route along the main road was opened. The coast tram (*Kusttram*) is now one of the most successful inter-urban metre gauge tramways in the world, and claimed by some to be the longest, at 67km (42 miles).

Chapter Twelve

Things to see and do now

There is less to see and do in the Ypres Sector than in the Arras and Somme Sectors of the Western Front. Much of the land is low, wet and agricultural. In this environment vestiges of former railways tend not to survive, especially of the light railways.

Belgium

In a series of changes between 1963 and 1993, Belgium reformed itself into a federal state consisting of a Dutch (Flemish) speaking region, a French speaking region, and a small German speaking region, with a bilingual capital territory (Brussels). The Province of West-Vlaanderen, capital Brugge (Bruges), which covers almost the whole of the Belgian area of this book, is in the Dutch speaking region with all place names and signage now in Dutch. Ieper (Ypres) is now classified as a city and is the administrative centre for the south-western part of the Province. The exception for this book is the small detached part of the Province of Hainaut, immediately north of the Leie (Lys) River from the frontier north of Armentières east to between Comines (Komen) and Wervik (Wervicq) - see Figure 3.1. Hainaut is French speaking. In this chapter we have used the contemporary Dutch names, with the French names (where different) in brackets, except for places in Hainaut where the Dutch names are in the brackets. For a listing of names in Dutch, French, and sometimes English, see Table 0.1 (page xiv).

France

In 1982, the *départements* of France were brought together again into 28 *régions*. The Nord and Pas-de-Calais *départements* formed the *région* Nord-Pas-de-Calais, capital Lille, *Préfecture* of the Nord *département*. From 1 January 2016 the total *régions* were reduced to 13, and this *région* was merged with Picardie to form the new *région* of Hauts de France, with the capital at Lille.

Railways

The closure of the line across the Belgian–French border between Halluin and Menin in 1939 was followed by the loss of all the other lines across the border by the early twenty-first century. The coast link from Bray-Dunes-Ghyvelde to Adinkerke (where the station is now called De Panne) was closed by 1960, but a service in France from Dunkerque to Bray-Dunes continued to 1994. Probably the line remained in use for occasional traffic, and it was reported to have been reopened from 1999 to 2003, when it finally closed. This leads to the situation now, in which there is no railway line across the border north and west of the Lille–Kortijk (Courtrai) line all the way to the North Sea. This is a length of border of more than 80km (50 miles).

The First World War

There are plenty of guides to the Ypres battlefields, and to the memorials and museum centres now available. Hooge Crater Museum is on the Meenseweg (Menin Road) at 50° 50' 47.17" N 2° 56' 36.52" E. There is a short section of twisted rusting 60cm gauge light track mounted on metal sleepers outside the museum. The Memorial Museum Passchendaele 1917 at Zonnebeke Château (50° 52' 14.50" N 2° 59' 19.16" E) has good dugout and trench reconstructions. There is no major light railway information, but a short length of very light 60cm track is on display, with a four wheel trolley. The trolley base is metal, with slots for stanchions, and we cannot identify this with any standard British Army type of wagon. However, it is known that many 'push' wagons on trench tramways were constructed in workshops near the front, using whatever materials were to hand. The restored Cloth Hall in the main square in Ieper (Ypres) houses the 'In Flanders Fields' Museum. All three are worth visiting for general information on the First World War but do not contribute to the specific railway history. Further south (in the Messines area) is the Plugstreet 14-18 experience, in Ploegsteert Wood, opened in 2013 and run by the Commune of Comines-Warneton (Komen-Waasten). Nearby is the Hyde Park Corner Cemetery and Memorial.

Graves and War Cemeteries

Railway work was never as hazardous as fighting in and from the front-line trenches. However, it was exposed to shell fire and aerial bombardment, and sometimes to machine gun fire as well. This was particularly the case for light railway and tramway work, often carried out near the front line, and even more so in the Ypres Salient, where so many fortifications, troops, artillery batteries, dumps and

Narrow Gauge in the Ypres Sector

12.1 Very light 60cm gauge track and a 'tramway' wagon on display at the Memorial Museum Passchendaele 1917 (Zonnebeke Château) in 2020. (*Authors*)

railways were packed together in a small area. Many railway troops died in action in this area, especially in 1917 and 1918. We have selected a few graves to visit, mostly in three cemeteries.

Poperinghe New Military Cemetery

This cemetery is at 50° 50' 50.65" N 2° 43' 59.28" E, on the N304 Deken de Bolaan, close to the ring road in south-east Poperinge (Poperinghe). The Commonwealth War Graves Commission (CWGC) continues to use the former version of the town name in the title of the cemetery rather than the present Dutch name.

There are many railway troops buried here, as the Cemetery Register shows. The 2CRT suffered especially building and repairing light railways of the B System in and around Ieper (Ypres). On 25 June 1917, a party had just reached the B4 line when a shell exploded immediately in their rear, killing one man outright. Two more died shortly after. The three, Sappers Shawcross, Tait and Swingler, were buried here the next day (Graves II A 25-27). On 11 August, Captain Hugh Anderson was in charge of a party relaying track on the B6 line in the ruins of Ieper (Ypres) when he was struck by a shell fragment and died soon after in the Ambulance Station in Ypres Prison Yard. He was buried here the next day (Grave II H 20). Major T.N. Elliott took a working party on 28 September to the Bedlington line near St-Jean, to repair shell breaks. At St-Jean there was heavy shelling, and they took shelter in a ditch at the side of the track. A shell fragment hit Major Elliott in the small of the back, killing him almost immediately. His body was brought back by light railway. The next day, his Company (A Company) all attended the funeral service at Battalion HQ, prior to his burial here (Grave II H 30). In all there are ten men of 2CRT, four of 9CRT, two of 5th New Zealand Light Railway Operating Company (LROC), and two of the ROD, standard gauge buried here.

Hagle Dump Cemetery

Hagle Dump was right by the path of the double track B12 line from the Westonhoek Yards, close to Triangle Junction where it met the B1 line. After April 1918 this very busy line was under observation by German troops on Mont Kemmel. The cemetery is located at 50° 51' 41.12" N 2° 46' 54.57" E, on the St-Pieterstraat. On 24 July 1918 a PE tractor of 12LROC (RE) was struck by a shell near the Culloden Locomotive Yard (B1 line) and the entire crew of three men were killed. The three (Sappers Devine, Gillett, and Coombes) were buried here (Graves II E 3-5). Also buried here are another man of 12LROC, one of 14LROC, five of the 4ATC, four of 5ATC, and one of the 2APB. The cemetery was started in April 1918 during the Battle of the Lys (Fourth Ypres), and all of these railway burials are in the period April to August 1918.

Coxyde Military Cemetery

Our third cemetery is at Koksijde (Coxyde) on the Belgian Coast, with burials from the rather forgotten period when the British Fourth Army was deployed here from June

12.2 The grave of Sapper C. Devine, of 12th Light Railway Operating Company RE, at Hagle Dump Cemetery, in February 2020. He was killed on 24 July 1918 when a shell struck his petrol electric tractor near Culloden Locomotive Yard on the B1 line. Sappers W.A. Gillett and J. Coombes, who died with him, are buried in adjacent graves. (*Authors*)

to November 1917. The cemetery is located at 51° 06' 13.30" N 2° 38' 35.65" E, on the north side of the N396. From 20 June to 8 August 1917 C and D Companies of 2nd Battalion Canadian Railway Troops (2CRT) were here while A and B Companies were at Ypres (Ieper). On 2 July a light railway working party were hit by shelling. One man was killed and another died of wounds. Corporal Rushworth was buried on 2 July (Grave I C 26) and Sapper Dillon on 3 July (Grave I C 25). On 29 July Sapper Bigras was fatally wounded by shelling during night work near Oost-Dunkerque (Grave I L 27).

There are also four men of 10CRT, who replaced 2CRT, buried here in September 1917. 4CRT are also represented here. Three Sergeants (Gilpin, Trudell and Pengelly), and one Private (McAuley) were killed by bombing at their camp at Coxyde on 14 July, and are buried at I H 6, I H 7, I G 26 and I H 5 respectively. The main operating company were the 13Can LROC. On 1 September 1917 a shell struck their dugout at Oost-Dunkerke, killing three men (Sergeant McAuley and Sappers Robertson and Fraser). The three men were buried in the afternoon at Coxyde (Graves III D 23-25). 32LROC replaced the Canadians; Sappers Doyle and Larner were killed in October (Graves IV F 14 and IV H 2).

Other cemeteries

It is worth noting that the 17(3)Aus LROC suffered high casualties operating from their base at Mimico, especially in October 1917, with eight deaths. Four men of this unit are buried at Lijssenthoek Military Cemetery, south-west of Poperinge (50° 49' 45.07.30" N 2° 41' 59.84" E). Also at Lijssenthoek are buried two men of 33LROC, who died on 25 April 1918 when a shell struck their party at Railhoek. Five others died that day, and two of wounds next day; four are buried at Nine Elms British Cemetery, west of Poperinge (50° 51' 03.20" N 2° 41' 49.35" E) and three at Haringhe (Bandaghem) Military Cemetery, at Haringe near Roesbrugge (50° 54' 05.74" N 2° 36' 53.54" E). Also buried at Nine Elms is Sapper H.C. Stone, of the 85Can ECC, killed at Poperinghe by a shell fragment on 9 April 1918, the first day of the German offensive, while the Company were operating metre gauge services from Bollezeele (Grave XIV B 20).

The Track

Most track of all gauges which was built for military use has disappeared. There is little left of the metre gauge civilian tracks.

Standard Gauge

The three main sidings of the Oakhanger (Westonhoek) RE Yard can be seen on satellite images stretching south from the N333 Elverdingseweg at 50° 51' 40" N 2° 44' 57" E, and one of the tracks curving north-east into the Edwaarthoek part of the yard can be picked out, but traces of the rest of these yards have disappeared (see Figure 7.5). In France the triangular junction for the Watten to Socx line, where it joined the Calais to St-Omer main line near St-Momelin, can still be clearly seen on satellite images at 50° 47' 52" N 2° 13' 47" E but the rest of this line has disappeared.

Belgian metre gauge tramways

A ride on the Kusttram (coastal tram), at least between Nieuwpoort (Nieuport) and Adinkerke, is recommended. The service is fast, frequent and comfortable; details and links to timings are on www.dekusttram.be. At the western terminus at Adinkerke (De Panne) station, the adjacent Plopsaland theme park is on the site of the Adinkerke Chaussée sidings and 60cm gauge transhipment, which in the summer of 1917 was the British Fourth Army workshops and Y Corps Dump.

The only other surviving SNCV/NMVB metre gauge lines are short lengths which are now part of the Charleroi Metro, and the Tramway des Grottes de Han. This latter is situated in Namur Province in the south of Belgium (50° 07' 34" N 5° 11' 16" E), and is run as a heritage line serving the caves. It is 3.3km (2 miles) long, and is operated using five diesel railcars and open-sided trailers (*balladeuses*). We do not know of any connection of the rolling stock with the area of this book.

Depots

The main depot for the coastal tramway at the west end is at the terminus at Adinkerke (De Panne) station (51° 04' 42" N 2° 36' 01" E). The former depot at De Panne (51° 05' 42" N 2° 35' 03" E) is off the N34 just south-east of the roundabout with Westhoeklaan, the modern route of the line to Adinkerke station. This is now a tram museum, the *Stelplaats Historische Trams*. Historical electric tram vehicles are on display.

The main standard gauge station at Veurne (Furnes), on the line from Adinkerke to Gent (Gand), is at 51° 04' 25" N 2° 40' 12" E. The Vicinal lines 2, 29 and 115 used the Staatiplaats (*Place de la Gare*) outside the station. East of the main building is the old goods hall. Beyond that at the east end of the square is the former SNCV/NMVB steam depot, with a large locomotive and/or carriage shed, a yard and other buildings.

Stations

The stations on the Belgian rural *Vicinal* lines are often difficult to pick out. Many have been modified, and some were rebuilt after the war. However many in the area of this book have a Dutch style mansard roof. Also of note is the general absence of goods buildings. Some stations of the SNCV/NMVB network in Belgium are shown in Table 12.1, with the metre gauge stations

12.3 The former station building at Watou, on line 115 from Poperinge (Poperinghe) to Veurne (Furnes), in February 2020. (*Authors*)

12.4 The former station building at De Seule (Le Seau) in November 2021. This was on line 75 from Ieper (Ypres) to Steenwerck, and from 1932 was the end of the line, just inside Belgium by the French frontier. (*Authors*)

Table 12.1 Location and present status of metre gauge stations in Belgian West Flanders and in French Flanders
The list of Belgian stations is not complete

Name	Type	Location	(Latitude, Longitude)	Present status & use, notes
Belgium				
Line 115 Veurne (Furnes) - Poperinge (Poperinghe)				
Veurne (Furnes)	NMBS/SNCB	in the Staatiplaats outside the station	51° 04' 25" N 2° 40' 12" E	in use for SG - old MG depot east end of square
Leisele (Leysele)		Stationsplein	50° 59' 09" N 2° 37' 22" E	station building not identified
Watou		west side of Moenaardestraat 100m north of the church.	50° 51' 36" N 2° 37' 13" E	private house
Poperinge (Poperinghe)	NMBS/SNCB	N308 outside the station	50° 51' 17" N 2° 44' 09" E	in use for SG
Line 29 Ieper (Ypres) - Veurne (Furnes)				
Veurne (Furnes)	(see Line 115 above)			
Elzendamme (Elsendamme)		in Elzendamme, corner of Stavelestraat	50° 56' 59" N 2° 43' 18" E	
Elverdinge (Elverdinghe)		south side of Boezingestraat, 100m east of the junction with Steenstraat.	50° 53' 11" N 2° 49' 09" E	private house
Brielen		on the N8 Veurnseweg	50° 52' 04" N 2° 50' 54" E	private house
Ieper (Ypres)	NMBS/SNCB	N37b, station forecourt	50° 50' 52" N 2° 52' 38" E	in use for SG
Line 75 Ieper (Ypres) - Steenwerck (F)				
Ieper (Ypres)	(see line 29 above)			
Kemmel		Kemmelstraat, N331	50° 46' 55" N 2° 49' 49" E	
De Seule (Le Seau)		N331 Seulestraat 130m north of the French border	50° 43' 18" N 2° 48' 19" E	private house
Steenwerck (La Crèche) (F)	SNCF	Rue des Ajoncs, north side of station	50° 42' 48" N 2° 47' 10" E	in use for SG Hazebrouck - Armentières
Line 75 branch Kemmel - Warneton (Waasten)				
Kemmel	(see line 75 above)			
Wijtschate (Wytschaete)		Schoolstraat	50° 47' 06" N 2° 52' 66" E	
Mesen (Messines)		Rijselstraat	50° 45' 52" N 2° 54' 01" E	
Warneton (Waasten)		Clos des Mountches	50° 45' 09" N 2° 56' 59" E	Café Tramstatie
Line 107 Diksmuide (Dixmude) - Poperinge (Poperinghe)				
Merkem (Merckem)		Stationsplein, Stationstraat	50° 57' 14" N 2° 51' 20" E	replaced with modern building 'Stationshuis'
Poperinge (Poperinghe)	(see line 115 above)			
Line 107 branch Merkem (Merckem) - Elverdinge (Elverdinghe)				
Merkem (Merckem)	(see line 1017 above)			
Bikschote (Bixschoote)		N369, by junction with Bikschotestraat	50° 55' 44" N 2° 51' 13" E	
Steenstraat		Steenstraat centre	50° 54' 56" N 2° 50' 13" E	
Zuidschote (Zuydschoote)		in Zuischtestraat	50° 54' 47" N 2° 49' 51" E	
Elverdinge (Elverdinghe)	(see line 29 above)			
Line 121 Ieper (Ypres) - Geluwe (Gheluwe)				
Ieper (Ypres)	(see line 29 above)			
Beselare (Becelaere)		N303 near village centre	50° 50' 51" N 3° 01' 27" E	
Geluwe (Gheluwe)		20m south of Iperstraat N8 270m north-west of junction with N311	50° 48' 36" N 3° 04' 19" E	also line 41, depot now a bus station (De Lijn)
France				
Hazebrouck to Hondschoote, and Rexpoëde to Bergues (CF)				
Hazebrouck to Hondschoote				
Hazebrouck	CdN/SNCF	D53 Rue de Vieux Berquin	50° 43' 30" N 2° 32' 29" E	in use by SNCF for SG
Hondeghem	standard	off D53 just north of D161 roundabout	50° 45' 15" N 2° 31' 24" E	private house
St-Sylvestre-Cappel*	standard	Rue de la Concorde, off D916	50° 46' 31" N 2° 33' 25" E	private house
Steenvoorde	standard	D18A, by the Mairie	50° 48' 34" N 2° 34' 39" E	private house
Winnezeele	standard	Rue de la Gare	50° 50' 17" N 2° 33' 00" E	private house
Herzeele	standard	Rue du Petit Train	50° 53' 08" N 2° 32' 09" E	private house

Things to see and do now

Bambecque		standard	Rue de la Gare, off the D4	50° 54' 11" N 2° 32' 39" E	private house
Rexpoëde		standard	Rue des Moeres D79	50° 56' 32" N 2° 32' 12" E	private house
Killem		standard	Rue des Acacias, off D55	50° 57' 26" N 2° 33' 32" E	private house
Hondschoote		standard	in *impasse* off D55 Rue de la Libération	50° 58' 38" N 2° 34' 43" E	private house

Rexpoëde to Bergues

Rexpoëde		(see Hazebrouck to Hondscoote)			
Warhem		standard	off Route d'Ypres D916	50° 57' 50" N 2° 29' 11" E	demolished
Bergues		CdN-SNCF	by D916 just west of old town	50° 58' 01" N 2° 25' 39" E	in use by SNCF for SG

Hondschoote - Bray-Dunes-Plage (NF)

Hondschoote		(see Hazebrouck to Hondscoote)			
Les Moëres		standard	Off the Rue St-Antoine	51° 00' 57" N 2° 33' 14" E	private house
Ghyvelde-Ville		NK, prob standard	off Rue de la Frontière	51° 03' 11" N 2° 31' 50" E	demolished
Ghyvelde-Bray-Dunes		CdN-SNCF	Place de la Gare, off Rue Roger Salengro	51° 04' 14" N 2° 31' 04" E	SG closed finally 2003, now retail premises
Bray-Dunes-Plage		Special	Boulevard Georges Pompidou, Rue de l'Ancienne Gare on east side	51° 04' 41" N 2° 30' 53" E	*Office de Tourisme* (Tourist Office)

Herzeele - St-Momelin, Bergues - Bollezeele (SE)

Herzeele - St-Momelin

Herzeele		standard (Lambert)	Rue du Petit Train	50° 53' 08" N 2° 32' 09" E	private house
Wormhoudt		standard	Rue de la Gare, off D916	50° 53' 06" N 2° 27' 54" E	private house
Esquelbecq halt			by D417 Rue de Bergues	50° 53' 22" N 2° 26' 04" E	uncertain if there was a building
Esquelbecq		CdN-SNCF	off Rue de la Gare	50° 53' 23" N 2° 24' 45" E	in use by SNCF for SG
Zeggers-Cappel*		standard	off Rue de la Poste by church	50° 53' 13" N 2° 23' 11" E	private house
Bollezeele*		standard		50° 51' 59" N 2° 19' 31" E	private house
Volkerinchove				50° 50' 07" N 2° 18' 40" E	private house
Lederzeele		standard	Rue de Bergues	50° 49' 13" N 2° 18' 11" E	no building, probably demolished
St-Momelin halt				50° 47' 50" N 2° 17' 01" E	private house, track side seen from Rue Neuve
St-Momelin*		standard	Rue de la Mairie	50° 47' 38" N 2° 15' 01" E	

Bergues – Bollezeele

Bergues		CdN/SNCF	by D916 just west of old town	50° 58' 06" N 2° 25' 32" E	in use by SNCF for SG
Bierne		standard	just off D352	50° 57' 39" N 2° 24' 16" E	private house
Steene (Gd. Mille Brugghe)		standard	off D52 south of Grand Mille Brigghe	50° 57' 33" N 2° 21' 00" E	private house
Pitgam Halt		demolished		50° 56' 15" N 2° 19' 12" E	
Pitgam		standard	Rue de la Gare	50° 55' 39" N 2° 19' 36" E	*Maison des Associations* (communal premises)
Drincham*		standard	on D11	50° 54' 25" N 2° 15' 24" E	private house
Bollezeele		(see Herzeele - St-Momelin)			

Armentières - Halluin (CEN)

Armentières		CdN/SNCF	Boulevard Faidherbe	50° 40' 51" N 2° 52' 39" E	In use by SNCF for SG
Houplines		Cdn/SNCF	just north of Rue Victor Hugo	50° 41' 20" N 2° 54' 11" E	SG closed 1988, building as reconstructed 1922
Frelinghien		station 1a	Rue du Pont Rouge	50° 42' 57" N 2° 56' 04" E	private house
Deûlémont		station 1a	Rue de Verdun	50° 44' 02" N 2° 56' 53" E	private house
Comines (France)		CdN/SNCF	Avenue de Versailles	50° 45' 39" N 3° 00' 35" E	SG closed 14 December 2019
Wervicq-Sud (France)		station 1a	Rue de l'Abbé Bonpain	50° 46' 19" N 3° 02' 28" E	demolished, date unknown, now a public garden
Bousbecque		station 1a	Rue de Wervicq	50° 46' 21" N 3° 04' 30" E	demolished 9 June 1937, now an open area
Halluin		CdN/SNCF	Rue Marthe Nollet	50° 46' 52" N 3° 07' 11" E	SG closed 1971 (passengers) 1990s (goods and line)

NF	*Compagnie des Chemins de fer d'Intérêt local du Nord de la France*
CF	*Compagnie des Chemins de fer des Flandres*
SE	*Société générale des Chemins de fer Economiques*
CEN	*Société des Chemins de fer Économiques du Nord*
CdN	*Compagnie du Nord*, standard gauge stations, now SNCF if still in use
NMBS	*Nationale Maatschappij der Belgische Spoorwegen* (in French SNCB)
SNCB	*Société Nationale des Chemins de Fer Belges* (in Dutch NMBS)
SNCF	*Société Nationale des Chemins de Fer Français*
*	Station or Halt has name still visible on building (not noted for CdN/SNCF stations)
SG	standard gauge
MG	metre gauge

of French Flanders. De Seule (Le Seau) was on the frontier, at the end of line 75, from 1932, when the extension to Steenwerck Station in France was closed.

Track

Little evidence remains of the track in most places. Sometimes the old metre gauge path was separated from the road and has survived as a separated cycle track. This can, for instance, be seen on the N308 between Roesbrugge (Rousbrugge) and Proven, with long sections of cycle track on the south-west side of the road, and some of the original brick culverts can be seen. In some places the metre gauge tramway had its own path. A good example in the area of this book is at Nieuwkerke (Neuve-Église), where the tramway ran north of the village, while climbing the hill. This is now a footpath, running from 50° 45' 05" N' 2° 49' 36" E at 44m to 50° 44' 99" N' 2° 44' 19" E at 67m.

Metre gauge lines in France

All of the metre gauge lines in French Flanders and in the Pas-de-Calais *département* had closed by 1954.

Track

There is a short section of track left at Bergues, where the former line from Rexpoëde crossed a ditch on a small *tablier métallique* bridge with brick abutments, at 50° 57' 50" N 2° 26' 06" E.

Tournehem to St-Momelin

This was a major metre gauge development of 1918, associated with the British Army, and there is still something to see. At Watten-Éperlecques station (50° 49' 28' N' 2° 12' 33" E) there is a derelict area on the east side leading down beside the standard gauge main line, which was at least in part metre gauge sidings and transhipment sidings. There is a concrete shelter by the

12.5 Path of the former *Vicinal* line 75, just after it has left the N331 road from Kemmel, to climb out of the Douve Valley away from the road to Nieuwkerke (Neuve-Église) village on the hill in the distance. November 2021. (*Authors*)

12.6 Track of the line from Rexpoëde to Bergues, south-east of the Porte de Cassel at Bergues, crossing a ditch on a small *tablier métallique* bridge with brick abutments. July 2022. (*Authors*)

12.7 Bridge for the 1918 Tournehem - St-Momelin military metre gauge line, near the Aa Canal just south of Watten, in June 2006. (*Authors*)

station and an individual shelter in the wood at the far end of this area, both from the Second World War.

The most interesting surviving artefact is the bridge, probably built for the line, over a wide drainage ditch between Watten station and the Aa canal (50° 49' 17.30" N 2° 12' 45.46" E). This is up a little side road on the D213 nearly half a mile south of its junction with the D207 at Watten. The side road is the path of the railway and 150 yards up it is barred and private, but the bridge is just beyond and is easily visible. This bridge is constructed of standard gauge sleepers laid across steel girders, which are themselves laid across the channel between concrete abutments, and is in the pattern of First World War military construction. Beyond the bridge the line went on through the woods to the standard gauge main line. The D213 runs down the west bank of the Aa canal and its junction with the side road is the site of the 1918 bridge over this canal, which was then much narrower. On the other side of the canal at this point, in the area called l'Ermitage (50° 49' 21.74" N 2° 12' 54.02" E), the line can be seen going into woods, with some old culverts, but the woods are private.

Terracing near the eastern end of this line, just before it crossed the standard gauge Watten-Socx line, can be seen in a field, looking south-east, 400 yards down the minor road which leaves the D326 half a mile north-east of St-Momelin village (50° 48' 15" N 2° 15' 39" E); see picture on page 178. The junction with the Anvin-Calais line was probably in the area just north of Zouafques (50° 49' 04" N 2° 03' 04" E), but apart from some rough ground by the road there is nothing to see there now.

Stations and halts, and other buildings

Many station buildings remain. The locations and current status of the metre gauge stations and halts of French Flanders are shown in Table 12.1. In many cases, the road side of the station is easier to see than the track side, which is often now the garden or private side of a house or business. The name of the station was posted on the track side.

Hazebrouck to Hondschoote, and Rexpoëde to Bergues

These lines were built and initially operated by the *Compagnie des Chemins de fer des Flandres* (CF), a Company associated with the entrepreneur Alfred Lambert. The stations are similar to those on other

12.8 The track side of the former station at St-Sylvestre-Cappel, in November 2021. This was a typical 'Lambert' type station on the line from Hazebrouck to Hondschoote. (*Authors*)

lines associated with him – the Aire to Berck line in the Pas-de-Calais *département*, and the lines from Noyon to Ham and to Montdidier, and from Milly to Formerie, in the Somme and Oise *départements*. Many examples can be seen. We recommend those at St-Sylvestre-Cappel, where the road approach is now on the track side, and the name can be seen: at Hondeghem: and at Herzeele.

Hondschoote to Bray-Dunes-Plage

This short-lived line was built and operated by a one-off Company, although associated with the NF. The situation was not regularised until after the line had closed, and the stations are unusual to match this. The section from Ghyvelde-Bray-Dunes, on the standard gauge line from Dunkerque to Adinkerke, to Bray-Dunes-Plage was only used from 1903 to 1914, and the grandiose terminus at Bray-Dunes-Plages was therefore something of a white elephant. This station is now in use as the local tourist information office, and the former track area is a large car park. The only 'ordinary' station we have found on this line is at Les Moëres.

Herzeele to St-Momelin, and Bergues to Bollezeele

Although these lines were owned and operated by the SE, the stations are quite different from those of the Somme lines of the same Company. The Somme lines had been constructed much earlier (opened 1888–90), whereas these lines were opened between 1910 and 1914. There may also have been pressure from the Nord *département* to have stations which were more similar to those of the existing lines in the area, opened 1894 to 1903. The main difference is that the ridge of the main passenger building is aligned parallel with the track, rather than at right angles as on the Somme lines. The stations are constructed of pale brick with red brick embellishments. We consider these the most attractive metre gauge stations in French Flanders. Good examples can be seen at Wormhoudt, Zeggers-Cappel, Pitgam, and Drincham. The station at Bollezeele is a mess, the brickwork covered in peeling white paint, but if you go round the corner, you can see the station name and some of the old depot buildings. The *Café de la Gare* across the road is much smarter.

12.9 The track side of the grandiose special terminus station at Bray-Dunes-Plage, for the former line from Hondschoote, in February 2020. (*Authors*)

12.10 The track side of the former SE station at St-Momelin in March 2007. Note the pale brick and the design, which differs from that for the SE Somme lines. (*Authors*)

Armentières to Halluin

This steam tramway was associated with the CEN, which was linked to the Belgian entrepreneur Baron Empain. It formed part of the inter-urban network which extended along the frontier between the conurbations of Valenciennes, Lille-Roubaix-Tourcoing, and Armentières. The lines from Lens to Frévent, and from Boulogne to Bonningues, both in the Pas-de-Calais *département*, were also operated by this Company, and the stations on this line are similar to those of all these others. The station buildings on this line were all of CEN type 1a, that is the standard two storey building without a goods hall. Only two tramway stations survive, at Frelinghien and Deûlémont.

At 50° 47' 11" N 3° 07' 37" E the tramway negotiated the right angle bend at the Belgian frontier. At the house on the facing corner the sign for the *Douane Française* is still displayed, but there are no customs posts here now; this is the European Union (see picture page 24). Close by on the Belgian side ran the *Vicinal* line 85 between Menen (Menin) and Kortrijk (Courtrai), but the two lines were never connected.

Bridges
Hazebrouck to Hondschoote, and Rexpoëde to Bergues

The line crossed the Yser river just south of Bambecque on a *tablier métallique* bridge. This is still in place at 50° 53' 56" N 3° 32' 27" E but is in woods on private land.

Herzeele to St-Momelin, Bergues to Bollezeele

On the section from Bergues to Bollezeele, abutments can be seen for a bridge across the Nouveau Bieren Dyck at 50° 57' 42" N 2° 25' 08" E. These are over what is now a side arm of the dyke, which appears to have been straightened since the line was dismantled. At Pitgam halt, at 50° 56' 15" N 2° 19' 12" E there is a brick arch bridge over a small ditch on the path of the line. Presumably there used to be a halt building here also, but no trace remains.

12.11 The CEN Type 1a station at Frelinghien, now a private house, in November 2021. (*Authors*)

12.12 The brick arched bridge of the Bergues to Bollezeele line over a drainage canal at Pitgam Halt in November 2021. This also shows the waterlogged nature of this land, 1m (3ft 3 inches) above mean sea level, after heavy rain. (*Authors*)

French, British and Belgian 60cm gauge railways

Almost nothing has survived of the 60cm gauge railways in this area. At 50° 54' 42" N 2° 44' 10" E there is a vague circular mound in the field. This is not high enough to be visible on satellite images, and it is difficult to photograph. However, it is exactly on the site of the structure shown on a British Army railway map of June 1918 as 'Circle Dump', part of the Swiss Cottage yards. This was a circle, apparently of light railways, with spurs of track inside. We have no idea what this was, but the circle on the 1918 map is about the right size for the mound, about 80 yards across.

Rolling Stock

Surviving metre gauge and 60cm gauge motive power units in Belgium, France and the UK are shown in Table 12.2. Further details of all the locations and organisations can be found on their websites.

Metre gauge locomotives and other rolling stock

Metre gauge rolling stock is only mentioned in this section if it was, or might have been, in use in the area of this book.

Belgium
Tramsite Schepdaal

This site just west of Brussels in the municipality of Dilbeek was originally the depot of one of the first

12.14 The manufacturers plate of SNCV/NMVB No. 979, May 2022, showing the Hawthorn Leslie manufacturer's No. 3228 (*Siegrid Vereertbrugghen, Tramsite Schepdaal*)

metre gauge *Vicinal* lines, from Ninove to Brussels, opened in 1887. In 1962, the site became the national Vicinal Tramway Museum, and it was reopened after refurbishment in 2009. The electric trams preserved here operated in the Brussels area. They have the only preserved Type 19 steam locomotive, which were all ex-British Army. This example, works number 3228, was manufactured by R & W Hawthorn, Leslie & Co of Newcastle upon Tyne in 1917. It was given the Railway Operating Division (ROD) number 244 and is known to have been at the depot at Bollezeele in March 1918, when the 85Can ECC were operating the line from there to Poperinghe and Crombeke

12.13 NMVB/SNCV Type 19 No. 979 at Tramsite Schepdaal in April 2021. This locomotive was at Bollezeele in March 1918, ROD number 244. (*Robert Jeanfils, Tramsite Schepdaal*)

Things to see and do now

Table 12.2 Present location of relevant metre or 60cm gauge motive power rolling stock

Item of rolling stock	Organisation	Country	Year	Type	Status
Metre gauge locomotives - Belgium					
Type 7 (SNCV/NMVB)	ASVi	Belgium	1888	steam	in working order
Type 18 (SNCV/NMVB) (JJ Gilain)	TTA	Belgium	1915	steam	not in use
Type 19 (SNCV/NMVB) (Hawthorn Leslie) (1)	STM	Belgium	1917	steam	in working order
Type 18 (SNCV/NMVB) (Grand Hornu)	TTA	Belgium	1920	steam	not in use
Type 18 (SNCV/NMVB) (Grand Hornu)	TTA	Belgium	1920	steam	not in use
Type 18 (SNCV/NMVB) (Haine-Saint-Pierre)	STM	Belgium	1920	steam	in working order
Metre Gauge tractors and railcars - France					
Locomotives 351 & 352 (ex-VFIL Flandres)	CFBS	France	1948	diesel	in use
Railcar 401 (M-41) (ex-VFIL Flandres)	CFBS	France	1936	diesel	derelict
60cm gauge locomotives and tractors					
Alco-Cooke 2-6-2T WD No.1265 (2)	FR	Wales	1916	steam	awaiting overhaul
Alco-Cooke 2-6-2T WD No. 1257	Tacot des Lacs	France	1916	steam	in steam
Hunslet 4-6-0T WD No. 303	WOLT	England	1916	steam	in steam
Baldwin 4-6-0T WD No. 778	LBNGRS	England	1917	steam	in use
Baldwin 4-6-0T WD No. 794	IWM/WHHR	Wales	1917	steam	in restoration
Baldwin 4-6-0T WD No. 608	F&WHR	Wales	1917	steam	in restoration
Baldwin 4-6-0T WD No. 779	Statfold Barn	England	1916	steam	in restoration
Alco-Cooke 2-6-2T WD No. 1240	AMTP	France	1916	steam	in static display
Hudswell Clarke 0-6-0WT Works No. 1238 (3)	MRT	England	1916	steam	in use
Decauville 0-6-0T Works number 1652 (4)	CFCD	France	1916	steam	in use
Kerr-Stuart 'Joffre' 0-6-0T Works No. 2405 (5)	WLLR	England	1915	steam	in use
Kerr-Stuart 'Joffre' 0-6-0T Works No. 2451 (6)	L&BR	England	1915	steam	in use
Kerr-Stuart 'Joffre' 0-6-0T Works No. 3014	MRT	England	1916	steam	in use
Henschel 0-8-0T DFB Works No. 15968 HF 1091	NGR	England	1918	steam	awaiting overhaul
Henschel 0-8-0T DFB Works No. 15311	AMTP	France	1917	steam	in use
Henschel 0-8-0T DFB Works No 3562	Tacot des Lacs	France	1917	steam	in static display
Hartmann DFB 0-8-0T Works No 4126	AMTP	France	1918	steam	in static display
Borsig 0-8-0T DFB	Tacot des Lacs	France	1917	steam	awaiting restoration
Krauss DFB 0-8-0T Works No. 7373 HF 1599	CFCD	France	1917	steam	in static display
Orenstein & Koppel DFB 0-8-0T Works No. 8627 HF 2199	CFCD	France	1918	steam	in static display
Borsig DFB 0-8-0 tender Works No. 10334 (7)	CFCD	France	1918	steam	in use
Orenstein & Koppel DFB 0-6-0T Works No. 8083	CFCD	France	1915	steam	awaiting restoration
Orenstein & Koppel DFB 0-10-0T Works No. 8285 HF 2085	CFCD	France	1917	steam	in static display
Baldwin 2-6-2T Works number 46828 (8)	Tacot des Lacs	France	1917	steam	in use
Dick, Kerr PE	Tacot des Lacs	France	1917	PE	in static display
Simplex 20hp WD No. LR 264	WHHR	Wales	1916	petrol	in working order
Simplex 20hp WD No. LR 2832	MRT	England	1918	petrol	in working order
Simplex 20hp WD No. LR 2593	AMHC	England	1918	petrol	in static display
Simplex 40hp 'protected' WD No. LR 3098	LBNGRS	England	1918	petrol	in working order
Simplex 40hp 'protected' WD No. LR 3090	MRT	England	1918	petrol	in working order
Simplex 40hp 'protected' WD No. LR 3101 (9)	AMHC	England	1918	petrol	in working order
Simplex 40hp 'armoured' WD No. LR 2182	LBNGRS	England	1917	petrol	awaiting restoration
Simplex 40hp 'open' LR 2228 (10)	FR	Wales	1917	petrol	in working order
McEwan-Pratt (Baguley) Works No. 736	WHHR	Wales	1916	diesel	in working order
Baldwin (US Army)	Tacot des Lacs	France	1917	diesel	in working order

(1) Formerly British Army Railway Operating Division, ROD No. 244, at Bollezeele in March 1918
(2) Mountaineer
(3) Ashanti
(4) Nord-Est
(5) Joffre
(6) Axe
(7) Geneviève
(8) Felin Hen
(9) restored as 'open' type
(10) Mary Ann

ASVi	*Association pour la sauvegarde du Vicinal*, Lobbes-Thuin, Belgium	L&BR	Lynton and Barnstaple Railway, North Devon
AMHC	Amberley Museum and Heritage Centre, West Sussex	MRT	Moseley Railway Trust, Apedale, Staffs
AMTP	*Association du Musée des Transports de Pithiviers*	NGR	North Gloucestershire Railway
CFCD	*Chemin de Fer Cappy Dompierre (P'tit train de la Haute Somme)*	STM	Schepdaal Tram Museum (*Tramsite Schepdaal*), Belgium
CFBS	*Chemin de Fer de la Baie de Somme*	TTA	*The Tramway Touristique de l'Aisne*, Erezée, Belgium
DFB	*Deutsche Feldbahnen* (German field railways)	VFIL	*Compagnie Générale des Voie Ferrées d'Intérêt Local*
FR	Ffestiniog Railway	WHHR	Welsh Highland Heritage Railway
F&WHR	Ffestiniog and Welsh Highland Railway	WLLR	West Lancashire Light Railway
HF	*Heeresfeldbahn* (Army field railway)	WOLT	War Office Locomotive Trust
LBNGRS	Leighton Buzzard Narrow Gauge Railway Society Ltd		

(see Chapter Seven). It was given the SNCV/NMVB number 979 which it still carries. Retired in 1950, it has since been restored to working order. Also preserved there is Type 18 locomotive number 1066, constructed by Haine-Saint-Pierre in 1920, which is in working order and in use. This type, first manufactured in 1910, was the type on which the British Army locomotives were modelled.

Association pour la sauvegarde du Vicinal Lobbes–Thuin

The HQ is at the *Centre de Découverte du Chemin de Fer Vicinal Tramway historique Lobbes-Thuin*. There is a museum here, and a metre gauge line, partly based on a former SNCV line and partly on disused re-gauged standard gauge line, with partial electrification. The centre is on the River Sambre south of Charleroi. The centre has a Type 7 steam locomotive of 1888, constructed by *Les Ateliers Métallurgique* (Tubize) numbered HL303, works number 704. This is in working order and is in use, and is the oldest preserved SNCV locomotive. There are also two diesel railcars of 1934 and 1947, 17 electric power cars, and a considerable number of carriages, passenger trailers, and wagons. Of the electric tramcars, from 1901 to 1956, ten are in working order, including the earliest.

The Tramway Touristique de l'Aisne

This is in the southern Belgian Province of Luxembourg. The Tramway operates 11.2km of a former SNCV line, using four railcars of 1933 to 1935. The museum also has three Type 18 SNCV/NMVB steam locomotives, not in working order.

France
Steam locomotives

As far as we know, no steam locomotives from the metre gauge lines of French Flanders have survived. During the war, a considerable number of additional locomotives were brought in from other parts of France, but none of those known to have worked on the Flanders lines has survived. Some Belgian locomotives almost certainly operated on French metre gauge lines during hostilities, but we cannot identify any of these. The French ordered 50 Baldwin 0-6-2T locomotives, and some may well have operated on the Flanders lines. After the war there is no record of these locomotives being used on these lines. There is no doubt that the British used many of the 0-6-0T *bicabine* locomotives constructed in 1916 and 1917 on the French Flanders lines, with 42 of these based at Bollezeele from January 1918, serving the line into Belgium to Poperinghe and Crombeke. Hawthorn Leslie no. 3228 now at Schepdaal was one of these.

Diesel locomotives and railcars

After the closure of the VFIL lines in 1954, bogie railcar No. 401 (M41) and diesel locomotives Nos 351 and 352 were sold in 1957 to SE for the Baie de Somme network. They are still at the *Chemins de Fer de la Baie de Somme*. The diesel locomotives are in use, but we

12.15 Diesel locomotive no. 352 in use at St-Valéry Port (CFBS) in July 2006. Built at the VFIL-CGL works at Lumbres and delivered to the VFIL Flanders lines in 1951, it was transferred to the Baie de Somme lines in 1957. (*Authors*)

believe that the railcar is derelict and at present not scheduled for restoration.

Passenger carriages

The six passenger carriages sold by SE from the Herzeele–Bergues–St-Momelin lines in 1936 to the Paris-Orléans-Midi Company for the Cerdagne line were renovated by SNCF in 1964-67 and 1985-88. They are still in service on the '*petit train jaune*' of the Cerdagne.

Light Railway (60cm gauge) locomotives and other rolling stock

There are numerous surviving locomotives from the British Army light railways (WDLR) of 1914-18. A comprehensive list is provided in *Narrow Gauge at War* (Plateway Press, revised 2008). There are also surviving examples of wagons. In respect of Motor Rail Simplex petrol tractors, the situation is complicated by the fact that the Simplex 20 continued to be produced until 1932, and the Simplex 40 until 1930, in identical or similar form to those used in the war. Very similar copies of the Simplex 20 were produced by other firms until 1951.

Chemin de Fer Cappy Dompierre (CFCD) – Le P'tit Train de la Haute Somme

We have not given details of the other locations where rolling stock is now, but CFCD is undoubtedly the most important contemporary site for 60cm gauge railways in northern France, and especially for those of the First World War. As far as we know this is the only preserved 60cm gauge line on the site of one built for military use in that war.

The history of the area and of this line has been told in our previous book '*Narrow Gauge in the Somme Sector*' (Pen & Sword Transport, 2019). There is a 60cm gauge line 7km long, and an extensive museum.

British Army 60cm gauge rolling stock

For British Army petrol tractors and steam locomotives, for the most part it is known that they were shipped to the western front and used, but not where and how. However, for the Ypres Sector we have two examples of locomotives which were definitely used in the area of this book, and which survive. These are Alco-Cooke 2-6-2T locomotives of WD numbers 1257 and 1265. Both served with the 13Can LROC, who were with the British Fourth Army on the Belgian Coast between 21 June and 11 September 1917. That we know this is thanks to the unusual thoroughness of this Company, who recorded the numbers of locomotives involved in incidents, most usually derailments. From this we know that they had sixteen Alco-Cookes (of a total production for the British Army of 100), but no other types of steam locomotives are recorded. 1257 and 1265 both derailed on 28 August 1917, in different parts of the network (see Chapter Seven). When 13Can LROC left the area 32LROC may have taken over, and in any case, we do not think that the Canadians took their rolling stock to Savy-Berlette, where they arrived on 12 September. However, when the Fourth Army left the area in November 1917, handing it back to the French and Belgians, the rolling stock was transferred south, probably to the Fifth Army area. This would explain why both these locomotives later arrived at Pithiviers.

Alco-Cooke WD 1265

Works number 57156 was built by the American Locomotive Company (Alco) of New Jersey for the British Army in 1917. After the war, it was used by the *Ministère des Régions Libérées* (Ministry for the Liberated Regions), probably on the Soissons-Laon network in the *département* of Aisne. It was acquired by the Vis-en-Artois system in 1926. In 1935 it was sold to the Tramway de Pithiviers à Toury (TPT) and given the number 3-23. After the closure of the TPT in 1964 3-23 was acquired by John Ransom and moved to a site in west London. In 1967 the locomotive was given to the Ffestiniog Railway on indefinite loan and allocated the name *Mountaineer*, after the original of that name built in 1863 and scrapped in 1879. The locomotive was significantly modified, most notably the profile of the cab was altered to accommodate the Ffestiniog loading gauge, and it was converted from coal to oil burning during 1971. Last operated in 2006, it is currently awaiting overhaul.

Alco-Cooke WD 1257

Works number 57148 was also built for the British Army in 1917. This was also previously owned and run by the TPT as their No. 3-20. Formerly at Froissy (CFCD), this locomotive is now at Tacot des Lacs. It has been refurbished to working condition.

Other Steam Locomotives

For the remainder we have indicated where examples of the main locomotives types can now be seen in Belgium, northern France and Great Britain (see Table 12.2).

Hunslet 4-6-0T

Hunslet 4-6-0T of 1916, works number 1215, WD No. 303, was used in France and there is a photograph at Boisleux-au-Mont, south of Arras, in September 1917, with the 14th US Engineers. Repatriated from Australia, it has been restored by the War Office

12.16 Alco-Cooke 2-6-2T WD No. 1257 on display at the Gare de l'Est in Paris on 10 November 1986. This locomotive used to be at the *Chemin de Fer Cappy-Dompierre*, but is now at *Tacot des Lacs*. It saw service with the 13th (Canadian) Light Railway Operating Company at Coxyde (Koksijde), near Nieuport (Nieuwpoort) in North Belgium, in August 1917, with WD No. 1265 of the same type, now 'Mountaineer' on the Ffestiniog Railway. (*Photo & collection Bernard Rozé*)

Locomotive Trust and is now in steam. The locomotive is normally kept at the Moseley Railway Trust (MRT) (Apedale Valley Light Railway).

Baldwin 4-6-0T

All four of these in the UK served on the Western Front. That at Leighton Buzzard Narrow Gauge railway (LBNGRS) and the associated Greensand Railway Museum, works number 44656 of 1917, WD No. 778, is regularly in steam. In October 1917 this locomotive was rostered with others by the 14th US Engineers at Boisleux-au-Mont.

Baldwin works number 44699 WD No. 794 went to India after serving on the Western Front, and worked on a sugar plantation. Repatriated in 1985 with No 778, it is now owned by the Imperial War Museum and is under restoration at the Vale of Rheidol Railway workshops. When restored it will be operated on the Welsh Highland Heritage Railway (WHHR), where it will carry the No. 590. The original Baldwin of this number operated on the Welsh Highland Railway from 1923 but was scrapped in 1942.

Two further Baldwins were repatriated from India in 2013. Works number 45190 of 1917, WD No. 608, has been restored and since October 2019 has been in use on the Ffestiniog and Welsh Highland Railway. Works number 44657 of 1916, WD No. 779, was at Statfold Barn Railway, but has now been transferred to Killamarsh (near Sheffield) for restoration.

Alco-Cooke 2-6-2T

In addition to WD Nos 1257 and 1265 (see above), the former TPT had a third Alco-Cooke 2-6-2T, works No. 57131, WD No. 1240. This is preserved in static display at the *Association du Musée des Transports de Pithiviers* (AMPT) at Pithiviers.

Hudswell Clarke 0-6-0WT

The MRT and Apedale Valley Light Railway have a Hudswell Clarke 0-6-0 well tank manufactured in 1916, works No. 1238, now known as *Ashanti*. This did not see war service but is identical with those which did. After restoration off site, it has been fully operational from autumn 2014.

Decauville 0-6-0T
Large numbers of these were produced for French Army Light Railways. CFCD operate one (No. 5, Nord-Est), works number 1652 of 1916, which after the war operated at the *sucrerie* at Toury, and was restored to working order in 2006.

Kerr-Stuart 'Joffre' class 0-6-0T
Seventy of these were produced by Kerr, Stuart of Stoke-on-Trent for French Army Light Railways in 1915 and 1916, based on the Decauville design. In 1956, five of the class were found derelict in France, and three are now in working order in England. Works No. 2405 of 1915 was acquired by the West Lancashire Light Railway in 1974. Now named *Joffre*, this locomotive has been fully restored. Works No. 2451 of 1915 was acquired by the Lynton and Barnstaple Railway in 1983. Restored and named *Axe*, it is in regular use based at Woody Bay station. Works No. 3014, of 1916 was probably used by the French Army around Verdun. This locomotive is now operational at MRT (Apedale).

Deutsche Feldbahn (DFB) locomotives
The most common DFB (German Field Railways) locomotives were the 0-8-0T type, of which about 2,500 were constructed by 14 different manufacturers between 1904 and 1919. AMTP operate a Henschel of this type, built in 1917, and have a Hartmann of 1918 in static display. CFCD operate a Borsig 0-8-0 of 1918, with a tender added later in Poland. They also have a Krauss of 1917 and an Orenstein and Koppel of 1918 in the museum at Froissy, with an 0-10-0T of 1917, in static display. An Orenstein and Koppel 0-6-0T of 1915 is at CFCD awaiting restoration. A Borsig 0-8-0T at *Tacots des Lacs* is awaiting restoration. The Hesnchel 0-8-0T at the North Gloucestershire Railway is awaiting overhaul.

US Army
From US First World War light railways *Tacots des Lacs* have 2-6-2T Baldwin number 5104, works number 46828, now named *Felin Hen*, in working order.

Petrol and diesel locomotives (tractors)
For these we have mostly mentioned those which can be seen in England and Wales.

Dick, Kerr and British Westinghouse petrol electric (PE) tractors
The only examples we know of which are accessible to the public are at *Tacot des Lacs*, where there is a Dick, Kerr PE tractor, built for the British Army in 1917, which presumably saw service in France. There is no number. Another Dick, Kerr and a British Westinghouse PE tractor, which are reported to be awaiting restoration, have been converted to standard gauge with new cabins, but there are no numbers.

Motor Rail Simplex 20 tractor
We have listed three in England and Wales related to the First World War. Simplex 20 WD No. 264 of 1916, is in private ownership, but is kept and displayed, in working order, at WHHR. At MRT (Apedale) works No. 1111, WD No. LR 2832, built in 1918 but too late for service, was used at Naburn Sewage Works near York, before arriving at MRT in 1992. It is in fully operational condition. Simplex 20 works No. 872, WD No. 2593, is on loan to Amberley Museum and Heritage Centre (AMHC). Built in 1918, this probably served in France. It was rebuilt by Motor Rail in 1925 and is in static display.

Motor Rail Simplex 40 tractor
We list here five in England and Wales related to the First World War. Three of these were of the 'protected' type and were delivered in 1918, too late to be sent to France. Works No. 1377, WD No. LR3098, is operated at LBNGRS, on long term loan from the National Railway Museum. Works No. 1369, WD No. LR 3090, has been rebuilt and is in working order at MRT (Apedale). Works No. 1381, WD No. LR 3101, was donated to AMHC. It was of 'protected' type but has now been restored as an 'open' example. It is in operational condition.

The National Army Museum has given LBNGRS 'armoured' 40hp Simplex WD No. LR2182, probably works No. 461, manufactured in 1917. This is thought to be the only 'armoured' Simplex in original mechanical condition, and it is awaiting restoration. In 1923 the Ffestiniog Railway purchased a reconstructed Simplex 40, probably WD No. LR 2228 of 'open' type. It was the first locomotive on the reopened railway in 1955, and in 1960 was fitted with a diesel engine. Called *Mary Ann*, she has now been put back in original condition with a rebuilt petrol engine.

McEwan-Pratt (Baguley) 10hp tractor
The only example of this type which saw service in the war is at WHHR, works number 760, WD No 297, restored with a diesel engine.

US Baldwin diesel tractor
From US First World War light railways *Tacot des Lacs* have a US Army Baldwin 0-4-0 diesel of 1917, number 2053, works number 48606, which is in working order.

Walks
There are few places in this area where remains of the metre gauge or light railway (60cm gauge) lines in this book can be seen, but we have included a walk in

Ypres. Also part of the former standard gauge single track line from Ieper (Ypres) to Roeselare (Roulers) is worth walking, between Zonnebeke (50° 52' 29" N 2° 59' 05" E) and Passendale (Passchendaele) station (50° 53' 42" N 3° 02' 10" E). This passes very close to Tyne Cot Cemetery (50° 53' 15" N 3° 00' 03" E), which can be visited as a small diversion.

A walk in Ieper (Ypres)

This walk is about 1 mile (1.6km) long, or 1½ miles (2.5km) to complete the circuit through the town centre. However, we suggest allowing up to 2 hours to look at everything, including the Menenpoort (Menin Gate) and the memorials on the nearby ramparts. The walk is shown in Figure 12.1.

The walk is mainly to follow some of the light railways built in the ruins of Ieper (Ypres) and around the ramparts and the moat, but it also allows an appreciation of part of these fortifications. We have described this walk starting at the Menin Gate (Menenpoort). It is easier to park across the gardens from the modern station than at the Menin Gate, and you can then walk through the town past the Cloth Hall to begin the walk. Near the Menin Gate there is some parking (with payment) at the Sint-Jacob (St Jacques) church just along the Bollingstraat from the Gate.

The B6 line in Ypres was started by early June 1917 but was not completed through the Menin Gate, to join the Y1 line at the beginning of the Menin Road, until late September. Work was suspended from mid-July to 11 August 1917 because of intense shelling. When work resumed on 11 August to reconstruct the damaged line, Captain Anderson of the 2nd Battalion, Canadian Railway Troops was killed by a shell fragment (see Poperinghe New Military Cemetery). We have found no explanation for the circuitous route of the line through Ypres, but most likely it was to avoid the main square by the Cloth Hall, one of the most targeted areas for German artillery. The need to join up with the D11 line at the south-west of the town may also have been a factor.

12.17 The view from outside the Menin Gate Memorial, looking into the town, with the Cloth Hall in the distance. One of the Australian lions is on the left. (*Authors*)

The original gate here was destroyed by the town council in the nineteenth century to widen the road, and became just a gap in the ramparts. The site was a major target for German artillery and the road and light railway required frequent repair. A fictional scene here in Autumn 1917 is depicted on the jacket of this book. The Menin Gate is now one of two major memorials to the missing, inscribed with the names of 54,000 officers and men who have no known graves (the other is at Tyne Cot). The moving ceremony of the Last Post is held here every evening at 8pm, best visited when there are fewer others, on a cold winter evening. The lions outside the gate were here during the war and damaged by shell fire. The present restored lions were given by the Australian government.

Inside the gate, facing the town centre, turn left into Bollingstraat. The B6 line ran inside the ramparts down this road, taking the sharp turn by using a curve through the ruins of the house on the corner. Rather than following the road, take the ramp up onto the ramparts, where there are some interesting memorials. After a few hundred yards take the steps back down to the route of the railway at the Sint-Jacob church. The road along the inside of the ramparts is now the Aalmoezeniersstraat. To the right, back toward the Menin Gate, are casemates, one of which is now a restaurant.

Turn left along the Aalmoezeniersstraat, following the inside of the ramparts, until you reach a small roundabout. Immediately before this on the left is a postern tunnel through the ramparts leading out to a footbridge across the moat, with a good view. There was a footbridge here in the war. At the roundabout go straight across, second exit, into the Rijkeklarenstraat, which is in effect a half right taking you and the route of the line into the southern part of the town. Cross the Bukkersstraat, dodging slightly to the left into the Burggraafstraat, with Sint-Pieterskerk (St Peter's Church) on your left. There was a passing loop on the B6 line here. At the end of the church cross the

12.18 The causeway across the 'moat' which in 1917–18 carried the D11 light railway line from the Mimico Depot to join the B6 line at the right hand end. View from the ramparts path, approximately on the path of the B6 line towards the Menin Gate. (*Authors*)

Risjelstraat, the old main road to the Lille Gate and on to Lille, with a quick right and then left into Klaverstraat. At the end is a grassy square, which the railway went straight across, and into what is now private property, where it turned towards the right to reach the moat again.

We must turn right into the Arsenaalstraat and then after 150m left into a long thin car park, with cars parked only on the right, immediately past the Ter Vesten B&B. At the far end of the car park there is a way through which brings us back to the moat and to the line of the railway. Turn right along the path by the moat and follow it round to the left. The B6 line ran to our right a little further away from the moat, turning left near the corner of the then Infantry Barracks, a massive building giving relatively sheltered billets. This is now housing, with the barracks grounds now a car park and the Post Office behind a brick wall.

You will reach a small road, Esplanade, coming from the right, and this continues, as Eiland, as a short causeway across the moat, on the left. By this on the far side is a small island with the Pacific Eiland restaurant. This is now a very pleasant area of water, old ramparts and trees, and the destruction is difficult to imagine. The D11 line came in across this causeway from the major light railway depot at Mimico, 1,100 yards (1km) to the west. The evidence which we have seen suggests that this line was complete by the end of May 1917, about the time that construction of the B6 line began. Its initial usefulness was probably the link which it had to the south end of Ypres station yards from the west side. When this part of B6 was built the two lines joined here. We do not think that the B6 line was complete through the Menin Gate to join the Y1 line in the Menin Road until September 1917 but it may have been usable north to the B5 junction at least some of the time before that. The Mimico depot was certainly open and active from the move there of the 17(3)Aus LROC on 30 September 1917, and from then the way through Ypres using the D11, B6 and Y1 lines would have been an important route to Hell Fire Corner and the forward light railways.

From this junction the B6 line curved north through what is now the car park to the Colaertplein, but for us it is best to follow the path round closer to the moat, the Vestingsroute, to the same point. The modern standard gauge station is to the left, and the metre gauge lines used to leave from the forecourt, but there is nothing left. The light railway went straight on, in what becomes Maloulaan and then Haiglaan to the junction with the B5 line at what used to be Clapham Junction, near the B4 junction at White Pole Corner, although the former is named as the latter on some maps.

It is possible to follow the rest of the B6 line, and then the B5 line through northern Ypres, but there is little to see. However, we have shown the routes of the former lines in the rest of Ieper in Figure 12.1 for those who wish to explore further. We have also shown other sites to visit. We suggest turning right at De Stuersstraat and going straight back to the Menin Gate past the Cloth Hall and through the Grote Markt. Or, as already suggested, you can start with this, depending on where you have left your car.

And finally ...

The British and Dominion Armies played a special part in the defence of Ypres and the surrounding area during the First World War. Passchendaele, and the Third Battle of Ypres, rank with Verdun and the Somme in the catalogue of horrors of that war. It was also the area of the development of the densest railway network anywhere along the Western Front. We think that the history of narrow gauge railways in this Sector will add to the overall story.

Bibliography

ANGELIER, Maryse. *La France Ferroviaire en Cartes Postales. Nord-Pas-de-Calais.* Éditions La Vie du Rail, Paris, 2001.

Association du Musée des Transports Pithiviers (AMTP). *La voie de 60 et les chemins de fer militaires. Dés locomotives Péchot-Bourdon à la ligne Maginot.* Éditions AMTP, 1991.

AVES, William A.T. *The lines behind the front. The Railways in Support of the British Expeditionary Forces in the Great War.* Lightmoor Press, Lydney, Glos. 2016.

AVES, William A.T. *R-O-D The Railway Operating Division on the Western Front. The Royal Engineers in France and Belgium 1915-1919.* Shaun Tyas Publishing, Donington, Lincs, 2009.

BROWN, Malcolm. *The Imperial War Museum Book of 1918 Year of Victory.* First published 1998, this edition Pan Books, London, 1999. Especially Chapter 4, pp 83-114, on the Lys pocket, April, 1918.

DAVIES, W.J.K. *Minor Railways of France.* East Harling, Norfolk, Plateway Press, 2000.

DAVIES, W.J.K. *The Vicinal Story. Light Railways in Belgium 1885 - 1991.* Light Rail Transit Association, Scarborough. Second Edition, 2006.

DAVIES, W.J.K. *Light Railways of the First World War. A History of Tactical Rail Communications on the British Fronts 1914 – 18.* David & Charles, Newton Abbot, 1967.

DELLEAUX, René. 'L'évolution du matériel roulant de la SNCV'. *Voie Étroite* 1986, Vol. 95, pp 13-15.

DOMENGIE, H. and Banaudo, J. *Les petits train de jadis. Nord de la France.* Les Éditions du Cabri, Breil-sur-Roya, France, 1995.

FAREBROTHER, Martin & Joan. *Tortillards of Artois. The Metre Gauge Railways and Tramways of the Western Pas-de-Calais.* Oakwood Press, Usk, Mon., 2008.

FAREBROTHER, Martin J.B. & Joan S. *Narrow Gauge in the Arras Sector. Before, during and after the First World War.* Pen & Sword Transport, Barnsley, South Yorkshire, 2015.

FAREBROTHER, Martin J.B. & Joan S. *Narrow Gauge in the Somme Sector. Before, during and after the First World War.* Pen & Sword Transport, Barnsley, South Yorkshire, 2018.

GITTINS, Sandra. *Between the Coast and the Western Front.* The History Press, 2014.

HENNIKER, A.M. *Transportation on the Western Front 1914-1918.* London, 1937, reprinted by the Imperial War Museum, London and the Battery Press, Nashville, USA, 1992.

HERITAGE, T.R. *The Light Track From Arras.* First published 1931. 2nd edition Plateway Press, East Harling, Norfolk, 1999.

HUGHES, I.G. *Hunslet 1215. A War Veteran's Story.* Oakwood Press, Usk, Mon., 2010.

JOHNSON, J. H. *1918 The Unexpected Victory.* Arms and Armour Press, 1997.

JONES, Mervyn. *The Essential Guide to French Heritage and Tourist Railways.* Oakwood Press, Usk, Mon., 2006.

KUSEE, Wim. *Tramways in Belgium. NMVB/SNCV.* www.kusee.nl.

Le HENAFF and BORNECQUE, H. *Les Chemins de Fer Français et la Guerre.* Paris: Librairie Chapelot, 1922.

LINK, Roy C. *WDLR Album.* RCL Publications, 2014.

Nord Pas-de-Calais, Picardie. Paris: Le Guide Vert Michelin, 2008 (in French, but more detailed than the English equivalent below).

Northern France and the Paris Region. Michelin Green Guide, 2014 (in English).

ORGAN, John. *Northern France Narrow Gauge.* Middleton Press, Midhurst, West Sussex, 2002.

PRÉVOT, Aurélien. *Les chemins de fer français dans la Première Guerre Mondiale.* LR Presse, 2014.

RAWSON, Andrew. *The Passchendaele Campaign 1917 (British Expeditionary Force).* Pen & Sword Military, Barnsley, 2017.

REED, Paul. *Walking Ypres (Battleground Ypres).* Pen & Sword Military, Barnsley, South Yorkshire, 2017.

ROZÉ, Bernard. *L'Apport en matériel moteur des chemins de fer secondaires à voie métrique aux réseaux touchés par la guerre 1914-1918.* Chemins de Fer régionaux et Tramways 2017 Vol 382 pp 4-21.

TAYLOR, A.J.P. *The First World War. An illustrated history.* New York: Perigee Books, The Berkley Publishing Group, 1963 (original publishers Hamish Hamilton).

TAYLORSON, Keith. *Narrow Gauge at War 2.* Plateway Press, East Harling, Norfolk, 1996.

TAYLORSON, Keith. *Narrow Gauge at War.* Plateway Press, East Harling, Norfolk, 2nd edition, 2008 (first published 1987).

The Railway Gazette and Railway News. *Special War Transportation Number,* 1920. Reprinted by the Moseley Railway Trust and Railway Gazette International, 2013.

THOMAS, David. Coordination (parts 1 & 2). *The SNCF Society Journal* March 2006, No. 121, & June 2006, No. 122.

WAGNER, Claude. *Les petits trains de Ch'Nord.* Éditions LR Presse, Auray, 2011.

Index

For British Army places in the First World War, *see also* Depots, Transhipments, & Yards

Aa river & canal, 1, 2, 18, 72, 125, 174, 177, 178, 206, 230
Abeele, 47, 69, 76, 77, 89, 91, 165, 172, 188
Abraham Heights, 114, 129
Accidents & incidents, 83, 85, 93, 105, 180, 184, 196
    Ammunition explosions, 85, 94, 118, 134, 169, 181
Adinkerke, & station, 6, 31, 133, 136, 196, 211, 213, 218, 220, 221, 224, 231
Aircraft & air raids: 94, 96, 114, 184, 193
Albert I, King of the Belgians, 1, 188-189, 192
Alveringhem, 40, 46, 77, 213
Amberley Museum and Heritage Centre (AMHC), 239
American Army, *see* United States Army
Ammunition explosions, *see* Accidents & incidents
Ammunition Refill Points (ARP), 181, 185, 186, 187
Antwerp, 58, 60, 212, 220
Apedale Valley Light Railway, *see* Moseley Railway Trust
Armentières, xiii, 1, 6, 23, 24, 26, 28, 58, 59, 60, 64, 65, 72, 85, 87, 89, 91, 163, 172, 191, 192, 194, 209, 232
Armentières, Battle for (1914), 58, 60
Armistice, 11th November, 1918, 57, 192
Arras sector, xii, 1, 48, 49, 65, 104, 128, 129, 130, 131, 136, 221
Artillery, 81, 82, 94, 106, 108, 116, 117, 118, 119, 122, 133, 188, 189, 190
Association du Musée des Transports de Pithiviers (AMTP), 238, 239
Association pour la sauvegarde du Vicinal Lobbes-Thuin, 236
Atherley Junction, 81, 105, 106, 129
Australian Army, *see also* railway & related transport companies, 51, 116, 184
Australian & New Zealand Army Corps (Anzac), 84, 96, 114, 116
Autorails, *see* Railcars

Bailleul, 69, 96, 97, 185, 191, 194, 195, 200
Ballast, 51, 54, 69, 76, 116, 131, 169, 176, 179, 180, 184, 185
Battles, *see under the name of the Battle*
Belgian Army, 1, 45, 58, 62, 64, 65, 100, 104, 107, 123, 124, 132, 163, 188, 192
Belgian Coast, 48, 86, 104, 107, 115, 123, 131-136, 192
Belgian - French border - *see* French - Belgian border

Belgian Government, 45, 220, 221
Belgian railways, xii, 45, 58, 220
Belgium, xii, 45, 220, 221
Bergues, 4, 11, 12, 13, 18, 19, 62, 65, 69, 73, 84, 86, 108, 125, 126, 165, 172, 173, 195, 197, 203, 206, 207, 228, 232
Béthune, 58, 59, 64, 98, 163
Birdwood, Sir H.R. (General, Australian, Fifth Army), 184
Bixschoote (Bikschote), 43, 44, 73, 216
Boescheppe, 76, 130, 185
Boesinghe (Boezinge), 58, 69, 73, 80, 100, 104, 106, 107, 108, 123, 165, 191, 192, 200
Bollezeele, 18, 22, 71, 72, 73, 86, 125, 126, 172, 176, 207, 208, 231, 232, 234, 236
Boulogne, 1, 50, 60, 84, 125
Brandhoek, 82, 172, 188, 190, 191
Bray-Dunes(& -Plage), 11, 16, 69, 71, 86, 208, 221, 231
Bridges, 11, 16, 19, 20, 24, 30, 60, 69, 81, 82, 124, 128, 136, 191, 192
    Aa Canal & side canals, 174, 177-178, 206, 230
    Light railways, 81, 104, 107, 113, 115, 116, 128, 135, 185, 186, 187, 188
    Metre gauge, 108, 176, 177-178, 206, 228, 230, 232
    Standard gauge, 60, 104, 107, 163, 165, 172, 174, 191, 196
    Yser canal crossings & bridges, 81, 104, 107, 113, 128, 165, 172, 191
Brielen, 39, 104, 165, 172, 191
British & Dominion armies, xii, 45, 58, 100, 116, 163, 174, 188, 242
    British Expeditionary Force (BEF), 45, 46, 58, 59, 64, 192
    General Headquarters (GHQ), 45, 46, 59, 60, 65, 118
    First Army, 47, 57, 59, 60, 91, 97, 98, 124, 163, 169, 191
    Second Army, 2, 45, 47, 48, 57, 59, 60, 64, 65, 77, 80, 89, 91, 98, 104, 107, 113, 114, 116, 117, 123, 128, 132, 136, 163, 169, 173, 177, 181, 187, 189, 191, 192, 195
    II Corps, 113, 115, 116, 188
    Canadian Corps, 114, 116
    Third Army, 48, 49, 57, 136, 169, 195
    Fourth Army, 45, 48, 49, 86, 91, 107, 123, 128, 131, 132, 133, 136, 223, 237
    Fifth Army, 45, 48, 98, 104, 107, 108, 113, 114, 116, 118, 123, 132, 169, 189, 191, 197, 237

    II Corps, 197, 198, 199
    XIV Corps, 106, 107, 108, 113
    XVIII Corps, 106, 113, 118, 122
    XIX Corps, 106, 108, 114, 197, 198, 199, 200
    Railway companies: *see* Railway & related transport companies
British War Cemeteries & Memorials (CWGC), 189, 221-224, 240, 241
Broodseinde (& Ridge), 58, 93, 116, 117, 118
Brugge (Bruges), 42, 44, 217, 221
Buses, including competition & replacement, 202, 206-208, 209, 210, 211, 218, 220
Busseboom, 69, 81, 82, 192
Byng (General, Third Army), 57

Calais, 1, 2, 46, 51, 60, 69, 77, 169, 180
Canadian Army, 64, 114, 116
Railway companies: *see* railway & related transport companies
Canals, 2, 40, 69, 71, 128, 174, 177
    Comines Canal, 2, 89, 95, 104, 163
    Yser Canal, *see under* Yser
Carriages, *see* Passenger Carriages
Cassel, 2, 28, 64, 128, 136, 172, 209
Casualties (railway troops):
    Australian, 82, 94, 96, 120, 135, 168, 223, 224
    British, 94, 105, 113, 131, 135, 172, 185, 189 222, 223, 224
    Canadian, 95, 113, 114, 117, 122, 131, 135, 168, 177, 222, 224, 240
    New Zealand, 222
    United States, 191
Cerdagne, le petit train jaune (SNCF), 206, 237
Château Wood, 123
Chemin de Fer Cappy Dompierre (CFCD), 135, 237, 239
Chemin de Fer de la Baie de Somme (CFBS), 204, 236
Chemin de Fer Électrique d'Ostende–Blankenberghe et Extensions (OB), 38, 39, 40
Chief Engineers (British Army), 77, 84, 99, 134
Chief Railway Construction Engineer (CRCE, British Army), 45, 46, 77, 128, 180, 187, 192
Closures:
    Metre Gauge :
        Belgian, 210, 211, 217, 218
        French, 202, 203, 208, 209
    Standard gauge, 196, 202, 210, 221

Index

Coastal tramway (*Kusttram*, Belgium), 211, 213, 218, 220, 224
Compagnie des chemins de fer des Flandres (CF), 10, 24, 203, 230
Compagnie des Chemins de Fer d'intérêt local du Nord de la France (NF), 11, 16, 86, 208, 209, 231
Compagnie (des chemins de fer) du Nord (CdN), 8, 10, 60, 86, 192, 203, 207, 208
Companie des Chemins de Fer Économique du Nord (CEN) 10, 18, 24, 25, 26, 209, 232
Compagnie Générale des Voie Ferrées d'Intérêt Local (CGL-VFIL), 10, 203, 205
Connections, *see* Timetables
Construction & Maintenance (WW1):
    Light railways, 80, 81, 82, 84, 91, 94, 95, 104-105, 106, 108, 113-118, 128-130, 132-135, 181, 184-188, 192-194, 197, 198, 199, 200
    Metre gauge, 108, 174, 176-180
    Standard gauge, 108, 132, 192, 196, 197
Controls & control posts (WW1), *see* Traffic control
Cornwall, J.K. (Lt Colonel, 8CRT), 181
Courtrai (Kortrijk), 43, 44, 60, 194, 197, 211, 216, 217, 218, 221, 232
Coxyde (Koksijde), 32, 39, 62, 132, 133, 135, 212, 213, 217, 223
Coxyde-Bains (Koksijde-Bad), 32, 39, 62, 132, 213
Crombeke, 42, 71, 108, 125, 126, 176, 181, 188
Crossings on the level (railways), *see* Diamond crossings
CWGC (Commonwealth War Graves Commission): *see* British War Cemeteries & Memorials

De Panne, *see* La Panne
De Seule, *see* Le Seau
Demolitions, 19, 20, 124, 168, 172, 174, 176
Depots & workshops (railway), *see also* Transhipments *and* Yards & supply dumps, WW1:
    British Army Light Railways:
        A system, 130, 169, 190
            Byng (AB line), 184, 191, 198
            International Corner, 98, 105, 118, 189
            Pretoria (Swiss Cottage), 181, 190, 191
            Vox Vrie (AB line), 188, 189, 190, 198, 199, 200
        B & B9 systems, 130, 169, 199
            Culloden locomotive yard (B1 line), 169, 185, 223
            Westonhoek, 116, 168, 169, 184, 185, 197
            Beaurainville, 51, 54
        Belgian coast (1917), 135, 136, 224
            Angus locomotive yard (Coxyde), 134
        Busseboom & Ouderdom systems, 91, 93, 168
            Angus locomotive yard (Ellarsyde), 83, 96
        Mimico (D11 line), 120, 122, 131, 168, 190, 224, 242
        Quintin (Poperinghe), 82, 83, 105, 120, 130, 131, 184, 185, 189, 191, 193, 194, 195
        Vauxhall (Elzenwalle, V1 line), 93, 94, 168, 190, 200
        Douve Valley & Romarin systems, 96, 99, 100
        L system:
            Heidebeek, 168, 185, 186, 188, 189, 191
            Ledringhem locomotive yard (L8 line), 185, 187, 191
            Yser locomotive depot (L7 line), 181, 187, 189
        La Lacque, 49, 51, 54, 94
        Light railway repair trains, 54, 56, 83, 186
        Savy(-Berlette), inc. Forward Depot & Training School (First Army), 49, 93, 94, 98, 131, 135, 191, 192, 195, 237
        Watou training & rest camp (1917), 106, 115
    British Army standard gauge (ROD), 172, 173, 188, 197
        Audruicq, 47, 48, 65, 69, 77, 125, 169, 172, 177, 195
        Bergues Exchange, 107, 108, 125, 126, 165, 172, 177, 197
        Borre, 91, 165, 173
        Merris, 91, 126, 163, 173, 176
        Peselhoek, 107, 125, 165, 172, 190
    Metre gauge, Belgium, 34, 40, 42, 62, 72, 126, 224
    Metre gauge, France, 12, 15, 22, 25, 72, 203
        Bollezeele (Bergues & Herzeele to St-Momelin), 72, 86, 125, 126, 231, 236
        Ghyvelde (British), 71, 72, 176
    RE depots, UK, 46, 50, 59, 77, 80, 82, 83, 84, 85, 189
    Tramways of Dunkerque, 30, 210
Derailments: 82, 93, 96, 135, 177, 184, 190, 237
Diamond crossings:
    Metre gauge - light railway, 132, 134, 176, 185, 186, 187
    Standard gauge - light railway, 105, 185, 186, 187, 193
    Standard - metre gauge, 70, 108, 126, 176, 177, 196
Dickebusch (Dikkebus), 69, 82, 83, 89, 91, 163, 168
Director (Directorate) of Construction (British Army), 57, 169, 174, 181, 195
Director of Railway Transport (DRT or DR, British Army), & Assistants, 45, 59, 60, 65, 70, 76, 77, 195
Director-General of Transportation (DGT, British Army), & Assistants, 45, 49, 57, 65, 77, 98
Dixmude (Diksmuide), 35, 42, 44, 77, 100, 134, 192, 211, 216, 217, 218
Douve river & valley, 58, 65, 69, 84, 89, 91, 192, 194, 201

Dual gauge, 33, 38, 71, 126
Dunkerque, 1, 2, 5, 6, 29, 30, 39, 132, 134, 172, 197, 201, 209, 210
Dutch language, xiii, 216, 220, 221

Electric trams, *see* Tramcars
Electrification, 26, 28, 30, 33, 38, 211, 213, 216, 217, 218, 220
Elverdinghe (Elverdinge), 39, 42, 69, 71, 72, 77, 80, 104, 106, 107, 108, 125, 165, 172, 191, 193, 194, 198, 216, 217
Elzenwalle, 93, 163, 168, 200
Empain, Baron, 23, 24, 35, 38, 60, 62, 132, 209, 213, 232
Esquelbecq, 11, 71, 73, 86, 125, 126, 176, 177, 187, 206, 207, 208

Fares, 16, 18, 26, 29
Ffestiniog & Welsh Highland railways (Porthmadog, North Wales), 48, 135, 237, 238, 239
First World War, xii, 1, 45, 58, 211, 221, 242
Flanders, 1, 2, 6, 10, 65
Flemish language, *see* Dutch
Foch, *Maréchal* Ferdinand (formerly General), 58, 163
French Army, 45, 58, 59, 62, 64, 65, 100, 132, 165, 169, 192
    First Army, 73, 104, 107, 108, 123, 124, 132
French - Belgian border, xiii, 1, 2, 24, 42, 44, 203, 210, 213, 216, 217, 221, 232
French, Field-Marshal Sir John (previously General), 45, 64
French Government, 45, 221
French language, xiii, 216, 220, 221
Furnes (Veurne), 1, 5, 35, 38, 39, 40, 42, 61, 62, 71, 72, 132, 134, 213, 216, 217, 224

Gas & gas attacks:
    German, 64, 89, 104, 131, 135, 136, 168, 185
    British, 189-191
Geddes, Sir Eric (Lt Colonel, DGT), xii, 45, 49, 57, 77
Gent (Gand, Ghent), 2, 192, 220, 224
German Army, 48, 58, 64, 107, 124, 169, 181, 185, 191, 192, 196, 210, 223
    advance, Somme front (from 21 March 1918), 56, 57, 94, 98, 124, 136, 169
    advance, Lys front (from 9 April 1918), *see* Ypres, Fourth battle of
    Nieuport bridgehead (capture of, 1917), 131-132, 134
Gheluvelt, 193, 194, 199, 217
Gheluvelt plateau, 58, 64, 93, 117
Gheluwe, 43, 44, 60, 194, 216, 217, 218
Ghyvelde & Ghyvelde-Bray-Dunes, 16, 51, 69, 71, 72, 73, 86, 104, 132, 172, 176, 208, 209, 221, 231
Glencorse Wood, 116
Government, *see under* British, French *etc.*
Gravenstafel, 114, 116

Haig, Field-Marshal Sir Douglas (later Earl), 45, 65, 136, 163
Hainaut (Province of Belgium), xiii, 1, 2, 221

245

Halluin, 23, 44, 64, 209, 221, 232
Hancox, Lt Colonel S H, 106
Hazebrouck, 1, 8, 11, 12, 47, 60, 73, 86, 89, 91, 125, 163, 172, 173, 188, 191, 195, 201, 203
Hell Fire Corner, 44, 108, 115, 120, 163, 165, 168, 189, 190, 193, 199, 242
Henniker, Major A. M. (later Lt. Colonel), 45, 48, 57, 177
Herzeele, 11, 18, 62, 71, 73, 86, 108, 126, 176, 177, 180, 203, 206, 207, 208, 231
Hill, 60, 87, 89, 100, 108
Hondeghem, 12, 180, 187, 188, 191, 196, 231
Hondschoote, 11, 12, 16, 86, 203, 208, 209
Hooge Crater & Museum, 123, 221
Horse traction, 29, 31, 32, 77
Houplines, 24, 25, 72, 85, 209
Houthem, 62, 71, 72, 208
Houtkerque, 71, 125, 203, 207, 208
Hyde Park Corner, 85, 95, 98, 194, 221

Ieper (Ypres), *also see* Ypres, xiii, 221, 222, 240
IJzer river, *see* Yser river
In Flanders Fields Museum (Cloth Hall, Ieper), 221
Influenza, 176, 181, 184, 187

Joffre, General Joseph (French Army), 65

Kemmel, 62, 69, 72, 84, 97, 168, 192, 199, 200, 201, 216
King of the Belgians, *see* Albert
Kitchener's Wood, 113, 114
Koksijde, *see* Coxyde
Kortrijk, *see* Courtrai
Kusttram, *see* Coastal Tramway (Belgium)

La Clytte, 69, 91, 93, 134, 163
La Crèche, 30, 40, 85, 96, 97, 98, 168, 191, 194, 200
La Panne (De Panne), 30, 31, 39, 40, 62, 134, 211, 212, 216, 218, 220, 224
Lambert, Alfred (Lambert Group), 10, 12, 14, 16, 203, 230
Langemarck (Langemark), 42, 43, 44, 73, 104, 107, 108, 113, 114, 128, 169, 181, 216, 218
Le Seau (De Seule), 30, 40, 72, 85, 96, 210, 216, 228
Leighton Buzzard Narrow Gauge Railway Society Ltd. (LBNGRS), 238, 239
Light railways, *see also* contemporary heritage railways by individual name, xii, 45, 47, 48, 65, 70, 76, 77, 104, 124, 126, 180, 192, 221, 234, 237
  Belgian light railways, 77, 126, 136, 193, 200-201
  British & Dominion Armies (WDLR), 46, 49-50, 64, 76, 77, 98, 169, 181, 188, 196, 197, 201, 221, 234
    North-south 'main lines', 50, 57, 97-98, 168
    A system (Proven - Boesinghe), 80, 104, 105-106, 108, 113, 118-119, 128, 130-131, 169, 181, 188, 191, 197, 198, 199, 234
    B system (Westonhoek), 80, 97, 104, 105, 106, 108, 114-116, 118, 119, 128-130, 131, 169, 181, 184-185, 188, 189, 193, 197, 199, 222, 242
    B6 line, 81, 105, 114, 120, 169, 190, 222, 240, 241, 242
    B9 system, 80, 104, 105, 106, 108, 113-114, 118-119, 128, 130-131, 169, 185, 188, 197, 199
    Belgian Coast: 132, 133-135, 136, 223, 224
    Busseboom system, 80, 81-84, 104, 116-118, 120, 122, 130, 131, 168, 192
    C lines, 91, 93, 104, 105, 108, 117-118, 120, 122, 123, 130, 131, 168, 169, 184, 185, 189, 190, 193, 194, 197, 199
    D lines, 91, 93, 104, 105, 120, 130, 131, 168, 181, 184-185, 190, 191, 192, 194, 197
    D11 line, 81, 82, 83, 120, 168, 190, 240, 242
    F lines, 93, 168, 184, 188, 192, 197
    Y lines, 115, 120, 130, 168, 169, 190, 193, 199, 240, 242
    Douve Valley (Messines front), 84, 95, 96, 105, 168, 192, 193, 194, 197, 201
    Eastern extensions (autumn 1918), 192-194, 197, 199, 201
    Hooge Crater monorail, 123
    L system, 130, 168, 181, 185-188, 198
      LX line, 108, 181, 188, 191, 192
      L4 line & extension, 130, 181, 185, 187, 188
      L6 line, 168, 177, 181, 185-186, 187, 188, 191
      L7 line, 168, 172, 176, 181, 186, 187
      L8 line, 173, 176, 185, 186, 187, 191
    Ouderdom system, 80, 82, 84, 91, 93, 94, 105, 117, 122, 168, 181, 184, 192
      K lines, 91, 168, 192, 193, 194
      P lines, 93, 94, 168, 184, 185
      V lines, 91, 93, 94, 95, 168, 184, 192
    Romarin system, 83, 85, 95, 96, 97, 98, 100, 168
    Sectors (after WW1), 197-200
    Trench tramways, 47, 49, 70, 76, 77, 84, 85, 95, 122-123
      B system, 116, 129, 193
      B9 system, 113, 122
      Belgian Coast (Nieuport), 135-136
      C system, 117, 123, 131
      Douve Valley (Messines), 84, 95, 96, 98
      Romarin (Ploegsteert) area, 85, 96, 98-99, 168
  Boescheppe quarry funicular, 76, 185

Deutsche Feldbahn (DFB), *see* German Army
Dunkerque to St-Pol (tramway), 29, 30
French Army, 48, 49, 64, 126, 128, 133, 136, 169, 198, 239
French, after WW1, 201, 237
German Army, 48, 49, 64, 192, 193, 194, 197, 239
La Panne to Adinkerke (tramway), 31, 64, 134, 201, 211, 213
La Panne to St-Idesbald (tramway), 32, 39, 64
Tramway de Pithiviers à Toury (TPT), 237, 238
Lille (Risjel), 1, 6, 23, 24, 192, 194, 221, 232, 242
Lynton & Barnstaple Railway, 239
Lys, Battle of the River, *see* Ypres, Fourth battle of
Lys (Leie) river, xiii, 1, 24, 44, 168, 189, 191, 192, 193, 194, 199
Malo-les-Bains (Dunkerque), 29, 30, 131, 132, 136, 209, 210
Marne, First Battle of (1914), 58
Menin (Menen), 43, 44, 60, 192, 194, 199, 216, 217, 221, 232
Menin Gate (Menenpoort, Ypres), 44, 190, 217, 240, 241, 242
Menin Road, *see* Roads
Messines (Mesen), 40, 84, 87, 91, 163, 168, 191, 216
Messines, Battle of, 65, 77, 81, 83, 84, 85, 87, 89, 93, 98, 104, 108, 163
Messines front, 65, 87, 91, 93, 117, 120, 123, 130, 131, 136, 201, 221
Messines (Mesen) Ridge, 2, 65, 87, 89, 91, 104, 107, 163, 191, 192, 194
Metre gauge railways, *see also* contemporary heritage railways by individual name, xii, 1, 6, 45, 48, 59, 65, 70, 124, 125, 132, 165, 174-180, 192
  Belgian, 6, 32, 44, 45, 60, 62, 204, 211, 218, 220, 224, 228, 234, 242
  Coastal Tramway (*Kusttram*, Belgium), 213, 218, 220, 224
  Courtrai Group, 24, 38, 43-44, 60, 194, 216-217, 218, 232
    Ypres (Ieper) to Gheluwe (line 121), 44, 60, 123, 199, 217
  Coxyde to Coxyde-Bains (tramway), 32, 39
  Dixmude Group, 38, 39-40, 42-43, 62, 73, 213, 216, 217-218
    Dixmude to Poperinghe & Ypres (line 107), 42, 44, 62, 71, 72, 73, 108, 125, 176, 180, 181, 216, 217
    Poperinghe to La Panne via Furnes (Veurne) (line 115), 40, 42, 62, 71, 72, 73, 108, 125, 172, 176, 180, 185, 216, 224
    Ypres (Ieper) to Furnes (Veurne) (line 29), 39-40, 44, 62, 71, 72, 73, 108, 125, 126, 213, 218, 224
    Ypres (Ieper) to Steenwerck (F) & Warneton (line 75), 9, 30, 40, 62, 70, 72, 73, 91, 97, 210, 212, 213, 216, 228

246

Index

Lines not built, 44, 211
Littoral Group, 32, 38, 39, 62, 70, 211, 212-213, 217
   Ostende to Furnes (Veurne) (line 2), 38, 39, 61, 132, 213, 217, 224
   Swevezeele Group, 38, 44, 217
First World War, 48, 60-62, 64, 69, 70, 71, 72, 104, 125, 165, 211
   Bollezeele to Poperinghe & Crombeke, 125-126, 176-177, 224, 234, 236
   French lines supplying locomotives, 73, 76
   Furnes avoiding line (lines 29 & 115), 71, 211
   German Army use (SNCV lines), 60, 64, 73, 211
   Herzeele to Watou (1915), 62, 71, 73, 108, 125, 176, 180, 186, 203, 204
   Pont-aux-Cerfs to Houthem (1915), 62, 71, 72, 208
   Proven to Crombeke, 71, 73, 108, 125, 180, 211
   Tournehem to St-Momelin (1918), 19, 173, 174, 177-179, 180, 204-206, 228, 230
French Flanders, 6, 10, 62, 73, 108, 202, 228, 230-232, 236
   Armentières to Halluin (tramway) (CEN), 23-27, 40, 44, 64, 72, 209, 232
   Hazebrouck to Hondschoote, and Rexpoëde to Bergues (CF), 10-16, 18, 62, 71, 108, 126, 176, 180, 187, 203-204, 228, 230-231
   Herzeele to St-Momelin, Bergues to Bollezeele (SE), 11, 12, 18-23, 62, 72, 125, 173, 176, 179, 206-208, 231, 237
   Hondschoote to Bray-Dunes(-Plage) (NF), 11, 12, 16-18, 62, 71, 176, 208-209, 231
   Lines never built, 18, 19, 23, 206
   Tramway de Cassel, 28, 29, 64, 209
   Tramways of Armentières, 26, 28, 64, 72
Other French lines, 12, 14, 26, 73, 231
Pas-de-Calais *département*, 177, 204, 231
   Aire to Berck, 12, 14, 174, 179, 203, 204, 231
   Anvin to Calais, 11, 19, 23, 62, 69, 71, 177, 179, 180, 196, 203, 204
   Béthune to Estaires, 76, 203, 212
   Boulogne to Bonningues (CEN), 25, 179, 232
   Lens to Frévent (CEN), 25, 57, 98, 232
Somme *département* network (SE), 11, 21, 22, 203, 206, 208, 231, 236
Mission (junction), 81, 104, 106, 116, 188

Mons, & battle of (1914), 58, 60, 192
Mont Kemmel, 2, 57, 69, 163, 165, 184, 191, 200, 223
Montreuil (-sur-Mer), 45, 65, 177, 197
Moseley Railway Trust (MRT), 238, 239
Mule haulage, 49, 82, 83, 95, 98, 99, 123

Nationale Maatschappij der Belgische Spoorwegen (NMBS), *see Société Nationale des Chemins de Fer Belges* (SNCB)
Nationale Maatschappij van Buurtspoorwegen (NMVB), *see Société Nationale des Chemins de Fer Vicinaux* (SNCV)
Neuve-Église (Nieuwkerke), 30, 40, 70, 72, 91, 228
New Zealand Army, *see also* Australian & New Zealand Army Corps (Anzac), 84, 194
Nieuport (Nieuwpoort) & Bains (Nieuwpoort-Bad), 1, 2, 38, 58, 60, 61, 62, 65, 107, 131, 132, 135, 136, 192, 212, 213, 217, 220
Nieuwkerke, *see* Neuve-Église
NM voor de Uitbating der Buurtspoorwegen van den Omtrek Diksmuide - Ieper (ODI), 35, 40, 42, 213
Nord *département* (France), 1, 2, 201, 204, 221, 231
North Gloucestershire Railway, 239

Oost-Dunkerque (-Duinkerke) & Bains (-Duinkerke-Bad), 38, 132, 133, 136, 213, 217, 224
Operations (WW1):
   Light Railways, 56-57, 80, 82-83, 84, 91, 93, 94, 95-96, 97-98, 105, 118-120, 130-131, 135, 168, 169, 181, 184, 185, 186, 187, 188, 189-191, 193, 194, 195, 197, 198, 199, 200
   Metre gauge, 72, 126, 192
   Standard gauge, 107, 108, 192, 195, 196-197
   Trench tramways, 99-100, 135-136
Ostende (Oostende), 6, 38, 39, 43, 58, 73, 132, 213, 216, 217, 220
Ostvleteren (Oostvleteren), 39, 42, 62, 71, 72, 73, 125
Ouderdom, 69, 84, 89, 91, 93, 168, 172, 184, 185, 192

Pas-de-Calais *département* (France), 1, 6, 12, 19, 179, 221
Passchendaele (Passendale) & Ridge, 2, 58, 107, 114, 119, 130, 136, 192, 193, 240, 242
Passchendaele Memorial Museum (Zonnebeke Château), 221
Passenger carriages, 13, 14, 17, 22, 26, 29, 30, 31, 34, 62, 72, 76, 206, 209, 211, 212, 224, 236, 237
Petrol locomotives, *see* tractors
Pilckem & Pilckem Ridge, 107, 108, 113, 165, 169, 172
Plan Z, *see* Z
Ploegsteert & Ploegsteert Wood, xiii, 85, 87, 99, 191, 194, 221

Plugstreet 14-18 experience, 221
Plumer, General (British Second Army), 64, 91, 123
Poelcapelle, 64, 113
Polygon Wood, 116, 117, 131
Pont-aux-Cerfs, 16, 62, 71, 208, 209
Poperinghe (Poperinge), 6, 8, 40, 42, 46, 60, 62, 69, 72, 73, 80, 83, 118, 120, 124, 125, 126, 130, 132, 168, 176, 177, 180, 195, 200, 216, 222, 224, 234, 236
Potijze, 81, 105, 115, 169
Proven, 69, 71, 80, 124, 125, 126, 165, 168, 172, 176, 177, 180, 181, 185, 194, 198, 211, 228

Quarries, 69, 136, 169, 172

Railcars (diesel), 203, 204, 212, 218, 224, 236
Rails, *see* Track
Railway & related transport companies, 45
   Australian, 46, 50, 121, 131, 195, 197, 200
   Australian Broad Gauge Operating Companies (Aus BGROC), 50, 65, 94, 131, 172
      5th (formerly 59th) - 59(5), later 5(59), 107, 108, 125, 165, 172
      6th (formerly 60th) - 60(6), later 6(60), 107, 125, 126, 165, 172, 176, 177, 180, 184
   Australian Light Railway Operating Companies (Aus LROC), 50, 131
      1st (formerly 15th) - 15(1), later 1(15), 50, 83, 93, 94, 98, 105, 120, 122, 131, 195
      2nd (formerly 16th) - 16(2), later 2(16), 50, 93, 94, 131, 168, 181, 184, 185, 186, 187, 188, 189, 190, 191, 195, 199, 200
      3rd (formerly 17th) - 17(3), later 3(17) (later 3rd Aus LRFC), 50, 57, 93, 94, 120, 122, 131, 135, 168, 181, 187, 200, 224, 242
   Australian Pioneer Battalions, 72, 82, 85, 94, 95, 96, 97, 98, 99, 100, 118, 123, 168, 223
   Belgian, 61, 71, 77, 177, 178, 180, 191, 196
   Section Vicinale des Chemins de Fer en Campagne (SVCFC), 60, 61, 62, 72, 73, 76, 108, 132, 180
   British, *see also* Royal Engineers (below), 95
      17th Battalion Northumberland Fusiliers, 46, 65, 91, 104, 106, 113, 128
   Canadian railway units, 46, 197, 200
      13th (Canadian) LROC, 46, 50, 132, 135, 195, 224, 237
      58th (Canadian) Broad Gauge ROC, 46, 65, 91, 126, 199
      85th (Canadian) Engine Crew Company, 46, 65, 86, 125, 126, 172, 176, 177, 180, 224, 234

Canadian Overseas Railway Construction Companies (CORCC), 46, 65, 69, 71, 77, 82, 134, 136
Canadian Railway Troops (Battalions, CRT), 46, 65, 77, 197
  1st (1CRT), 104, 107, 126
  2nd (2CRT), 19, 20, 46, 81, 87, 104, 105, 106, 114, 116, 128-129, 132, 133, 135, 173, 174, 176, 177, 178, 179, 188, 222, 224, 240
  4th (4CRT), 80, 81, 104, 132, 134, 135, 224
  5th (5CRT), 117, 118, 122, 123
  7th (7CRT), 80, 81, 82, 83, 84, 91, 105, 108, 113, 114, 117, 128, 130
  8th (8CRT), 82, 85, 94, 95, 96, 97, 118, 168, 169, 181, 184, 185, 187, 191, 192, 193, 194
  9th (9CRT), 91, 104, 106, 108, 113, 114, 116, 117, 118, 120, 130, 222
  10th (10CRT), 91, 130, 132, 134, 168, 181, 184, 185, 186, 187, 188, 189, 190, 191, 192, 193, 194, 224
Tramway Companies, Canadian Engineers (TC CE), 46, 50, 51, 57
French companies (*Sections de Chemins de Fer de Campagne*, SCFC), 45, 48, 60, 71, 172, 187
  10ème *section*, 45, 62, 70, 71, 72, 73, 76, 86, 108, 125, 176, 177, 180
Labour forces and companies, 46, 98, 106, 114, 135, 200
Royal Engineers (RE), 45, 46, 196
  Army Tramway Companies (ATC), 49, 84, 95, 96, 135-136, 223
  Army Tramway & Forward Companies (AT&FC), 122, 123, 198
  Foreways Companies (FWC), 49, 190
  Group Railway Construction Companies or Engineeers (RCC/E), 46-48, 60, 77
    RCE II, 47, 48, 65, 71, 72, 73, 77, 80, 84, 91, 104, 105, 129, 165, 176, 180, 185, 189, 191, 192, 197
    RCE III, 47, 48, 60, 71, 83, 91, 174, 176, 178, 188, 192, 193, 194, 195, 197
    RCC IV, 48, 91, 195
    RCE VI, 48, 132, 136
    RCE Comms, 48, 197
  Light Railway Companies, 46, 50, 194
    Director(ate) of Light Railways, 45, 49, 65, 118
      Assistant Directors of Light Railways (ADLR), 47, 49, 57, 130, 135
        ADLR II, 77, 81, 84, 96, 116, 118, 128, 168, 181, 186
        ADLR V, 105, 197, 198, 199, 200
    Light Railway Construction Engineers (LRCE), 49, 81, 116
    Superintendents of Light Railways (SLR), 49, 189
    Light Railway Forward Companies (LRFC), 51, 57, 194, 195, 198, 199, 200
    Light Railway Operating Companies (LROCs), 46, 50-51, 57, 131
      1st (1LROC), 50, 97-98, 118
      2nd (2LROC), 105, 131
      10th (10LROC), 105, 118, 189
      12th (12LROC), 83, 85, 96, 97, 98, 105, 131, 168, 169, 184, 185, 187, 188, 189, 190, 191, 193, 194, 195, 223
      14th (14LROC), 84, 93, 94, 131, 168, 185, 189, 223
      15th, 16th & 17th (Aus), *see* 1st, 2nd & 3rd Australian LROCs
      29th (29LROC), 50, 105, 118, 169, 181, 186, 189, 191, 193, 194
      32nd (32LROC), 135, 224, 237
      33rd (33LROC), 50, 84, 93, 94, 105, 118, 119, 172, 189, 224
      34th (34LROC), 50, 199, 200
      35th (35LROC), 50, 181
      54th (54LROC), 50, 118, 119
    Light Railway Train Crews Companies (TCC), 51
      21st (later 21st LROC), 83, 93, 105, 131, 189, 200
      22nd (later 22nd LROC), 51, 105, 119, 120, 131, 189
  Railway Companies (RC), 46, 59, 65
    Royal Monmouth & Royal Anglesey, 46, 59
    8th (8RC), 46, 59, 60, 72
    10th (10RC), 46, 59, 60, 69, 71, 72, 84, 108, 176
    109th (109RC), 59, 60, 72, 85, 91, 128, 185, 186, 187
    112th (112RC), 59, 60, 69, 71, 72, 84, 85, 105, 107, 108, 187
    113th (113RC), 59, 60, 129, 176, 177, 180, 185, 186, 187, 193
    114th (114RC), 59, 116, 128, 129, 130, 132, 196
    120th (120RC), 93, 180, 181, 187, 196
    260th (260RC), 128, 129, 130
    263rd (263RC), 128, 129, 130
    277th (277RC), 80, 104, 105-106, 108, 113, 128, 187
    298th (298RC), 71, 178, 180, 196
    Bridging (287RC, 297RC), 46, 174, 178
    Other companies, 59, 60, 72, 104, 107, 108, 128, 176, 177, 178, 179, 185, 196
  Railway Operating Division (ROD), xii, 45, 46, 48, 65, 73, 76, 82, 86, 91, 93, 94, 108, 125, 126, 172, 176, 180, 192, 199, 222, 234
5th New Zealand Light Railway Operating Company (5NZ LROC), 46, 50, 82-84, 91, 93, 105, 120, 123, 168, 169, 181, 184, 185, 186, 187, 188, 189, 191, 193, 194, 195, 222
92nd South African Broad Gauge Operating Company (92SA BGROC), 131
South African Light Railway Operating Companies (LROC), 50, 96, 119
  7th (7SA LROC), 85, 96, 105, 119
United States Railway Engineers, 51, 237, 238

Rexpoëde, 11, 13, 126, 176, 187, 203, 230
Roads: 84, 96, 98, 108, 130, 196, 228
  Neuve-Église (Nieuwkerke) to Kemmel, 69, 84, 97, 201
  Ypres to Menin (The Menin Road, *Meenseweg*), 44, 60, 93, 115, 117, 123, 169, 199, 216, 217, 221
Rolling stock, *see* steam locomotives, tractors (petrol & diesel locomotives), diesel railcars, passenger carriages, wagons & tramcars (electric)
Roulers (Roeselare), 42, 43, 44, 58, 73, 192, 195, 216, 217, 218, 240
Rousbrugge (Roesbrugge), 42, 125, 165, 172, 185, 187, 196, 224, 228
Royal Engineers (RE), *see* Railway and related transport companies

St-Idesbald, 32, 39, 133, 135
St-Jean (Sint Jan), 81, 105, 106, 130, 131
St-Momelin, 18, 19, 23, 72, 86, 125, 126, 177, 178, 179, 204, 205, 206, 207, 208, 224, 230, 232
St-Omer, 1, 18, 19, 23, 45, 59, 60, 205, 206, 207, 208
Salvaging, 163, 165, 168, 169, 172, 176, 181, 185, 197-200
Sanctuary Wood, & junction, 117, 118, 193
Second World War, 206, 210, 217-218, 228, 230
Smith-Dorrien, General, 59, 64
Société Anonyme des Railways Economiques de Liège–Seriang (LS), 35, 38, 40
Société Anonyme Intercommunale Courtrai (IC), 35, 43
Société Anonyme pour l'Exploitation du CFV de Courtrai–Menin–Werwicq (CMV), 43
Société Anonyme pour l'Exploitation du CFV Thielt–Hooghlede, 44
Société Générale des Chemins de Fer Economique (SE), 11, 18, 21, 76, 86, 125, 126, 203, 205, 206, 208, 209, 211, 231, 236, 237
Société Générale des Transports Départementaux (SGTD), 209
Société Nationale des Chemins de Fer Belges (SNCB), 9, 196, 220
Société Nationale des Chemins de Fer Français (SNCF), 8, 202, 203, 206, 237
Société Nationale des Chemins de fer Vicinal (SNCV), 24, 30, 31, 32, 35, 38, 39, 125, 210, 211, 212, 218, 220, 224, 236
Société pour l'Exploitation des Lignes Vicinales d'Ostende et des Plages Belges (SELVOP), 213, 216, 218
Socx, 173, 176, 177, 178, 196, 224
Somme, battle of (1916), xii, 45, 49, 65, 77, 104, 242

Index

Somme Sector, xii, 1, 48, 49, 50, 96, 98, 104, 105, 123, 130, 134, 135, 163, 168, 172, 174, 181, 189, 191, 195, 199, 221
Standard gauge railways, xii, 1, 6, 45, 59, 65, 100, 104, 107, 124, 132, 163, 165, 169, 172-174, 192, 221, 224
   British Army lines & facilities (*see also* Yards & supply dumps, WW1), 69, 163
      Abeele to Ouderdom & Dickebusch (Ouderdom line), 69, 89, 91, 95, 163, 165, 172, 196
      Bergues to Proven, 69, 107, 108, 126, 172, 176, 180, 184, 185, 187
      Crombeke Road to Reigersburg (Great Midland Railway) & extension, 69, 104, 107, 125, 165, 172, 185, 188, 191, 196
      Douve Valley line, 69, 89, 91, 97, 163, 191, 192
      Elverdinghe loop, 67, 102, 196
      Hazebrouck avoiding line, 163, 187, 196
      Kemmel (later Brulooze) line, 69, 93, 163
      Poperinghe avoiding line, 71, 172, 188, 193
      Proven to Elverdinghe and Boesinghe (Northern Line), 69, 80, 104, 105, 107, 108, 165, 172, 180, 187, 191, 192, 193, 196
      Steenwerck to Neuve-Église, 70, 72, 91
      Steenwerck to Romarin & Petit-Pont, 85, 91, 163
      Watten to Socx, 173-174, 176, 177, 178, 179, 196, 224, 230
   Belgian lines, 9, 38, 44, 45, 58, 59, 196-197, 199, 224
      Adinkerke to Poperinghe via Proven, 6, 9, 72, 107, 185, 196
      Hazebrouck (F) to Ypres via Poperinghe, 1, 8, 47, 48, 59, 60, 65, 77, 82, 82, 86, 89, 130, 172, 176, 184, 186, 188, 192, 196, 197, 200
      Menin (Menen) to Roulers (Roeselare), 43, 44, 192, 193, 195
      Ypres (Ieper) to Courtrai (Kortrijk) via Comine (Komen), 58, 59, 87, 89, 104, 108, 165, 191, 192
      Ypres (Ieper) to Roulers (Roeselare), 44, 59, 104, 108, 114, 116, 117, 130, 165, 192, 193, 199, 240
      Ypres (Ieper) to Thourout (Torout) via Staden & Boesinghe, 59, 69, 104, 107, 108, 113, 165, 191, 192, 196
   French & Belgian lines, 1, 6, 8, 24, 221
      Armentières to Comines (Komen) & Menin (Menen), 24, 99, 192, 194
      Dunkerque to Furnes (Veurne) & Dixmude (Diksmuide), 1, 16, 30, 31, 59, 69, 72, 86, 132, 132, 208, 213, 218, 220, 221, 224, 231
      Lille to Courtrai (Kortijk), 44, 193, 221

French Army lines & facilities, 125
   Heidebeek to Noordhoek, Zuydschoote & Lizerne (Steenstraat) ('Far North' line), 107, 125, 126, 128, 165, 172, 180, 181
French lines of *Intérêt Général*, 6, 60. 69, 172, 177, 196-197
   Calais to St-Omer & Hazebrouck, 1, 59, 60, 69, 173, 177, 187, 206, 224
   Hazebrouck to Armentières & Lille, 1, 9, 30, 40, 59, 60, 89, 91, 123, 163, 172, 191, 192, 197, 210
   Hazebrouck to Dunkerque, 1, 18, 71, 104, 172-173, 187, 188
French lines of *Intérêt Local*, xii, 6, 202, 203
   Tramways of Dunkerque, 29-30, 209-210
Statfold Barn Railway, 238
Stations, 12, 16, 20, 21, 25, 33-34, 203, 206, 208, 224, 228, 230-232
Steam Locomotives, xii
   light railways (60 cm, and near Imperial gauges):
      British War Department, 49, 51, 85, 94, 98, 105, 116, 131, 168, 181, 200, 201, 237-239
      Alco-Cooke (Alco), 52, 93, 120, 122, 135, 237, 238
      Baldwin, 51, 83, 84, 93, 96, 184, 238
      Barclay, 83, 84, 93, 96
      Hudson (Hudswell-Clarke), 51, 83, 238
      Hunslet, 51, 83, 84, 93, 237
      French Army, 51, 52, 83, 239
      Kerr Stuart (Joffre class), 52, 239
      German Army, 51, 239
      La Panne to Adinkerke, 211
      US Army, Baldwin, 52, 239
   metre gauge, 14, 17, 22, 72, 91, 125, 203, 206, 208, 234, 236
      British War Department, 34, 48, 73, 76, 125, 126, 177, 180, 212, 234, 236
      French Army, 62, 73, 76, 236
      German Army (use of), 73
      Manufacturers, 14, 17, 22, 26, 34, 73, 76, 203, 206, 208, 236
      SNCV (Belgium), 34, 62, 73, 76, 125, 177, 211-212, 218, 236
      Type 19 (ex-British), 212, 234, 236
   standard gauge, 72, 105
Steenbeek stream, 107, 108, 113, 165, 169
Steenvoorde, 11, 176, 187
Steenwerck, 30, 40, 60, 62, 70, 72, 77, 85, 91, 96, 200, 210, 213, 216
Stelplaats Historische Trams (De Panne), 224
Stewart, Brigadier General J.W. (Australian), 45

Tacots des Lacs (Seine-et-Marne département), 135, 237, 239

Timetables & connections, xiii, 86
   French Flanders, 26, 29, 30
      Armentières to Halluin, 26, 209
      Herzeele to St-Momelin, Bergues to Bollezeele, 22, 23, 62, 86, 126, 206-208
      Hazebrouck to Hondschoote, and Rexpoëde to Bergues, 14, 15, 62, 86, 203
      Hondschoote to Bray-Dunes (-Plage), 18, 62, 86, 209
Tournehem, 18, 19, 23, 177, 179, 180, 204, 205
Track:
   Light railway (60cm gauge), 49, 50, 51, 64, 80, 82, 98, 104, 105, 106, 113, 115, 116, 118, 123, 130, 133, 136, 184, 194, 221
   Metre gauge, 11, 12, 19, 24, 28, 32, 33, 72, 125, 211, 220, 228
   Standard gauge, 185, 187
Tractors (petrol & diesel locomotives), 49, 211, 239
   British Army light railways, 49, 51, 52, 94, 98, 105, 168, 184, 190, 191, 200, 201, 239
      Petrol Mechanical (PM), 51, 57, 239
      Crewe tractor, 52, 84
      McEwen & Pratt, 49, 52, 83, 239
      Simplex, 52, 116, 131, 211, 237, 239
      Simplex 20hp, 52, 57, 81, 83, 84, 96, 131, 135, 200, 237, 239
      Simplex 40hp, 52, 54, 84, 120, 122, 135, 188, 200, 237, 239
      Petrol Electric (PE), 51, 52, 81, 84, 85, 116, 131, 135, 185, 200, 223, 239
   US Army light railways, 239
   Metre gauge (diesel), 203, 204, 219, 236
Traffic, WW1:
   Light railways, 83, 84, 93, 94, 97, 98, 105, 106, 118, 119, 128, 130, 131, 134, 135, 168, 191, 194
      Ammunition, 54, 56, 81, 82, 83, 84, 85, 91, 93, 94, 96, 105, 106, 108, 113, 116, 117, 118, 120, 131, 135, 168, 197
      Casualties, 93, 96, 118, 119, 120, 194, 222
      Other goods, 115, 134, 168, 189-191
      Rations & water, 96, 106, 131, 198
      Supplies (including RE), 96, 97, 98, 120, 131, 168
      Salvage, 197, 198, 199, 200
      Track & ballast, 105, 118, 131, 184
      Troops, 94, 98, 131, 194
   Metre gauge, 126, 176, 179, 180
   Standard gauge, 165, 197
Traffic control & control posts (WW1):
   Light railways & trench tramways, 57, 83, 93, 97, 99-100, 131, 168, 169, 184, 188, 189
   Standard & metre gauge, 72, 165
Tramcars (electric), 26, 28, 30, 35, 38, 72
   SNCV, 212, 218, 219, 220, 224, 234, 236

249

Tramsite Schepdaal (National Vicinal Tramway Museum), 126, 234, 236
Tramway des Grottes de Han, 224
Tramway Touristique de l'Aisne, 236
Tramways Electriques d'Ostende–Littoral (TEOL), 35, 38
Transhipments, *see also* Depots and workshops *and* Yards & supply dumps, WW1:
  standard gauge to light railway, British,
    A system & related lines, 69, 80, 191
      International Corner & St-Sixte, 69, 80, 98, 105, 106, 118, 180, 181, 189
      Swiss Cottage, 72, 80, 104, 168, 181, 190, 191, 234
    B & B9 systems, 129, 169, 185
      Westonhoek, 69, 72, 81, 98, 104, 105, 116, 128, 168, 169, 184, 185, 222, 224
    Belgian coast (1917), 133, 135, 136
    Busseboom & Ouderdom systems, 69, 82, 84, 93, 95, 185
      Ellarsyde (Busseboom), 69, 82, 83, 93, 94, 105, 120, 131, 165, 168, 184, 185
      Fuzeville, 69, 84, 94, 165, 172, 192
      Heksken, 69, 84, 165, 172, 188, 192
    Douve Valley & Romarin systems, 96, 97, 194, 200
      De Kennebak, 69, 95, 96, 97, 98, 201
      La Crèche, 85, 96, 98, 191, 194, 200
    L system, 172, 185, 187
      Watou, 172, 180, 186, 188, 198
  standard gauge to metre gauge, 18, 28, 71, 72, 73, 91, 180
    Heidebeek, 125, 176, 180, 211
    Swiss Cottage (British), 70, 71-72, 80, 211
    Watou, 73, 172, 180
    Watten (-Éperlecques) station, 177, 205, 228
    Watten - Socx line (British), 173, 174, 177, 178-179, 196
  standard gauge to metre gauge & light railway:
    Noordhoek, 126, 128, 165
  metre gauge to light railway:
    Wagenbruge, 176, 187
Trench tramways, *see* light railways
Twiss, Brigadier General J.H. (formerly Lt. Colonel) (DRT, British Army), 45, 60

United States Army, 1, 188, 191
United States Government, 65

Veurne, *see* Furnes
VFIL, *see* Compagnie Générale des Voie Ferrées d'Intérêt Local

Vicinal Tramway historique Lobbes-Thuin, 236
Vlamertinghe (Vlamertinge), 60, 82, 118, 129, 130, 172, 188, 191
Voormezeele, 91, 93, 95, 117, 190, 200

Wagons (trucks):
  Light railway (60cm gauge), 54, 57, 84, 93, 94, 96, 98, 105, 131, 181, 185, 188, 190, 191, 200
  Metre gauge, 14, 17, 22, 26, 29, 34-35, 62, 76, 91, 125, 126, 180, 206, 209, 212, 236
  Trench tramways, 49, 99-100, 221
War Office Locomotive Trust (WOLT), 237
Warneton (Waasten), 40, 62, 192, 200, 201, 216, 221
Water supply (locomotives), 82, 105, 120, 178
Watou, 40, 62, 71, 73, 125, 168, 172, 177, 180, 186, 188. 203, 216
Watten, 173, 177, 178, 196, 206, 224, 228, 230
Weather & flooding, 58, 65, 93, 94, 96, 107, 108, 114, 115, 116, 124, 128, 130, 131, 135, 163, 192, 193, 200
Welsh Highland Heritage Railway (WHHR) (Porthmadog, North Wales), 238, 239
West Flanders (*West-Vlaanderen*, Province of Belgium), xiii, 1, 2, 35, 59, 221
West Lancashire Light Railway (WLLR) (nr. Preston, Lancashire), 239
Westhoek (& Ridge), 117, 118, 120, 122, 123, 130, 131
White Pole Corner (Ypres), 81, 105, 106, 242
Wieltje, 107, 115, 116, 125, 129, 165, 191, 199
Workshops, *see* Depots
Wormhoudt (Wormhout), 18, 165, 201, 207, 208, 231
Wulverghem, 69, 96, 168, 191
Wytschaete (Wijtschate), 2, 87, 91, 95, 163, 168, 191, 194

Yards & supply dumps, WW1: *see also* Depots & workshops *and* Transhipments
  Belgian, standard gauge - Waayenberg, 107, 185
  British, light railways,
    A system, 106
    B & B9 systems:
      Hagle Dump (B1 line), 168, 169, 185, 222
      Triangle (B1 B12 jct), 81, 104, 105, 106, 169, 185, 222
      Manner's (B4, St-Jean), 116, 131, 189, 193, 198, 199, 200
    Belgian coast (1917), 134, 135
    Busseboom & Ouderdom systems, 82, 93, 118, 129, 168, 185, 189, 190, 191, 192, 193

    Birr Cross Roads (C2 line), 118, 120, 123, 193, 194, 195
    Pacific (D1 line), 82, 83, 168
    Douve Valley & Romarin systems, 85, 96, 105, 131
    L system, 185, 187
    Watou, 186, 188, 198
    Salvage dumps (1918-1919), 198, 199
  British, metre gauge, 71
    Ferme Bleue (St-Momelin), 173, 178, 179
    Pas-de-Calais lines, 70, 71, 179, 180
    St-Momelin Wharf, 72, 125, 177
  British, standard gauge, 60, 69, 72, 91, 107, 108, 126, 163, 165, 169, 172, 173, 174, 178, 185, 188, 192, 196, 201
    Edwaarthoek & Oakhanger, 69, 116, 128, 224
    Peselhoek, 69, 107, 128, 165, 172, 188, 190
    Railhoek, 71, 105, 172, 189, 224
    Reigersburg, 69, 165, 172, 191
    St-Jean, 107, 108, 165, 172, 185, 198, 199
    Vendroux (Calais), 60, 69, 70, 71, 126, 169, 177, 180, 196
    Wieltje, 107, 125, 129, 165, 191
  French, 176, 187
Ypres, battles of:
  First (1914), 58, 59, 60
  Second (1915), 60, 62, 64, 65
  Third ('Passchendaele', 1917), 1, 2, 49, 70, 77, 86, 87, 89, 94, 96, 100, 104, 105, 107, 108, 113, 114, 122, 123, 124, 128, 129, 132, 136, 163, 242
  Fourth (Battle of the Lys River) (1918), 1, 50, 54, 56, 57, 86, 87, 94, 96, 136, 163, 169, 185, 223, 224
  Fifth (Breakout from the Ypres Salient) (1918), 192
Ypres (Ieper), xiii, 1, 2, 3, 30, 35, 39, 40, 42, 44, 58, 59, 60, 64, 81, 104, 105, 107, 108, 114, 115, 120, 124, 163, 165, 169, 190, 191, 192, 200, 210, 212, 213, 217, 218, 221, 222, 224, 240-242
Ypres Salient, 1, 58, 59, 60, 64, 65, 69, 87, 91, 93, 97, 98, 100, 104, 107, 118, 123, 124, 132, 163, 192, 221
Ypres Sector, 1, 35, 48, 59, 189, 197, 212, 221, 237
Yser, Battle of the (1914), 58, 60
Yser Canal, 2, 71, 81, 104, 105, 107, 108, 113, 114, 118, 119, 129, 131, 163, 165, 169, 172, 192, 197, 198, 199
Yser (IJzer) river & estuary, 2, 11, 58, 59, 60, 131, 134, 135, 136, 187, 196, 232

Z - Plan Z, 169, 179
Zeggers-Cappel, 173, 174, 177, 196, 231
Zillebeke (& Lake), 117, 165, 199
Zonnebeke, 116, 118, 120, 123, 129, 130, 193, 221, 240